CAVITATION AND MULTIPHASE FLOW PHENOMENA

McGRAW-HILL
INTERNATIONAL
BOOK COMPANY

New York
St Louis
San Francisco
Auckland
Bogotá
Guatemala
Hamburg
Johannesburg
Lisbon
London
Madrid
Mexico
Montreal
New Delhi
Panama
Paris
San Juan
São Paulo
Singapore
Sydney
Tokyo
Toronto

FREDERICK G. HAMMITT
The University of Michigan
Department of Mechanical Engineering

Cavitation and Multiphase Flow Phenomena

This book was set in Times Roman

Library of Congress Cataloguing in Publication Data

Hammitt, F G
  Cavitation and multiphase flow phenomena
  1. Cavitation
  I. Title
  532'.0595   TC171   78-40929

ISBN 0-07-025907-0

CAVITATION AND MULTIPHASE FLOW PHENOMENA

1 2 3 4 5 MP MP 8 2 1 0

Printed and bound in the United States of America

This book is dedicated to my wife Barbara
and my children Frederick, Harry, and
Jane, without whose forbearance and patience
it could not have been written.

ACKNOWLEDGMENTS

The figures in this volume have come from a variety of sources. In each case the acknowledgment appears as a parenthetical reference in the figure caption. These are keyed by number to the list of references which appears at the end of each chapter, where the source is cited in full. Those which originate directly from our own previous project and technical reports are not always so referenced.

NOMENCLATURE AND UNITS

This book is written, for the most part, using the SI system of units, which no doubt will be used rather completely within the next 5 to 10 years, even in the United States. However, this is obviously not the case at present, particularly in the U.S. engineering community, and various special units are habitually used in certain applications. For this reason, these special units are used in the book where pertinent, as it is necessary to make the material meaningful to those accustomed to these units. In addition, much of the data here presented originate from earlier research papers, so that it often is not practical to convert these detailed numerical materials into SI units. Hence, the unit systems in which the original research results were reported has been continued in this book, as a complete conversion to SI units would be impractical. Where practical, double units including SI are given.

To assist the reader of any nationality to convert the results into terms familiar to him, the following Nomenclature Table is presented. In this connection a recent international hydraulic "dictionary" by A. T. Troskolanski, *Vocabulary of Hydraulic Machinery*, published by Elsevier Press, 1978, which lists accepted terminology in English, French, German, Polish, and Russian, may be helpful.

NOMENCLATURE AND UNITS

Mass   1 lbm = 0.4536 kg

Length   1 ft = 0.3048 m
             1 in = 2.540 cm = 0.0254 m
             1 mil = 10^{-3} in = 2.540×10^{-3} cm = 25.40 μm

Volume   1 ft^3 = 0.02832 m^3 = 28.32 liter
             1 gallon (US) = 0.003785 m^3 = 3.785 liter
                               = 0.8327 gallon (Imperial)

Force   1 lbf = 4.448 newton

Pressure   1 atm = 76.0 cm Hg at 32°F
                    = 1.033 kg/cm^2
                    = 14.696 lb/in^2
                    = 1.013 bar

          1 lbf/in^2 = 0.6895 N/cm^2 = 6895 N/m^2 (or Pascal, Pa) (lb/in^2 also signifies absolute pressure, and lb/in^2 gage is gage pressure, i.e., pressure above atmospheric)

          1 bar = 10^5 N/m^2 (or Pa) = 14.50 lb/in^2

Energy   1 Btu = 1.0551 kJ

Power   1 hp = 0.7457 kW

Flow rate 1 gal/min (US) = 0.22710 m^3/h = 3.785 l/min
             1 ft^3/s = 448.8 gal/min (US) = 101.92 m^3/h

CONTENTS

Number of institutes and laboratories reviewed

PREFACE

This book is closely related to the previous book *Cavitation* by R. T. Knapp, J. W. Daily, and the present author, published by McGraw-Hill Book Company, 1970, which is at present the only comprehensive book on that subject in the English language (reprinted 1979 by The Institute of Hydraulic Research, University of Iowa Press, Iowa City, Iowa). However, it is not simply an updating of that text. It includes the additional multiphase flow subjects of liquid-droplet impact phenomena and thin liquid film behavior in wet-steam flows such as en-countered, for example, in the low-pressure end of large steam turbines, which were not treated in *Cavitation*. It does not include some subjects, such as stationary cavity flows, treated there. The main heart of the cavitation subject, i.e., nucleation, bubble growth and collapse, and the resultant material erosion, are treated in both books. In general, the basic and presumably well-known portions of these subjects are not repeated in detail in the new work. Summarization and references to the previous work are utilized, so that in a sense the present work is a "companion piece" to the former.

While *Cavitation* summarized research results up to the late 1960s, the present book updates this subject to the late 1970s, since very substantial progress has been made during that period in these subjects. *Cavitation* tended to emphasize the research program at the California Institute of Technology, while the present work tends to emphasize that at the University of Michigan, the author's home institute. However, an effort has been made to provide as even-handed as possible a treatment of contributions from all sources. In addition, very copious reference lists from all sources, including Western and Eastern Europe and Japan, accompany each chapter.

Finally, the last chapter is a listing of the world research facilities engaged in the pertinent fields of research. Of course this cannot be entirely complete.

ONE

INTRODUCTION

The technical subjects considered in this book primarily involve vapor or gas bubbles in liquids, or liquid droplets in vapors or gases. Both of these flow regimes have many important features in common. These include the processes of (1) nucleation; (2) growth and collapse, or simply gradual diminuation; (3) effects on the overall flow regime due to the replacement of single-phase by multiphase flow, and the resultant effects upon machine performance and erosion.

Major technical applications to which the above considerations obviously are directly pertinent are: cavitation, wet-steam (or other vapor) flows, and rain erosion of high-speed aircraft or missiles. Except for the lack of major erosion problems, any boiling or condensation applications fit. With regard to erosion in cavitation or liquid impingement phenomena, it has been historically well recognized [1] that metallic erosion produced by either process is often of very similar appearance. In fact, starting in the 1930s and continuing today, experimental impact facilities [2, 3] with impact velocity in the range of 100 m/s have been used to evaluate the supposed "cavitation" resistance of materials used in the construction of hydraulic turbines, pumps, marine propellors, etc. This usage was based on some of the earliest cavitation damage research experiments [3] and upon the later-recognized close similarity in the damage produced by either process. Today this earlier-observed similarity is entirely consistent with theoretical expectations deriving from relatively recent and careful basic research using both ultra-high-speed photography [1, 4–8] and very complex computerized numerical analyses [1, 9–12]. It has been shown that, at least in common engineering flowing systems, cavitation bubble collapse in close enough proximity to a wall to be damaging is a highly asymmetrical process, resulting in the formation of a high-velocity liquid microjet, which strikes the wall and is at least a major contributor to the damage. Thus, according to present knowledge, cavitation damage includes an important contribution from liquid impact. Curiously enough, the direct liquid-drop (or jet) impact erosion processes as encountered with the aforementioned wet-vapor or raindrop impact flows may also include a secondary cavitation phenomenon within the impacting liquid slug.

Another flow regime involving important erosion, with or without cavitation, is that of slurry flows. In these cases the erosion probably includes a contribution from solid as well as liquid particle impact.

Some of the phenomena under consideration here are characterized by extremely high specific energy levels of the fluid. This must certainly be true, for example, of those processes such as cavitation and liquid impingement erosion which depend upon mechanical actions. In both cases the phenomena act as very strong "mechanical amplifiers" in the sense that the kinetic and pressure energy from a relatively large amount of fluid is concentrated in a much smaller amount. As a case in point, the original analysis of Rayleigh [13] for the symmetrical collapse of an empty sphere (bubble in an inviscid, incompressible liquid) shows the attainment of an infinite pressure and velocity at the center of collapse when the bubble radius becomes zero. Of course, this infinite specific energy is then concentrated in a zero mass of liquid. However, as the bubble radius *approaches* zero, the specific energy of the finite liquid mass surrounding the collapsing bubble becomes extremely high, as does the specific energy within any gas or non-condensing vapor trapped within the bubble [13–15]. The overall energy input for this process can be considered, as by Rayleigh, as originating from the pressure energy in the overall static liquid relative to the initial pressure level within the bubble (zero if this is void). While the Rayleigh ideal-fluid spherical-collapse model for cavitation bubbles is not considered to be entirely valid today, as previously pointed out, it is still true that, for one reason or another, an extremely large concentration of specific energy in the liquid must occur if mechanical damage is to result from the bubble collapse. The small size of the resultant pit [6, 8], where a 0.1-mm diameter, approximately circular crater results from the collapse of a spark-generated bubble of approximately 4 mm original diameter, also indicates that this is the case. Thus, the diameter-reduction ratio was ~ 40 and the volume-reduction ratio $\sim 64,000$. An even more striking indication of the high energy levels within a collapsing cavitation bubble is the phenomenon called "sonoluminescence," the faintly visible, usually bluish light resulting (most strongly at least) from "ultrasonic" cavitation, i.e., that induced in a static liquid by the imposition of relatively high-frequency pressure waves. It has more recently also been measured and studied in flowing systems such as a cavitating venturi [16]. The most widely accepted present theory for the origin of this light is the very high temperatures created in the gas trapped within a collapsing cavitation bubble [14–16] and the resultant luminescence, perhaps, of various impurities in the system such as traces of carbon, etc., as well as of the pure gases involved. Calculated temperatures are in the range of 10^3 to 10^4 K [14–16].

As previously indicated, according to present beliefs, cavitation damage results to a large extent from liquid impact, and hence can be considered along with such phenomena per se in regard to mechanical amplification of energy. In the case of liquid impact, again the concentration of the low specific energy levels of a relatively large quantity of liquid into very high specific energy levels of a much smaller quantity occurs, which is then capable of creating mechanical damage. In this particular case, as in the case of cavitation, the highly transitory nature of the process is instrumental in producing the result. This is made obvious by the experimental fact [2] that impacting water velocities of the order of 100 m/s rapidly erode materials as resistant as stainless steels in quite short exposure times. In this case, the stagnation pressure is only about 50 bar, and hence its imposition cannot account for the observed erosion. However, when the transient phenomenon known as "water hammer" is considered, the calculated pressure from this velocity is in the order of 3500 bar, and even greater pressures are possible due to special geometrical effects [17–19]. Hence, the observed erosion can be easily accounted for by these "water-hammer" pressures, even though the relatively steady-state stagnation pressure is obviously insufficient. In the case of water hammer, of course, a portion (depending upon the various detailed parameters describing the impact) of the entire kinetic energy of the impacting liquid drop is converted into a very high pressure and kinetic energy in a small portion of the drop which is in contact with the impacted surface during the collision. Thus, as with cavitation, the specific energy level of the liquid is greatly amplified by fluid-flow effects, but over only a very small time and space domain. Nevertheless, this is sufficient for the provocation of the highly important erosion effects associated with these processes.

Incidentally, light flashes are also observed with liquid impact [20, 21] if the velocity is sufficient ($\sim$ 500 to 1000 m/s). This is possibly due to the compression of the air film between the impacting drop and impacted surface, since these experiments, using a liquid-gun-type device, were conducted under ordinary atmospheric conditions. Again the light emission is indicative of the high specific energy levels involved.

High specific energies are no doubt also involved in many boiling and condensing phenomena, although not so predominently as in the impact and cavitation processes already discussed, and also in slurry and other solid-particle flows where erosion results. In the case of boiling, where high degrees of subcooling are involved, the process is almost indistinguishable from cavitation, and hence the energy levels of bubble collapse are similar. However, this bubble collapse usually occurs well within the fluid so that erosion of containing walls does not usually result. Saturated (non-subcooled) boiling does not result in extremely high specific energy levels.

REFERENCES

1. Knapp, R. T., J. W. Daily, and F. G. Hammitt: "Cavitation," McGraw-Hill, New York, 1970.
2. Ackeret, J., and P. DeHaller: Uber die Zerstorung von Werkstoffen durch Tropfenschlag und Kavitation, *Schweiz. Bauzeitung,* vol. 108, no. 5, September, 1936.
3. Parson, C. A., and S. S. Cook: Investigations into the Cause of Corrosion or Erosion of Propellors, *Trans. Inst. Nav. Arch.,* vol. 61, pp. 233–240, 1919.
4. Naudé, C. F., and A. T. Ellis: On the Mechanism of Cavitation Damage by Nonhemispherical Cavities Collapsing in Contact with a Solid Boundary, *Trans. ASME, J. Basic Engr.,* vol. 83, ser. D, pp. 648–656, 1961.
5. Shutler, N. D., and R. B. Mesler: A Photographic Study of the Dynamics and Damage Capabilities of Bubbles Collapsing near Solid Boundaries, *Trans. ASME, J. Basic Engr.,* vol. 87, ser. D, pp. 511–517, 1965.
6. Kling, C. L., and F. G. Hammitt: A Photographic Study of Spark-Induced Cavitation Bubble Collapse, *Trans. ASME, J. Basic Engr.,* vol. 94, ser. D, no. 4, pp. 825–833, 1972.
7. Kling, C. L.: "A High Speed Photographic Study of Cavitation Bubble Collapse," Ph.D. thesis, Nuclear Engr. Dept., University of Michigan, Ann Arbor, Mich., March, 1970.
8. Timm, E. E.: "An Experimental Photographic Study of Vapor Bubble Collapse and Liquid Jet Impingement," Ph.D. thesis, Chemical Engr. Dept., University of Michigan, 1974.
9. Plesset, M. S., and R. B. Chapman: Collapse of a Vapor Cavity in the Neighborhood of a Solid Wall, *J. Fluid Mech.,* vol. 47, no. 2, p. 238, May, 1971.
10. Mitchell, T. M., and F. G. Hammitt: Asymmetrical Cavitation Bubble Collapse, *Trans. ASME, J. Fluids Engr.,* vol. 95, no. 1, pp. 29–37, March, 1973.
11. Mitchell, T. M.: "Numerical Studies of Asymmetric and Thermodynamic Effects on Cavitation Bubble Collapse," Ph.D. thesis, Nuclear Engr. Dept., University of Michigan, 1972.
12. Mitchell, T. M., and F. G. Hammitt: Collapse of a Spherical Bubble in a Pressure Gradient, *1970 ASME Cavitation Forum,* pp. 44–47, 1970.
13. Lord Rayleigh (John William Strutt): On the Pressure Developed in a Liquid During the Collapse of a Spherical Cavity, *Phil. Mag.,* vol. 34, pp. 94–98, August, 1917.
14. Hickling, R.: Effects of Thermal Conduction in Sonoluminescence, *J. Acoustic Soc. Am.,* vol. 35, pp. 967–974, 1963.
15. Mitchell, T. M., and F. G. Hammitt: On the Effects of Heat Transfer upon Collapsing Bubbles, *Nucl. Sci. and Engr.,* vol. 53, no. 3, pp. 263–267, March, 1974.
16. Jarman, P. D., and K. J. Taylor: Light Emission from Cavitation Water, *Brit. J. Appl. Phys.,* vol. 15, pp. 321–322, 1964.
17. Heymann, F. J.: On the Shock Wave Velocity and Impact Pressure in High-Speed Liquid–Solid Impact, *Trans. ASME, J. Basic Engr.,* vol. 90, ser. D, pp. 400–402, 1968.
18. Hwang, J. B.: "The Impact Between a Liquid Drop and an Elastic Half-Space," Ph.D. thesis, Mech. Engr. Dept., University of Michigan, 1975.
19. Hwang, J. B., and F. G. Hammitt: High Speed Impact Between a Curved Liquid Surface and Rigid Flat Surface, ASME paper 76-WA/FE-34, *Trans. ASME, J. Basic Engr.,* 1977; ORA Rept. UMICH 012449-10-T, University of Michigan, October, 1975.
20. Hoff, G., and G. Langbein: Resistance of Materials Towards Various Types of Mechanical Stress, in A. Fyall and R. B. King (eds.), *Proc. Second Meersburg Conf. on Rain Erosion and Assoc. Phenomena,* 16–18 Aug. 1967, pp. 655–682, Royal Aircraft Est., Farnborough, England.
21. De Corso, S. M., and R. E. Kothmann: Light Flashes, *Material Research and Standards, ASTM,* p. 525, 1965.

TWO

BACKGROUND LITERATURE AND PAST RESEARCH

While background literature and past studies for the various multiphase-flow subjects considered here obviously have some features in common they must, of course, be considered in detail and, to some extent, separately. Past work will be dealt with in this way in the present chapter, while on-going treatments applying to several applications in question are considered in later chapters. Also, of course, under the subject of each application (such as cavitation, liquid impact, etc.), the reviews of pertinent background literature must be divided according to such typical subject divisions as nucleation, growth and collapse, erosion, effects upon overall flow regime, and machine performance.

2-1 CAVITATION

For more than half a century, cavitation erosion has been an important technological problem in a variety of fields utilizing fluid-handling machinery, and as a result a very large technical literature has grown up in this field. It is not by any means the purpose of this section to provide an exhaustive list of such literature. Rather, the main lines of research which have been followed will be indicated, along with the results which have been attained, particularly in relation to the present understanding of the subject. This will provide a guide to the literature adequate to allow more detailed study, for those presently not entirely familiar with the subject.

The overall subject includes many subdivisions. Very briefly, one must consider the following portions of the overall phenomenon. The bubbles, or voids of whatever sort, are created by the flow dynamics in a low-pressure region and then somehow transported to the collapse region where the liquid pressure is high enough to cause bubble collapse. The resulting elevated pressures and velocities created in the liquid by such collapse may interact with the material surface to cause erosion. This interaction is affected by many parameters, including those of the fluid, such as the local pressure, temperature, and velocity, and the viscosity, compressibility, and surface tension of the liquid, plus the mechanical properties of the eroded material, perhaps on a microscale and for extremely high rates of loading.

2-1-1 Nucleation of Bubbles and Voids

It is, of course, generally believed and agreed that bubbles or "voids" (actually containing vapor and/or gas) form in a region where the liquid pressure is sufficiently below that within the "microbubbles" (from which visible and audible bubbles form), to allow these to grow in spite of the restraining influence of surface tension. The growth process involves primarily the action of pressure forces, but these are the result of the interplay of surface tension, inertia, and viscosity, and also gas diffusion and evaporation. The microbubble "nuclei" may be carried to the low-pressure region by the stream velocity or they may grow in cracks or crevices in solid surfaces, primarily through the effects of gas diffusion, until they become large enough to become entrained in the liquid stream through the interaction of drag forces and surface tension. In other cases they may conceivably be formed by liquid-turbulence effects along the interface between the liquid and the vapor-gas mixture in a relatively stationary "cavity" formed along the surface of a solid obstacle in the flow such as a turbine or pump blade, etc. In any case, if damage is to be caused they must then be transported to a region of adequately high pressure to cause their collapse adjacent to a material surface. The details of nucleation, and the as-yet incompletely understood processes of nuclei stabilization in the liquid, will not be considered in detail here, but in Chap. 3. However, Refs. 1 to 4 provide good and recent surveys of the state of the art concerning both cavitation and boiling nucleation.

2-1-2 Overall Flow Regime

Once formed in a region of low pressure, if damage is to occur the bubbles or voids must then be transported to a region of high pressure in close proximity to a structure. The mode of such transport depends on the overall flow regime and no doubt differs importantly for different flow geometries, etc. The flow is almost always turbulent in cases of engineering interest. It is also more or less importantly affected by the presence of the voids or bubbles depending upon the situation concerned. Otherwise the details may differ greatly.

The flow may be essentially translational as along a streamlined object where there is no separation and where the bubbles are carried along at approximately stream velocity and are widely enough separated so that their influence upon each other and the flow is minimal. Many flows which appear frothy to the unaided eye are essentially of this type when viewed with a stroboscopic light.

Again, the flow may involve separation, as in the case of flow along a relatively blunt body, and a relatively "steady" cavity may appear along the body. High-speed photography often shows the substantial oscillations of the walls of such a cavity. However, in that case bubbles may be created by entrainment by the turbulent liquid flow along the interface, or may originate from entrained gas-vapor "nuclei" carried along by the stream and reaching the low-pressure region of the cavity. In either case, if they are carried by the stream to the region of higher pressure downstream of the cavity where the liquid flow reattaches to the body, an adequate pressure differential will be available to cause collapse and perhaps damage. If the region downstream of the cavity contains a sufficient concentration of bubbles, the two-phase nature of the flow will affect the local pressure and velocity distributions in the region of bubble collapse, i.e., a frothy region may exist. In this case, a form of condensation shock, such as has often been observed in cavitating venturis, for example, may be formed [5]. The possible importance of this type of flow in cavitating pumps has been previously discussed in the literature [6, 7].

Another type of cavitating flow, important from the viewpoint of cavitation damage, is vortex cavitation, such as is often observed downstream of the tip-to-casing clearance in unshrouded centrifugal or axial flow pumps, turbines, etc. A flow of this type can in some cases be extremely damaging [8], but the damage capability depends upon many apparently minor details of the flow situation, so that the likelihood of damage in this case is largely unpredictable.

It is clear from the above that the ability to predict the occurrence of damage or of overall performance effects from a knowledge of the population of "nuclei" available upstream of the cavitating region is not in general within the present state of the art, and in any particular case involves a detailed study of the flow geometries and regimes involved. In most cases, even with modern computer capabilities, it is not possible (or feasible) to analyze such complex, turbulent, multiphase, multidimensional flows in sufficient detail to allow the meaningful prediction of cavitation effects in a given case, even if the mechanisms of bubble collapse and material reaction were fully understood—which, of course, is also unfortunately not the case. However, it is hoped that certain relatively simplified mathematical models of the flow with as much general applicability as possible can be analyzed.

For the classical cases of "bubble cavitation" or subcooled boiling, i.e., nucleation, growth, and collapse of individual bubbles with a concentration sufficiently small so that their presence does not affect the overall flow significantly, there is the problem of the individual bubble dynamics between nucleation and collapse which is affected by the pressure and velocity fields to which it is exposed, and also the effects of gas and heat diffusion. In addition, there is also the problem of the trajectory of the bubble with respect to the liquid motion. For example, it is well known that negative "slip" will exist in a rising pressure gradient, positive slip for a falling pressure, and sidewise motion of the bubble relative to the liquid for pressure gradients normal to the liquid velocity. In essence, there is the problem of whether or not the bubble will arrive with the liquid at the damaged location. There are very few papers in the literature [9–11] where the trajectory problem has been analyzed for conditions applicable to cavitation.

There are numerous papers in the literature treating the problems of individual bubble growth and collapse. Their comprehensiveness in terms of the inclusion of "real-fluid effects" (inertial and heat transfer), viscosity, gas diffusion (including the effects of turbulence), surface tension, compressibility, etc., has increased with the years as the capability of computers for the treatment of complex problems has increased. Reference 2 treats this subject in considerable detail with a relatively comprehensive bibliography. The basic equations are found in many papers, one of the best of the earlier treatments being that by Gallant [12].

A problem far more complex than those discussed above (i.e., that of individual bubble dynamics between nucleation and collapse where the interactions between bubbles is minimal) is that of truly two-phase flows, boiling or cavitation, where the existence of the gas-vapor phase is important to the overall flow regime. No fully comprehensive analysis of such flows, even in very simple geometries, is yet available. The analyses to the present can be grouped as follows.

Lumped parameter approaches in straight pipes or ducts   Either the mixture must be considered as homogeneous (which is seldom an admissible assumption) or the conservation relations of mass, momentum, and energy must be written for each phase separately and then some form of connecting parameter (such as "slip," which may be measured empirically or assumed to depend upon other flow parameters as "void fraction," pressure gradient, etc.) assumed. One very important present difficulty with analyses of this flow situation is the fact that even the type of flow regime is not generally known a priori, i.e., "slug" flow, stratified flow, bubble flow, annular flow, etc. The determination of type of flow in this sense depends upon many parameters such as Reynolds number, orientation with respect to gravity, geometry, void fraction, pressure gradients, etc. One well-known representation of such flow regimes is the "Baker plot" [13], discussed in the book by Tong [14]. This book and that by Wallis [15] provide summaries of the state of the art with regard to the behavior of such ducted two-phase flows.

In classical "cavity flow" analyses around profiles, etc., the flow is assumed to separate from a convex wall due to the insufficiency of the normal pressure gradient, since the fluid pressure is limited to values above approximate vapor pressure if no cavity is to form. Such analyses usually assume laminar flow and no interfacial complications such as entrainment of gas, heat, or mass transfer, etc. Nevertheless, solutions exist for only very simple cases. The present state of the art for this type of problem is discussed in detail in Ref. 2. A relatively comprehensive recent treatment is also provided by Wu [16]. Most analyses consider only ideal, inviscid, laminar flow and require linearizing assumptions. The inclusion of nonsteady effects is still relatively rare. Such analyses are summarized in further detail in the 1968 book by Wu [17]. Needless to say, even if one uses ideal-fluid assumptions and simplified geometries, these are problems of very great mathematical complexity. Some simplification is often obtained by the assumption of a "supercavity," i.e., one which terminates only at infinite distance. However, such a cavity is clearly not pertinent to the cavitation-damage problem, since damage would be expected to occur at the downstream end of the cavity only where the liquid flow, with entrained bubbles, would reattach to the wall. This model, for an important class of cavitation-damage flows, was suggested by Knapp [2, 18].

A relatively simplified model for a cavity attached to pump inlet blades (e.g., where the effects of evaporation, entrainment, and disentrainment were included in an approximate fashion) has been investigated by a group at the NASA Lewis Research Laboratory [19]. With the addition of some empiricism, this approach succeeded to some extent in predicting pump-head decrease with cavitation. However, no application to damage has been made.

2-1-3 Cavitation Bubble Collapse and Damage Mechanisms

Much of the past research on cavitation damage, and the area in which the most significant progress has been made, has been that associated with bubble collapse and the material-surface reaction to the resultant pressures and velocities. The problems involved are extremely complex so that even in this rather limited area no very definite answers are yet forthcoming. A relatively recent article [20] by the writer provides a state of the art review of the subject, as does Ref. 2. This portion of the overall subject is considered in the following.

Single-bubble studies (theoretical and experimental) Historically, Rayleigh [21], publishing in 1917, is usually credited with "fathering" the concept of surface-material damage through the collapse of individual cavitation bubbles. His was an "ideal-fluid" analysis of the collapse of an initially spherical bubble, either empty or filled with vapor, which remains at constant pressure during the collapse. He showed that under these conditions the liquid velocity and pressure in the immediate vicinity of the center of collapse approach infinity, i.e., a bubble collapse of this type is an excellent mechanical amplifier.

A solution of the same problem by a different method had earlier been published by Besant [22] in 1859 (of which Rayleigh was perhaps unaware). Besant drew no conclusion regarding the production of material damage through this phenomenon, and in fact the existence of cavitation damage was unknown until approximately 1900.

Because Rayleigh's analysis (and also that of Besant) included no non-symmetrical parameters (such as the effect of a nearby wall, pressure gradients, relative velocity of surrounding liquid, gravity, etc.), the resultant bubble collapse (and rebound) were of necessity spherically symmetric. Since for an "ideal fluid" infinite pressure and velocity would exist at the center of collapse if such symmetry were maintained, at least the possibility of very high pressures and velocities (and also high temperatures within the bubble due to the compression during collapse of the gas-vapor mixture within) around the mathematical point of collapse exists for real fluids. Though these extreme pressures, temperatures, and velocities are indeed very local and very transient, there is the possibility of resultant damage to a nearby surface.

While it appears unlikely at present that high temperature† or velocities due to bubble collapse are important in cavitation damage, the radiation of pressure "shock waves" through the liquid with an intensity high enough to damage most materials certainly appeared to be a possibility. However, much more comprehensive and complex analyses would be necessary to demonstrate that this possibility is in fact a reality.

With this objective, numerous theoretical studies of increasing complexity have indeed been undertaken in the years since Rayleigh, and some experimental measures have also been obtained. However, the analyses even for a single spherically symmetric bubble collapse generally defy solution without the aid of modern computers, and for this reason the analyses up to the era of the 1950s are of mainly historical interest today. They are well recounted in Ref. 2 and will not be reviewed in detail here. If asymmetric effects are included, which in fact are of extreme importance in practical cases, only a relatively small part of the desired studies have yet been made, and these generally only for "ideal fluids." For the present generation of computers it does not appear feasible to make a complete solution of a nonsymmetrical collapse including all pertinent real-fluid effects. However, for spherical symmetry the major portion of analyses fruitful from the viewpoint of cavitation damage appear already complete.

Spherically symmetric bubble collapse If no asymmetrical parameters are included in the analysis, clearly the calculated collapse (and possible "rebound") will be spherically symmetric. However, the existing evidence from high-speed photography indicates that this condition is in fact very difficult to approximate. The following effects are roughly in order of difficulty for inclusion:

1. Surface tension
2. Viscosity, primarily of liquid rather than bubble contents
3. Liquid compressibility
4. Heat transfer to and from the bubble ("thermodynamic effects" according to cavitation literature)

Surface tension The effect of the inclusion in the equations of a term to represent surface tension is relatively trival, although the ability to achieve a simple analytical solution such as that of Rayleigh [21] or Besant [22] is then lost. However, relatively recent numerical studies from the writer's laboratory [26–28] and elsewhere [29] show that the surface-tension term is almost never of importance in the collapse process. While its magnitude grows inversely with bubble radius, in general the inertial term grows more rapidly as the collapse proceeds, so that the

† Their existence is confirmed by the experimental observation of sonoluminescence [23, 24, 25].

surface-tension effects remain overshadowed. Shima and Nakaj [30] suggest otherwise. Surface tension is, of course, very important in nucleation and the early phases of bubble growth, in that it controls directly the size of entrained gas "nuclei" in the flow. Thus it also indirectly affects cavitation damage by influencing strongly the number and size of bubbles available for collapse. In the limit, if the surface tension of the liquid were sufficiently great in a given case, cavitation itself (and hence damage) would not exist. Thus it would appear that a decrease of surface tension would tend to increase damage by increasing the number of bubbles.

Viscosity After surface tension, the least-difficult effect to include in the governing equations is that of liquid and/or vapor-gas viscosity. An interesting paradox arises in this connection, since the term involving viscosity disappears from the governing Navier-Stokes equations for irrotational flows such as the source or sink flow of spherically symmetric bubble growth or collapse, thus implying that viscosity has no effect on bubble dynamics. However, as originally shown by Poritsky [31] in 1952, the viscous effect comes in through the boundary condition at the bubble wall when the normal principle stresses in liquid and gas-vapor mixture are equated there (including the correction for surface tension). This question will be discussed in more detail in Chap. 4. However, the effects of viscosity appear not to be great, unless the viscosity is in the order of that of heavy oil.

Liquid compressibility More important than viscous effects upon cavitation damage are probably those of liquid compressibility. However, their inclusion involves a far greater complication of the analysis than inclusion of viscosity. The pioneering analysis of compressibility effects was that by Gilmore [32] in 1952, who utilized the approximation suggested in 1942 by Kirkwood and Bethe [33] in World War II work involving underwater explosions. Details of this treatment are given in Ref. 2.

Gilmore's essentially precomputer analysis did not provide much numerical data applicable to the cavitation-damage case. However, it has been the basis of many subsequent studies, including that of Ivany in the writer's laboratory [26–28] and of Hickling and Plesset at Cal Tech [25, 34, 35]. Reference 25 is a very comprehensive numerical study made during approximately the same period as Ivany's work. Hickling's study included a direct numerical treatment of the Navier-Stokes equations as well as a solution following Gilmore [32] and Kirkwood and Bethe [33]. He found the error due to the Kirkwood-Bethe approximation to be negligible. Hickling's studies [25, 34, 35] on the temperature profiles within the bubble are the most precise of those presently published on these subjects. Their basis differs from those of Ivany [26–28] in that they did not include viscosity and surface tension as did Ivany.

The investigations of Ivany [26–28] and Hickling [25, 34, 35] show that pressures radiated into the liquid during bubble collapse are probably not sufficient to explain observed cavitation damage unless the collapse center moves very

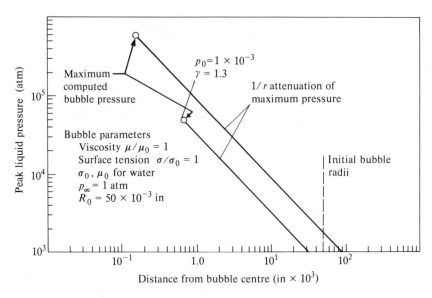

Figure 2-1 Peak liquid pressure on rebounding pressure wave vs. distance from bubble center, assuming $1/r$ attentuation (Ivany [26–28]).

strongly toward the wall during collapse (which is a theoretical possibility). However, in that case the assumption of spherical symmetry is obviously not usable. Hickling also showed the possibility of higher liquid pressures than those during collapse if the bubble contains a minute quantity of permanent gas (or vapor which behaves as permanent gas toward the end of collapse where time for condensation and heat transfer is not afforded), so that a "rebound" follows collapse. Figure 2-1 shows pressure applied to a wall during rebound for a "typical" cavitation bubble in water (assumed to remain spherical throughout), as calculated by Ivany [26–28]. The pressure at a distance from the collapse center (assumed to remain stationary) equal to the initial bubble radius is indicated by the dotted lines. This hypothesized collapse is as close to the wall as is conceivable, even though such a collapse at such close distance certainly could not remain spherical as required by the calculation. Nevertheless (Fig. 2-1), the maximum pressure upon the wall for an internal gas pressure of 10^{-4} bar would be about 2000 bar, damaging only, for example, to weaker metallic alloys such as aluminum. For a smaller internal gas pressure the wall pressure would be higher, but such numerical results are only of hypothetical interest because of the impossibility of spherical symmetry for such a case.

Figure 2-2 shows the pressures in the liquid during collapse as calculated by Ivany [26–28] for a typical bubble in water (internal gas pressure = 10^{-3} bar, $R_0 = 1.27$ mm, $p_\infty = 1$ bar). Close to the bubble wall, these pressures reach a maximum of almost 10^5 bar just before rebound starts. However, the pressure at a distance from a (stationary) center of collapse at this time is very small. The maximum pressure (Fig. 2-2) at a distance equal to the initial bubble radius

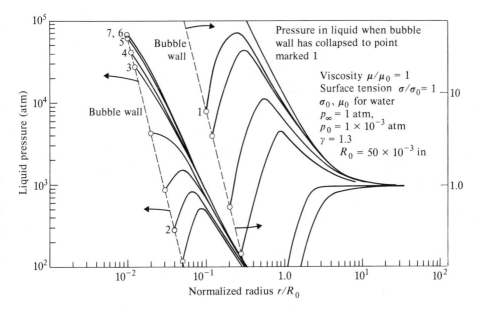

Figure 2-2 Liquid pressure vs. normalized radius for bubble containing gas (Ivany [26–28]).

from the collapse center is attained when the radius has reached about one-tenth of its initial value, and the pressure upon the wall at this time is only about 10 bar. Thus, even if spherical symmetry were possible, if the collapse center is substantially stationary, damage is unlikely during bubble collapse. It seems probable that the radiated pressures would be largest for a spherically symmetric collapse through a given volume ratio.

The calculations discussed above are unrealistic because of two opposing assumptions: that of spherical symmetry and that of a stationary center of collapse. They have also neglected thermal effects. It is not possible to say, in the light of presently available evidence, which of these errors is of greater importance, and hence whether or not the resultant numerical values are, or are not, conservative. In any case, it is clear that little meaningful additional information on cavitation-damage mechanisms can be obtained from calculations which neglect both asymmetries and the effects of motion of the bubble centroid during growth and collapse, as well as the thermal effects. Further consideration of the effects of these additional complicating factors will be made later in this chapter and in Chap. 4. Nevertheless, spherical analyses such as those of Ivany allow evaluation of the various real-fluid effects and thus add significantly to our knowledge.

Figure 2-3 summarizes many of the results of the Ivany calculations [26–28], showing the effects of viscosity and compressibility individually upon a "typical" cavitation bubble collapse in water if spherical symmetry were maintained. Bubble wall velocity and liquid Mach number are plotted against radius ratio and compared with the results of the incompressible, inviscid Rayleigh solution, for which velocity goes to infinity as radius approaches zero. However, the inclusion

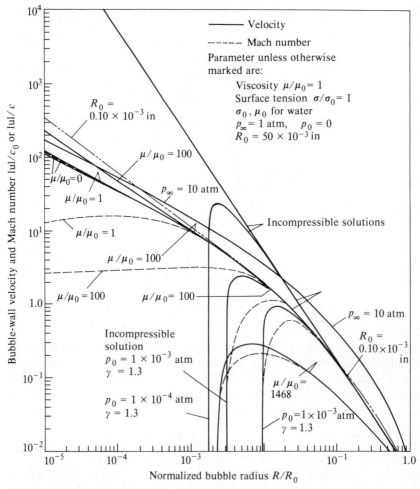

Figure 2-3 Bubble-wall velocity and Mach number vs. normalized bubble radius for reference bubble parameters except where noted otherwise (Ivany [26–28]).

of viscosity and/or compressibility reduces the wall velocity and wall Mach number to finite values, as was also shown by Hickling's calculations [25, 34, 35]. Figure 2-3 shows, for example, that a viscosity as great as 100 times that of cold water has little effect upon collapse velocities. In fact, an anomaly in the numerical calculations [26–28], possibly due to certain other conflicting assumptions in the analysis, indicates somewhat-increased collapse velocity for viscosities in this range. However, Fig. 2-3 does show that a viscosity of 1468 times that of cold water (which is the threshold viscosity predicted by Poritsky [31]) reduced collapse velocity almost to zero. Figure 2-3 indicates the strong effect of compressibility, with or without viscosity, in reducing wall velocities and Mach numbers substantially below the Rayleigh prediction for radius ratios in excess of about 10^2, as was also confirmed by Hickling [25, 34, 35]. The complete arresting of collapse for cases where an internal gas content exists is shown.

Thermodynamic effects and heat transfer   It has been well recognized since the mid-1950s [36, 37], because of observed performance of various types of fluid-handling machinery, that cavitation bubble-collapse intensity was significantly reduced, as compared, for example, with cold water, by thermodynamic and heat-transfer restraints involving the rate of transport of the latent heat and the heat of compression of the bubble contents away from a collapsing bubble.

These effects become important when vapor density, latent heat, and gas content of the bubble are relatively high. The first two conditions are realized, e.g., in hot water as compared with cold water, in cryogenic fluids, in certain organic liquids, etc. However, until recently many cavitation bubble analyses have completely ignored these effects. Their inclusion in detail has proved very difficult, even when the calculation is restricted to spherically symmetric collapses, so that no fully comprehensive results are as yet available.

There are, however, many studies in the literature oriented toward heat transfer which consider bubble growth in the absence of inertial effects and many oriented toward cavitation which include inertial effects but neglect the thermal restraints. It is only relatively recently that studies, utilizing present-day computer capabilities, have appeared which attempt to consider fully both thermal and inertial restraints.

Numerical studies by Hickling [25, 34, 35], Canavelis [29], and Mitchell [38] are representative. However, numerical results remain somewhat uncertain because of the uncertainty of some of the physical parameters involved; this will be discussed in further detail in later chapters. These calculations show in general that if the collapse remained spherical, temperatures within the bubble toward the end of collapse would rise very substantially, since toward the conclusion of collapse insufficient time and surface area are available for heat transfer from the bubble. This phenomenon probably explains the experimental observation of "sonoluminescence" [23, 24], as well as light flashes which are sometimes reported in tail races, etc. However, there is little connection with damage. Some investigators [39] in the past have suggested the possibility of damage due to surface heating, based upon metallographic observations of damaged surfaces. However, this damage mechanism appears improbable because the total heat capacity of the bubble contents is extremely small, and there is no mechanism for adequate heat transfer to the damaged surface. Also, the thermal conductivity of such surfaces is extremely high compared to that of the bubble contents, so that the attainment of a high temperature of the material surface is unlikely.

Another general result from numerical calculations of this type is that bubble collapse can, in those cases where vapor density and latent heat are high, be substantially retarded by the "thermodynamic" effects. This result is also consistent with much experimental information both on field and test devices. The importance of the mechanism depends upon the volume ratio of collapse, which of course depends upon the duration of symmetry, which is generally unknown. It appears likely that thermodynamic effects will be of less importance in highly asymmetric cases. Still, the experimental evidence of the importance of these effects is undeniable.

Various past studies have been made to determine a priori those cases where thermal (or inertial) effects are overriding. A pioneering study of this sort which is largely theoretical was that by Florschuetz and Chao [40] in 1965. Later work by Bonnin [41] was directed more specifically toward the cavitation application.

Field experience with pumps in particular has indicated for years the reduced damaging capability of those fluids for which thermodynamic restraints would be expected. These observations formed a portion of the background information, suggesting the earlier studies [36, 37] already mentioned. Confirmatory information now exists from damage tests with vibratory cavitation-damage facilities. It is certainly true with water [42–45], and apparently with all liquids yet tested, that the damage rate falls off very substantially as temperature is either increased, or decreased, from a value about midway between boiling and melting points at the test pressure. This general behavior appears to be true for all pressures also [44, 45]. The decrease of damage toward the low-temperature end is not yet well explained and is not of as great magnitude as that at high temperature. The decrease toward high temperature certainly depends upon the thermodynamic effect, since it occurs [45] even though the test is conducted at constant suppression pressure (rather than absolute pressure). These effects will be discussed in more detail in the appropriate chapter.

Asymmetric bubble collapse   Photographic records of bubble collapse in cavitation-damage situations indicate that spherically symmetrical collapse as envisioned by Rayleigh [21] is not generally a reality. There are now many good references on this subject, many of which are reviewed in Ref. 2. One of the few studies applied to flowing rather than static systems was that by Kling [46–48], where an actual one-to-one correspondence between the number of spark-generated bubbles and craters in a soft aluminum sample downstream were observed and correlated with high-speed motion pictures of the bubble collapse. Another recent study by Gibson [49] demonstrated that spark bubbles are indeed an adequate model for natural cavitation bubbles in an experiment such as that of Kling [46–48].

Numerical analyses, consistent with the available high-speed photographs, have now been completed for cases of wall proximity, liquid velocity relative to the bubble, pressure gradient, and the effect of a neighboring identical bubble. All of the above cases have recently been analyzed numerically by Mitchell [38, 50], using a program in which viscosity was included. The case of wall proximity was also analyzed by Chapman [51, 52], using a different numerical program for an inviscid fluid. In neither case were compressibility, surface tension, or other real-fluid effects included. Comparison of the results of Chapman with those of Mitchell shows that the neglect of viscosity is of little importance. It is possible that the other real-fluid parameters are also not of great importance for this type of collapse, since the bubble-volume ratio during collapse is not sufficient, judging from the spherical bubble-collapse studies already mentioned [25–28, 34, 35].

Previous theoretical studies are reported and discussed in Ref. 2. These showed many of the same results, but with more limiting assumptions, since they were performed prior to the availability of the necessary computer capacity.

Numerical estimates of microjet velocity resulting from asymmetrical bubble collapse from both photograph and analyses [46–53] cover the range from 100 to 600 m/s. The upper value is probably sufficient to produce permanent deformation in a single impact for most materials of interest. Since most cavitation damage appears to be of the nature of low-cycle fatigue failure, jet velocities, even at the lower value, could also contribute to damage production.

Typically, "water-hammer" pressure has been assumed to be a good estimate of the pressure under an impacting drop or jet, giving ~ 8400 bars for 600 m/s and 1400 bars for 100 m/s for cold water. Various past analyses [53–56] have been made, tending generally to confirm the rough validity of the water-hammer estimate. This very complicated subject will be discussed in detail in Chaps. 5 and 6.

Actual damage production in a cavitating field   Studies of single-bubble behavior as described in the foregoing have comprised much of the basic research in the cavitation-damage field over the past several years, because this area of study represents the limit of feasible mathematical analysis, and also of experimentation of a type wherein the pertinent variables can be well controlled. However, cavitation damage as encountered in the field is the result of a mixture of phenomena far more complex than those involved with single-bubble collapse under laboratory conditions. Some of the special conditions involved with field cavitation damage and some of the pertinent research is discussed next.

Phenomena involved

1. *Corrosion.* Cavitation damage under most field conditions is in most cases admittedly the result of a mixture of corrosive and mechanical attack. The resultant deterioration of the material surface usually occurs far more rapidly than would be the case if similar corrosive and mechanical attacks were encountered separately. However, the relative importance of corrosion and mechanical attack in a given case is not yet well known. This highly complex subject will be discussed in Chap. 5. Corrosion can be suppressed to a large extent by "cathodic protection," first suggested (to the writer's knowledge) by Petracchi [57] and later investigated by Plesset [58, 59].
2. *Multibubble and other effects.* Almost all previous bubble-collapse studies have considered single bubbles only. There are a few recent rather limited studies of some of the effects of one bubble upon a nearby bubble [60], although no comprehensive study of such effects appears to exist. A fully comprehensive study of such a type is probably not feasible, although more limited studies are certainly possible, as discussed in Chap. 4.

Resistance of materials and correlations with mechanical properties

General problem   One major goal of cavitation-damage research has always been the attainment of an ability to relate the cavitation-resistance capability of an untested material to that of materials already tested, in terms of known and easily measurable properties of the material. If such a relationship were known, a material could be "designed" to obtain a desired level of cavitation resistance without the intermediate step of cavitation testing. So far, the attainment of this goal has proved elusive for a combination of the following reasons.

1. The conventional mechanical properties of most materials are measured under semistatic conditions as compared with the very rapid rate of loading in cavitation attack. Also, they reflect average properties for a relatively large section, while cavitation attack is upon a microscale.
2. Material failure in conventional mechanical-property tests does not include the effects of corrosion, which are involved in cavitation attack. No test other than a true cavitation test has yet been devised which provides the proper mix of corrosion and mechanical effects. For this same reason, a given type of accelerated cavitation test does not provide data applicable to many (or most) field conditions. Cavitation tests could conceivably be conducted in a suitably adjusted corrosion environment to match a given field condition. However, the complexities of corrosion attack are such that this might not be a practical alternative. In an attempt to further understand the relationship between corrosive and mechanical effects in cavitation damage, Plesset [58, 59] has made tests with a pulsed cavitation field in a standard vibratory facility. The period between cavitation bursts is designed to allow an increased corrosive action.

It is conceivable that the only fully meaningful test for cavitation-damage resistance is an actual cavitation test. Even so, there is then the difficulty of providing a proper mixture of mechanical and corrosive effects.

Correlations between material properties and cavitation damage   Much information on cavitation-damage rates for various materials in various sorts of cavitation-damage tests have been published over the past half century, although relatively little well-documented information from field devices exists. Also, many more or less systematic and well-documented attempts have been made to correlate erosion rates with mechanical properties of materials. The general subject is well reviewed in Ref. 2 up to the late 1960s. A few summarizing comments will be made here.

In any given cavitation-damage (or impact-erosion) test, rate is time dependent. In general, the damage rate follows an S-shaped curve, starting with little or no measurable damage rate ("incubation period"), passing through a maximum rate portion of the test, and finally into a declining rate. Hence, it is necessary to compare the corresponding parts of the curves. This was first emphasized in the publications of Thiruvengadam [61, 62].

Data from various types of damage tests can be compared and utilized in a final correlation if the rates are normalized to that obtained in one type of test for a material common to the different tests. This has been done in recent quite-comprehensive correlations, one from the writer's laboratory [63] and one by Heymann [64]. In general, it appears that a correlation with a predicting ability better than by a factor of at least 2 does not exist, and that Brinell hardness (which has long been commonly used for this purpose) is about as useful as more complex parameters. Its use is also favored because of its extreme ease of measurement compared to more complex properties. Strain energy to failure for the material, i.e., the energy under a stress-strain curve, has been suggested in recent years by Thiruvengadam [62]. More recently, Hobbs [65] and others, including those at the author's laboratory [43, 63], have found that "ultimate resilience" (i.e., energy to failure, if failure is brittle in nature) is a better correlation parameter. This highly complex subject will be discussed in detail in Chap. 5.

Another factor of possible interest in the correlation of cavitation-damage resistance is the state of applied stress under which the component might be working. A study of this effect [66] in the author's laboratory showed that important effects existed with some materials and not with others.

Effects of fluid parameters   While most past cavitation-damage testing has been performed in cold water at room temperature and atmospheric pressure, the effects upon damage pattern and damage rate due to variations in the fluid parameters are of considerable importance. While most applications involving cavitation damage occur in ordinary cold water, it is very likely that the pressure in the damage zone in many cases is not atmospheric. Also, there is a growing number of applications at present where elevated temperatures are involved, often with fluids other than fresh water, i.e., sea water, organic fluids, liquid metals, cryogenics, etc. Hence, it is necessary to consider the effects of variations in the pertinent fluid parameters.

Temperature   The temperature is important both through the effect of thermodynamic and heat-transfer restraints on bubble collapse previously mentioned and through its effect on corrosion intensities and upon material mechanical properties, if the variation in temperature is large. However, the reduction of damage for temperatures approaching the boiling point is very strong [42–45], so that it is questionable whether or not cavitation damage is an important consideration for applications such as, for example, the primary circuit of a pressurized water nuclear reactor. Obviously, cavitation (and damage) would disappear completely at or above the critical temperature of the fluid, since no distinction between the phases would then exist, or at the boiling point, since no pressure differential to cause bubble collapse would then exist.

Pressure   At present the effect of pressure is less well documented than that of temperature. In general, it is clear that two opposing mechanisms exist for a given case.

1. An increase in pressure reduces the number and average size of bubbles and the extent of the cavitating region, thus tending to decrease damage.
2. An increase in pressure increases the pressure differential for bubble collapse, increasing collapse velocities and resultant damage per bubble. Thus, in an arbitrary case, an increase in pressure may result in either an increase or decrease of damage, depending upon the specific application. It is clear, for example, that if the pressure is increased sufficiently damage (and cavitation) will disappear completely. On the other hand, tests with vibratory facilities at the writer's laboratory, at NASA, and at NEL have shown that a moderate increase of pressure (from $\frac{1}{2}$ to 4 bar) increases the damage rate much more rapidly than in inverse proportion to the pressure over that range [45, 67, 68].

Gas content   The effects of gas content upon cavitation damage have been recently discussed by the writer in a survey report on that subject [1]. Generally, similar opposing trends exist in this case also. As gas content is increased, the number of bubbles and the extent of the cavitating region also increases, but the collapse violence of individual bubbles decreases due to a "cushioning" effect. If the gas content is reduced to nearly zero, so that the number of entrained "nuclei" to cause cavitation under ordinary conditions of suppression pressure becomes very small, cavitation may cease to exist, and the liquid may exhibit a strong tensile capacity.

Other fluid physical parameters   Various other fluid physical parameters, such as density, viscosity, compressibility, and surface tension, can influence cavitation-damage capability. Since these are in general not as important as those already discussed, their effects will be considered in later chapters.

2-2 DROPLET AND PARTICLE IMPINGEMENT

Liquid- or solid-particle impact upon solid surfaces, and the erosion thereby caused, is a phenomenon which has been of considerable technological interest for many years. Basically, these are quite similar in mechanisms and appearance to cavitation erosion. Perhaps they first achieved importance in connection with erosion of the blades of steam turbines operating with "wet" steam, i.e., steam which has been expanded, usually from a condition of original superheat into the saturated region as the pressure is reduced upon its passage through a multistaged turbine. This application is still of importance today because of the very high velocities of turbine blades in the lower stages of the very large turbines now being used. It is especially a problem with turbines for first-generation nuclear plants because of the large moisture contents at the low-pressure end due to originally saturated steam conditions and to large flow rates thus required (giving very large diameter, high-velocity stages).

Another technological application where droplet-impingement erosion has been an important problem for many years (as it is today) is that of impulse

(Pelton) hydraulic turbines. With the very high and increasing fluid velocities used, as heads are continuously increased, direct impact of the water stream on the buckets of the wheel would no doubt lead to an entirely prohibitive erosion rate. This is, of course, avoided by the proper design of the bucket contours. However, any roughening of the surface from whatever cause (as perhaps from cavitation, in some cases, where flow separation may be encountered) will allow a direct impact upon asperities thus created. The very high "water-hammer" pressures" engendered by such impact will then no doubt lead to a very rapidly increasing and prohibitive erosion rate.

The creation of very high pressures locally with particle impact is common to both droplet or solid-particle impact. The above example of Pelton wheel erosion may often be the result of both liquid- and solid-particle impact, since in many cases the impinging liquid stream contains sand particles in suspension. However, various important examples of solid-particle impact alone also exist. One of considerable present importance is that of helicopter blades under conditions of landing or take-off, where the blades often operate in a sandstorm of their own making in certain types of dry or desert terrain. Under such conditions sand-erosion problems also exist for the first-stage compressor blades of the rotor-drive gas turbine. A similar problem also exists for the compressor blades of fixed-wing aircraft gas turbines, and can be important for the propellor blades of propellor-driven aircraft, where this type of problem was first noted during World War II. Another case where solid-particle impact is involved is that of micro-meteorite bombardment of spacecraft. This is a case of hypersonic impact where the kinetic energies of impact are extremely large. A case of low-velocity impact (< 100 m/s) is encountered in coal-liquification plants, and has also proved to be highly damaging.

The previously discussed examples of liquid- and solid-particle erosion are cases where the effects are harmful and research is warranted for their avoidance. However, there are other cases where the effects of the erosion may be useful, so that research aimed at its enhancement is desirable. An application of this type that is quite old is that of sand-blasting or shot-peening of castings to remove surface impurities and improve surface properties. Another application presently being researched, which is only now achieving full practicality, is that of high-velocity water jets for cutting materials such as rock to provide an improved method for tunneling [69].

It is the major purpose of this section to review the state of the art relative to particle impact—most particularly, but not exclusively, liquid-particle impact (i.e., droplet impact). However, an additional and more subtle point is the analogy between droplet-impact erosion and cavitation erosion. The basic mechanisms of cavitation and droplet-impingement erosion are quite similar, differing primarily only by a factor of scale, i.e., cavitation attack is on a relatively micro scale compared to impingement erosion. The study of impingement erosion thus offers a possible tool of cavitation research, and vice versa. The close connection between these forms of erosion has been recognized for a very long time [70], but it is only quite recently that the details of cavitation attack have become sufficiently clarified (involving microjet liquid impingement generated in asymmetric bubble collapse near a wall, etc.) to allow the closeness of the analogy to become fully evident. This situation is well reviewed in Ref. 2. The analogy between these apparently differing forms of erosion became so well recognized that Committee G-2 of the American Society for Testing and Materials (ASTM) on Erosion by Cavitation and Impingement† was formed about 15 years ago to promulgate standards in this area and encourage research.

In order to further review the state of the art in these fields it is useful first to introduce the types of experimental devices employed, since this material has not, to my knowledge, previously appeared in book form. This will be done in the following sections. On the other hand, the cavitation devices are well described in Ref. 2.

2-2-1 Test Devices

Rotating-wheel facility   The earliest form of a droplet-impact erosion facility was apparently the rotating-wheel apparatus, where specimens of the material to be tested are whirled at high velocity though a relatively slowly moving jet of liquid [70]. Since this type of facility is still in wide use (as well as modern versions of the original concept with much higher velocity) it will be discussed at this point. Facilities of this type are now (or have recently been) in use in France [71, 72], the United States [73], the United Kingdom [74], Russia [75, 76], Poland [77], Czechoslovakia, Yugoslavia, etc. [78]. The velocity of impact in the older facilities was about 100 m/s. Experience shows that this is adequate to provoke rapid erosion in materials such as stainless steel and even Stellite [71, 72]. In these cases the erosion process may be aided by local cavitation induced in the collision [79–82] and by the irregular geometry of the impact surfaces.

Rotating-arm facilities   Test facilities more recently developed have often been of the whirling-arm type (rather than wheel), where even higher speeds may be attained. The practical limitations are the stress in the whirling member and the power required to drive the device. The former, as well as centrifugal loading of the test specimen itself, can be minimized if a single arm of large diameter is used, rather than a complete wheel of relatively small diameter. The centrifugal stress on the specimen, which is proportional to V^2/R, is thus reduced. This may be a necessity to provide proper test conditions. The device resembles to some extent an aircraft propellor designed for zero thrust. Devices of this type have been developed primarily for the study of rain (and dust or hail [83]) erosion of high-speed aircraft (primarily supersonic) components such as radomes, wing-leading edges, etc. Another erosion application for this type of device is that for helicopter blades and compressor blades for aircraft and helicopter gas turbines. The earliest facilities of this type were built in the later 1940s and 1950s for the study of propellor-tip rain and dust erosion.

---

† Committee name has recently been changed to "Erosion and Wear."

The velocity ranges produced are also generally applicable to the study of the wet-steam droplet erosion of the low-pressure stages of large steam turbines. This latter problem has been studied in various countries, including the United States, Britain, Germany, France, Russia, and the other countries of Eastern Europe [84]. Much of the recent work of this type is well covered in the special publications of the ASTM [85–88], and it is expected that this series may continue at intervals of a few years.

Propellor-arm facilities have ranged from the earlier units with speeds in the range of Mach† 0.5 up to Mach 3 at the Bell Supersonic Aircraft Company, Buffalo, N.Y. [89], which went into operation about 1970. In all these units artificial rain- and/or duststorms are provided. They have the advantage of relatively rapid testing in highly realistic conditions, but are not particularly suited to basic research into the damage mechanism as produced by a few impacts only. They do have the advantage of providing impact with relatively spherical drops (as with natural rain). The impact velocity and drop size are fairly well determinable. A disadvantage is the high centrifugal field produced.

Rocket-sled facilities As will be noted, the two types of devices previously described are capable of producing impacts with droplets which are approximately spherical at the time of impact. However, this has not usually been the case with the older-type impact wheels where collision was with cylindrical liquid slugs cut from a continuous low-velocity jet by the rotating impacting specimen. However, a relatively recent type developed by Ripken [73] does provide impact with spherical drops. This feature is highly desirable if aircraft rain erosion is the application under study, since in that case collision is with the relatively undisturbed drops falling at relatively low speed. However, for impingement in wet-vapor turbines, droplet shape is irregular.

The newer whirling-arm facilities, developed particularly for the study of rain erosion, utilize an artificial falling-rain field which provides approximately spherical droplets covering the size range of natural rain. However, these devices are limited in speed by centrifugal stresses in the arm itself, and also in the test specimens. It is true that the limiting speed is quite high in the case of the newest facilities, such as that at Bell Aircraft [89], which provides operation up to ~ Mach 3, but in such cases the test device becomes a major and very expensive facility. An alternative type of device, available before such high-speed propellor arms, is the rocket sled. This can also provide extremely high-speed collisions with spherical drops in the size range of natural rain, but is again an extremely expensive device.

In the United States a rocket-sled facility,‡ which has been used by the Air Force for rain-erosion tests [90] but which is used for many other purposes also, exists at Holloman Air Force Base. Speeds up to air Mach 5 have been utilized. Track length in an artificial rain field is about 1800 m, providing specimen exposure

---

† Mach numbers refer to air at STP, unless otherwise stated.

‡ Total track length is 50,000 ft (15 km). Mach numbers refer to air at STP, unless otherwise stated.

in a given run of only a few seconds. However, at such high speeds a single run is capable of destroying many otherwise suitable radome-type materials. An advantage of this type of system, compared with a whirling arm, is that up to 80 specimens can be tested simultaneously under essentially identical rain conditions, whereas the whirling arms can usually handle at most only two specimens at a time. A disadvantage of the rocket-sled device is that test duration is fixed by the track length and intermediate observations are not possible. This is particularly disadvantageous for tests of those materials which are completely destroyed in a single pass through the rain field. Another disadvantage is that the expense of a single run is very large.

Moving specimen; additional multiple-impact devices Moving-specimen devices are advantageous in that collisions can be with essentially spherical drops. Another alternative to whirling-arm (or wheel) devices or rocket-sled devices is a gun device, wherein a vehicle carrying the specimen (or specimens) is fired through an artificial rain field. An obvious difficulty here is the provision of adequate "arresting gear," so that the specimens can be recovered for later examination without prohibitive damage from the necessary high acceleration and deceleration. Another obvious difficulty is the fact that unless the gun is of very large caliber only one specimen can be tested in a given run, so that comparison between specimens under closely identical conditions may be difficult. To the writer's knowledge, only one such device has been built and utilized (by Saab Aircraft in Sweden).

A somewhat similar device is that developed by Fyall and others [91, 92] at the Royal Aircraft Establishment, Farnborough, England. This is a gun apparatus (about 25-mm caliber), but provides only single impact in a given run. Somewhat similar devices have since been built elsewhere. A spherical liquid droplet is affixed to a network of fine wires suspended in the path of the "bullet" which carries the test material. Arresting gear is provided to slow the specimen assembly after impact so that the target material can be recovered for examination. Since the single droplet is in a fixed position, it is possible to obtain excellent photographs of the collision, the camera being triggered by the rupture of a fine wire placed across the trajectory just ahead of the assembly from which the droplet is suspended. Excellent photographic results have been obtained with this apparatus which have been very important in advancing basic understanding of the collision process. However, the device is obviously not suited to studying the resistance of materials to repeated random impacts such as would be encountered in an actual rain field, since loading and firing is an extremely tedious process.

Stationary-target moving-droplet devices Stationary-target moving-droplet devices often have the advantage of simplicity and economy of operation, but so far have suffered from the virtual impossibility of obtaining droplets of the diameter encountered in actual applications such as natural rain or wet-vapor turbines, at realistic velocity, which remain approximately spherical up to impact. The difficulty is the break-up and/or severe distortion of large drops by aerodynamic forces or by the large forces necessary to accelerate the drop to the desired velocity in the test apparatus. It has been suggested in the past that a critical Weber number for good droplet stability is about 23, where $We = \rho V^2 d_c/\sigma$, ρ is air density, and σ is liquid surface tension. If this criterion is correct, the maximum velocity for droplets of the size range of raindrops is only about 1 m/s. However, a study of droplets separating from steam-turbine cascades indicates the possibility of spherical droplets with diameters as large as 0.4 mm at about 500 m/s [93], but the average raindrop diameter is in the order of 1.0 to 1.5 mm. This is also the probable size range of droplets causing steam-turbine erosion. It is clear at the moment that the precise method of formation and acceleration of the droplet is extremely important to its ability to retain the spherical shape up to impact; insufficient precise information is available to allow any definitive statements in this regard. However, it is also clear that up to the moment no device has been devised which can produce spherical high-velocity drops in the size and speed range suitable for the study of either the rain-erosion or the wet-vapor-turbine problem.

Many of the present devices in use produce an elongated jet rather than a droplet. However, as already discussed, a liquid slug of this shape may be reasonably useful for the study of material damage, provided the leading edge is a rough spheroid, since according to theoretical considerations [94] only the first instant of impact may actually be of importance in the damaging process.

Momentum-exchange gun devices (single impact) A momentum-exchange device using an air gun as the driver was originally developed by Bowden and Brunton [94] at Cambridge. They have been used at several other laboratories, including our own at the University of Michigan, and at Dornier Aircraft [95], where a water-jet velocity of 2000 m/s has been attained using a standard small-caliber rifle as driver. All of these devices use a small momentum-exchange chamber containing water (Fig. 2-4), sealed at the rear by a flexible diaphragm (perhaps

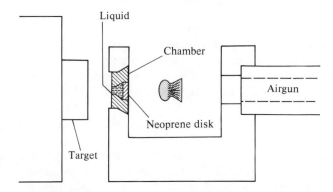

Figure 2-4 Brunton water gun—schematic [117].

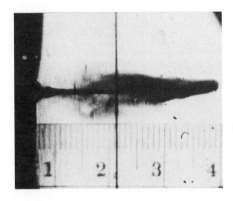

Figure 2-5 Droplet ejected from University of Michigan water gun at 550 m/s.

rubber) and with a small orifice at the front from which the jet is ejected. The bullet or lead slug from the driver gun impacts the flexible diaphragm, exchanges momentum with the water in the chamber, and ejects a small high-velocity jet toward the target material. A velocity ratio as high as 1:5 between the impacting bullet and the water jet may be attained [95]. The jet characteristics are well described in Ref. 94, and depend upon the water-meniscus position within the chamber. If the filling is such that a concave meniscus is attained, a "shaped-charge" effect exists which produces a small precursor ("Monroe") jet, with a velocity perhaps of 1.5 to 2 times the main jet velocity. For another meniscus position, a leading edge of somewhat greater diameter than the main jet is produced, which at least roughly approximates a hemispherical shape (Fig. 2-5—from the author's laboratory). Though the velocity so attained is somewhat less, this type of jet is probably better adapted to the study of erosion damage. This type of device obviously produces only a single impact, and is well adapted to the basic study of the impact process and to the provision of detailed high-speed photographs of the actual impact [94]. However, it is not suitable for the study of material resistance to multiple random droplet bombardment because of the limitation to an individual impact for each loading and the difficulty of producing a realistically random impact pattern if repeated loadings, though tedious, were used.

Momentum-exchange device (multiple impact) A multiple-impact device, which is economical and practical on a laboratory scale and is quite similar in principle to the gun device discussed in the preceding section, was developed and reported (1967) by Kenyon [96] at Associated Electrical Industries Ltd. in England for the testing of materials for steam-turbine blading. A somewhat modified arrangement was constructed later at our laboratory [97–99], shown schematically in Fig. 6-39.

These devices produce jets very similar to those of the Brunton-type gun [94], with velocities up to about 600 m/s. Thus the jet shown in Fig. 2-5 could actually be from either a single-shot device of the Brunton type or from a repetitive device of the Kenyon type. The automation is easily provided using a cam-operated device to withdraw and release the bolt to be driven by a heavy spring against the metallic sealing diaphragm. Fatigue of the diaphragm and spring appear to limit the useful impact velocity to about 600 m/s. The transfer of momentum and shock waves across the diaphragm into the water produces the desired jet, through an orifice similar to that used in the Brunton device. The same care must be taken in filling and obtaining the required meniscus configuration as for the Brunton device, but these features can also be easily automated. Jet-velocity repeatability to the order of ±2 percent is obtained [97]. The repeatability in the writer's laboratory of this automated device is superior to that of our single-shot device, probably because of the more precise guidance given to the driving member (by a machined cylinder) as opposed to the bullet impact in the gun device which appears to have more variation. The automated device is capable of about 30 shots per minute. It is thus well adapted to both basic research into the mechanism of single-drop impact (as is the Brunton gun) and also to material-resistance tests for repeated impact. It is clear that this is potentially a device fully capable of relatively realistic screening tests of materials for droplet-impact resistance such as can be obtained with the much more expensive devices such as whirling arms (if a similar velocity range is to be attained) or rocket sleds. However, complete realism is lacking with regard to the somewhat irregular droplet shape, which is approximately hemispherical with long trailing jet (Fig. 2-5), and to the difficulty of achieving a truly random impact pattern. This latter effect could be attained with somewhat greater complexity using some form of cam drive to provide a wobble motion in the plane perpendicular to the jet axis.

The device is obviously well adapted to the study of the effects of velocity, angle of impact, number of impacts, diameter of jet, fluid conditions, ambient pressure, target temperature, etc.

Wind-tunnel tests   It is clearly possible to use a test method wherein the target is suspended in a high-velocity wind tunnel, with artificial rain (or dust particles) injected into the airstream. Such tests have been conducted at various laboratories. The obvious disadvantage is the fact that large drops are almost certainly fragmented by injection into the airstream, so that the actual rain conditions are not well modeled. However, good realism can be attained with respect to target geometry and scale, as well as air temperature and pressure if the tunnel provides such capabilities. For example, a full wind-shield assembly from a large aircraft was tested in this fashion at ONERA, Modane, France [100]. An apparent disadvantage is the high cost, particularly if erosion tests are the objective, since operation of the full wind-tunnel complex is required.

2-2-2 Theoretical Considerations and Research (Droplet Impact)

General considerations pertinent to droplet impact   Essentially, droplet impact is a non-steady-state process in which the nonsteady aspect is of major importance. A pressure differential approximating the "water-hammer pressure," i.e., $\Delta P = \rho V C$, where V is velocity of impact, C is sonic velocity in liquid, and ρ is liquid density, is generated at the first instant of impact. Shock waves then traverse the drop originating from the solid surface of impact. Research to date has indicated the following general features of the phenomenon. The approximate water-hammer pressure persists until reflection waves or other pressure release mechanisms return to the surface of impact. The major release mechanism other than reflected waves is the generation of a liquid velocity parallel to the surface, with a stagnation pressure approximately equal to that of the central impact region from which it originated. Thus this radial velocity can be several times that of the impacting velocity, and may be important in the damage mechanism.

The general model of the phenomenon is particularly obvious if one considers a droplet shaped as a disc with its axis parallel to the impact velocity. This case approximates the classical water-hammer problem found in pipe flows if the disc periphery is relatively distant from the axis, so that there is little "edge effect." Then the disc is in fact confined within a rigid geometry as in the usual water-hammer case. Of course, the situation is much more complicated for spherical drops or short cylindrical drops, but the essential features are much the same. In all cases, an analysis requires some inclusion of liquid-compressibility effects (as in the classical water-hammer model), since otherwise infinite pressure would be generated at the instant of impact with a rigid solid. Since the velocity of sound in a completely incompressible liquid would be infinite, reflection waves would also return instantly, so that the infinite pressure portion of the impact would persist for zero time. In brief, no meaningful analysis is possible without consideration of compressibility of the liquid. If the target material is also considered compressible, the realism of the analysis is of course improved. However, the solid compressibility is not as important as that of the liquid, since the degree of solid compressibility is normally much less.

While the droplet-impact problem is essentially non-steady-state during the important period while the water-hammer type of pressure persists, this time is actually extremely short. For a spherical water drop of 1 mm diameter the time necessary for a pressure wave to traverse the drop from the surface of impact to the free surface and return is about 1 μs. It is thus during this very short time only that extremely high pressures can exist for most velocities and during which the very high radial velocities can be generated. Conservation of energy, of course, implies that an average relative velocity can be no more than the impact velocity, and an average pressure against the surface is the normal stagnation pressure. However, during the first few microseconds of impact these values can be greatly exceeded. For example, a cold water drop with 100 m/s velocity impacting a rigid surface would generate a "water-hammer" pressure of about 2000 bar, whereas the actual stagnation pressure for such a case is only ~50 bar. However, the total duration of the collision would be about 10 μs for a 1-mm diameter drop, during most of which the approximate stagnation pressure would exist, as opposed to the first 1 μs when the very high initial pressure would apply. Since experience shows that impact velocities of this order can damage materials such as stainless steel [71], it seems likely that the essential portion of the phenomenon from the viewpoint of damage is that during the first microsecond.

Recent comprehensive numerical calculations [101–103] confirm the above rough estimates. The last reference [103] also includes the effect of target material elasticity. Thus instrumentation (and high-speed photographic techniques) useful for basic research into the droplet-impact phenomenon must provide information during this initial microsecond.

The foregoing discussion indicates that at least for cases where the impact velocity is well below the sonic velocity of the liquid ($\sim$1500 m/s for cold water), the existence of a long trailing jet behind the striking portion (as in the water guns previously discussed) may not be important to the damage phenomenon, since the high pressures and velocities are a function only of the initial striking portion of the drop. The shape of the striking portion is thus important, but the relatively long trailing portion of the liquid slug merely assures the continuance of stagnation pressure and radial velocity of the same order as the impacting velocity for some additional microseconds. Since these velocities and pressures are often not sufficient to cause damage in themselves (as in the 100 m/s impact case previously discussed), they probably do not make an important contribution to the damage phenomenon. Thus devices such as the water guns may indeed model the damage phenomenon more closely than might be initially supposed, since the striking edge of the liquid slug can be approximately spherical.

Pertinent basic theoretical and experimental past results The "one-dimensional" problem, i.e., a large liquid slab or disc impacting a rigid solid, could be analyzed easily using hand calculations and classical water-hammer theory as applied to flow in completely rigid pipes.† In fact, the cases are entirely analogous. Such a simplified model, as compared to the actual process, is still sufficient to indicate many of the essential features of the actual phenomenon discussed below. However, this is not sufficient for obtaining an improved detailed understanding of the erosion phenomenon, since it occurs with finite drops of complex shape. Most experimental evidence [104] indicates a strong increase in damage rate (at a given impact velocity) with increased drop size. Water-hammer pressure is, of course, independent of drop size for a given velocity, but its duration is proportional to drop diameter. Thus at least a two-dimensional analysis is necessary to achieve further realism in this regard. Such an axially symmetric analysis could incorporate the cases of perpendicular impact with spherical or cylindrical drops (or a combination of shapes), but it could not handle the case of oblique impact, for example.

Until the last few years only very approximate analyses of the impact phenomenon have been made; for the most part these were modifications of the one-dimensional water-hammer analysis to include some correction at least for such effects as droplet shape, pressure dependence of sonic velocity of liquid, compressibility of solid, etc. The last point is most important for impact with soft coating materials (e.g., rubbers) or in the case of hypersonic impact velocity. In these cases, if the deformation of the solid is substantial, the flow regime may be

---

† See any elementary fluids text.

substantially altered. Recent photographs taken in the writer's laboratory [97, 98] indicate that this is the case with rubber coatings in the velocity range of 250 m/s. The effect of surface flexibility appears to be even more important for cavitation erosion, since here the overall collapse behavior of the bubble is strongly affected in such a way that a resultant jet may be oriented away from the surface, rather than toward it as in the case of a rigid surface. Recent photographs from the writer's laboratory [105, 106], as well as from Cambridge University [107, 108], indicate that this is the case. Such behavior would be expected on theoretical grounds (discussed in a later chapter) and is consistent with field experience on cavitation damage.

The important effect of pressure upon liquid sonic velocity (and also density), and hence upon water-hammer pressure, appears to have been recognized only very recently. It has been pointed out by Smith, Kent, and Armstrong [93] that the calculated (one-dimensional analysis) pressure can be increased by about 60 percent for an impact velocity of about 600 m/s if one takes account of this effect. Heymann [109, 110] presents a very simple relation for the computation of the modified sonic velocity:

$$C = C_0 + KM_{liq}$$

where $K \cong 2.0$ for cold water, if C is in feet per second, and M_{liq} is the liquid Mach number. This question will be discussed in further detail in Chap. 6.

Comprehensive numerical analyses It is clear from the foregoing that only comprehensive numerical analyses of the problem, or extremely careful and sophisticated experimentation, can advance the theoretical understanding beyond that attained in the light of modifications of the essentially one-dimensional analysis. Work of this type will be discussed in a later chapter.

Huang [101, 102], at the University of Michigan, analyzed numerically a spherical drop, a cylindrical drop of $L/D = 1$, and a combined spherical-cylindrical drop (i.e., "rounded corners") for cold water and a completely rigid flat material surface. This is the first comprehensive numerical analysis of the problem which appears to have been made. The results are independent of scale. They depend only on the initial-impact liquid Mach number (referred to the velocity of sound in the undisturbed liquid) and droplet shape. The liquid Mach numbers investigated were 0.2 and 0.5 (about 300 and 750 m/s for water).

All results are given in nondimensional terms. Figures 2-6 and 2-7 provide an example for impact velocity of liquid Mach 0.2 for a spherical drop of cold water impacting a fully rigid flat plane. A free-slip boundary condition was used, thus neglecting the major part of the viscosity effect.† Pressure (vertical axis in Fig. 2-7) is given in multiples of the classical water-hammer pressure for the undisturbed fluid, and the position (horizontal axis) is radius divided by initial

---

† An "artificial viscosity" was sometimes used to stabilize the computer calculation. However, comparison of free-slip and zero-slip boundary conditions showed the effect of viscosity on the results to be small.

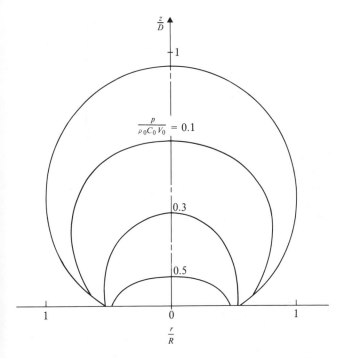

Figure 2-6 Isobar distribution in an initially spherical droplet at time $(Ct/D) = 0.5$, for impact Mach number of 0.2, and for free-slip boundary condition.

drop radius. The times after impact are given as a nondimensional time t', defined as $t' = Ct/D$, where C is sonic velocity in the liquid, t is actual time, and D is initial drop radius. The numerical time values shown become actual values in microseconds if the ratio $t/D \cong 10^6$. Since $C \cong 1500$ m/s for water, this would be the case for a droplet with an initial diameter of 1.5 mm (close to the drop size of interest in many cases). For such a drop at M_{liq} 0.2 impact velocity, peak pressure on the solid wall is reached after ~ 0.25 μs. Thereafter, it falls rapidly to an eventual steady-state value of about 0.10 times the water-hammer pressure (i.e., stagnation pressure), being only slightly above this value after 2.5 μs.

This brief presentation of numerical results has been made here only to show the essential features of the impact phenomena. Further results from this and other analyses are discussed in Chap. 6.

A result later than Huang by Hwang [103, 111, 112] rigorously also included the effect of elasticity of the target surface. As expected, this somewhat reduces the impact pressures and also allows the calculation of the stress in the target material. Plexiglas and aluminum were investigated.

Photographic evidence roughly tends to confirm the results of the Huang and Hwang analyses. Such studies include those by Fyall [91, 92, 113] and his co-workers at the Royal Aircraft Establishment at Farnborough, England, and Brunton and his colleagues at the University of Cambridge [53, 80, 81, 114–117].

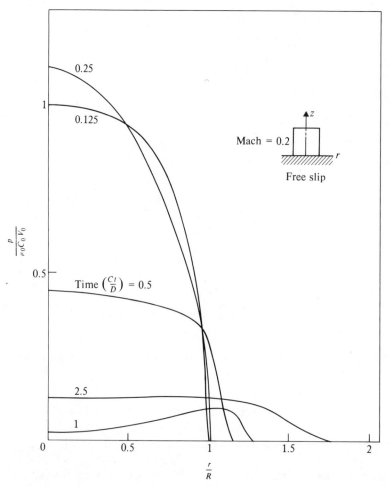

Figure 2-7 Pressure-time history at liquid-solid interface ($z = 0$) of an initially cylindrical droplet with $L/D = 1$, for impact Mach number of 0.2 and for free-slip boundary condition [101].

2-2-3 Solid-Particle Impingement

While most of the past work on particle-impact erosion has no doubt been concentrated upon liquid-droplet impact, some has also been devoted to solid-particle problems. Dust erosion of rotating blades as for helicopters, gas-turbine compressors for helicopter drive, etc., are motivating applications. Solid-particle erosion work to date has been largely experimental, and has been carried out mostly over the last decade. Theoretical work is far less advanced than that with liquid droplets, because of the much more complex nature of treating the "flow" of solid particles of irregular shape and complex mechanical properties as compared to that of a liquid.

Test results so far have been quite limited [118–120], and have often not followed the trends exhibited by liquid-droplet impact, i.e., the very strong increase of damage rate with increased velocity and rapid reduction of damage as the impact angle is decreased from perpendicular. Also, in general, the material mechanical-property correlations developed for liquid impact do not apply for solid impact.

Experimental information [119] shows substantial erosion of materials as strong as 304 ~ SS by solid dust particles with flow velocities smaller than 100 m/s. This would probably not be the case for liquid droplets under the same conditions. The difference is probably due to the very pronounced particle-geometry effects with hard solid particles, making the water-hammer model essentially inapplicable, even though the sonic velocity of the solid particles is much greater than that of water. Calculations provided to the author by Adler [121] based upon the hertzian contact model [122, 123] indicate contact pressures in excess of 10^5 bar for an impact velocity of 60 m/s. This indicates a multiplying factor of ~ 11 times that computed by the water-hammer model. While the hertzian model is based upon quasi-steady-state assumptions, trial calculations [121] show that it is surprisingly accurate for the impact problem.

An interesting limiting case for both solid- and liquid-impact erosion is that of hypersonic impact. Here the two otherwise diverse processes become essentially identical, since the stresses developed during such a very high velocity impact much exceed the ultimate failure stress of even most solids. Hence, in this case, it is usually considered that the solid behaves as a fluid exhibiting typical fluid properties, as does also the liquid. A suitable equation of state is thus required, perhaps of the form originally proposed by Tait [124]. This has also been used in our previously mentioned liquid-impact analyses [101–103, 111, 112]. For cases of extremely high velocity impact, of either liquid or solid particle, a gaseous equation of state has sometimes been used for the "particles," since the stress intensity during impact in these cases exceeds the binding energy (or even chemical energy) of the material, whatever it may be. Thus a collision at such velocities with a stone or a block of TNT would have approximately equal effect.

2-2-4 Wet-Steam and Thin Liquid-Film Flows

Closely related to the droplet-erosion problem as it occurs in the low-pressure end of large steam turbines is the problem of two-phase low-pressure flows, wherein the liquid phase deposited on the stator blades is in the form of a thin film, driven by shear with the high-velocity steam flow. Erosion of the next moving row results when large droplets are shed from this thin liquid film by the shear action of the high-velocity steam. Since the study of this flow in particular bears a very intimate relation with the droplet-erosion problem, it is also included as a portion of Chap. 7.

The present state of the art is reviewed briefly here. Other very closely related two-phase flows will also be included to some extent. Other variations of the liquid-film flows which are particularly applicable to the steam-turbine problem are cases where the film is partially or entirely driven by gravity and where the effects of heat transfer are important. Such flows occur in boiling liquid-cooled reactor cores, e.g., where one of the important problems is that of film "dry-out." These flows involve not only the study of the steady-state behavior but also film stability and breakdown aspects. These aspects are not only paramount, particularly in the case of steam-turbine erosion, but also in such problems as nuclear reactor core dry-out and other boiling heat-transfer cases.

Steam- or gas-driven thin liquid films   The primary case under consideration here is that of a thin water film (~ 10 to 100 μm thickness) deposited upon the stationary blading, primarily in the low-pressure end of large steam turbines. The film is driven by surface shear due to the high-velocity main steam flow. While the steam flow is no doubt turbulent in this case, that of the liquid film is laminar (classical Couette flow). If the film flow is "fully developed," the restraining force, just sufficient to prevent acceleration, is shear with the stator-blade surface. In this condition, the velocity distribution for the film is linear. If gravity plays an important role (not usually the case for the steam-turbine application), a parabolic distribution would be superimposed on the linear uniform-shear flow. Since steam and liquid shears at the surface are equal and opposite, the liquid shear could be computed by measuring the axial pressure drop of the steam flow if it is sufficiently great for accurate measurement (not usually the case). Otherwise, the shear must be estimated by estimating a "friction factor" for the steam flow. With present knowledge, such an "effective" friction factor is not subject to precise estimation, since little data exist for shear between gas (or vapor) and liquid flows.

If the rate of mass flow into the film is known (e.g., from condensation of the steam), the film thickness can be computed from the considerations of Couette flow [125]. Its measurement is also possible (as discussed later).

The preceding discussion assumes a steady laminar-flow condition in the film. However, much recent photographic and other evidence indicated the existence of a complex wave pattern [125] on the surface of such a film. Waves can run in either axial or transverse direction, or in some combination thereof [125]. If the waves achieve sufficient amplitude, film breakdown will occur. In some cases this appears to be in the form of axial rivulets with intervening dry areas. In other cases, irregular dry spots may occur [126, 127].

Several theoretical analyses to predict film breakdown have been made over recent years [128–136], though no good agreement with experimental observations appears yet to have been achieved. These questions are treated in greater detail in Chap. 7.

There is an obvious analogy between the thin-film flow problem as it exists in turbines and that of shallow-sea wind-driven surface waves. This latter problem has been treated in some detail [137] so that it may be possible to apply any analytical procedures, which may have been developed for this quite different application, to the thin-film problem existing in large steam turbines or to problems of nuclear reactor high-void fraction dry-out, which has also been treated in considerable detail [14, 15, 138–141]. Of course, this latter situation differs from the others mentioned since heat-transfer effects are important, so that it cannot be considered as even approximately adiabatic.

Another aspect of the turbine problem, upon which considerable research has been done, is that of the kinetics of the droplet motion after they have been shed from the stator-blade trailing edge, and their subsequent impact at high relative velocity upon the downstream rotor blades [77, 84, 93] and the consequent erosion of the rotor blades.

Still another aspect of this overall problem is that of the mass transfer between the main wet-steam flow and the liquid film upon the stator blading. This process involves both the deposit of some of the primary droplets in the steam and also other forms of steam condensation. Since the primary droplets in the steam flow are in the size range 0.01 to 0.1 μm [142–144], their measurement and count is extremely difficult.

Suffice it to say at this point that no easily practicable method for measuring the number and size distribution of the primary droplets has yet been devised, but it is a subject of active research in several laboratories throughout the world, primarily, at the time of writing, in Russia and Europe, both East and West.

Another phenomenon under continuing and active research, primarily in the same laboratories, is that of the deposit of the liquid film from the primary droplets onto the stator blades. The primary objective is to determine how much of the water condensed in the main steam flow is actually deposited upon the stator blades, since this stator film gives rise to the much larger secondary droplets in the stator wake which create the erosion problem. In principle, this can involve the effects of centrifugal force, since stator passages are curved and the flow usually has a large circumferential velocity component at this point, turbulent diffusion, and brownian motion at the least.

The measurement of secondary droplet population and size spectra as they are shed from the stator film is a considerably easier problem than that of the primary droplets, since they are much larger, being in the size range 10 to 1000 μm. This problem has been attacked primarily photographically [127, 145], but other methods such as light scattering [146–149] are possible due to the larger droplet size.

While most of the work to date has been concentrated upon the low-pressure stages of large turbines, there is also an erosion problem in the high-pressure stages which is now receiving attention [142, 143, 150, 151]. This results from the lack of adequate superheat and/or reheat in most of the nuclear power plants.

Film-thickness measurement can be attempted in several ways, but the most practical at present seems to be through electrical conductivity gages [125, 145, 152].

Choked flows Supersonic wet-steam flows such as exist in steam turbines, for example (and are thus involved in the droplet-impact erosion and thin liquid-film flow phenomena discussed earlier), are essentially "choked" flows in the sense of gas dynamics and can thus give rise to shock waves, i.e., sudden changes of pressure in the flow direction, due to vapor condensation, i.e., "condensation" shocks. In the steam-turbine case, the location of such a "shock" has been traditionally called the "Wilson line" [153]. In general, the axial pressure gradients are not as severe as is usual for supersonic gas flows passing through a conventional normal shock. It can be either positive or negative [151] depending upon the upstream conditions. These flows have in recent years been studied in some detail [154–158].

The condensation-shock and choked-flow behavior of supersonic wet-vapor flows is somewhat analogous to the "condensation" shock in cavitation flows, such as in a cavitating venturi [5] where the local velocity of the two-phase medium is sonic. This somewhat surprising situation is possible in many real situations, since the sonic velocity in vapor-liquid (or gas-liquid) mixtures can be extremely low because they combine high compressibility with high density as opposed to single-phase fluids, either liquid or gaseous.

Choked flows of either the wet-steam or cavitation type, i.e., either high or low "void fraction," are also of importance in the analysis of the sudden venting, as in pipe rupture, of a pressurized vessel containing a saturated vapor-liquid mixture. Applications of this type are of present substantial interest in the safety analyses of liquid-cooled nuclear reactor systems.

2-2-5 Primary Sources of Information on Droplet and Cavitation Erosion

Numerous key references have been made in this chapter, although the list is by no means exhaustive. However, there are certain key sources of information, which will be briefly reviewed below.

American Society of Mechanical Engineers (New York, N.Y.) The Fluids Engineering Division of ASME holds two meetings each year (normally November and May) at which technical papers on impact and cavitation, among many other subjects, are presented. Those deemed worthy of permanent record are later published in *Transactions of ASME, Journal of Basic Engineering.*† In the spring meeting of the Fluids Engineering Division of ASME a "Cavitation Forum" (recently "Cavitation and Polyphase Flow Forum")‡ is held, and a booklet composed of the various short papers and discussions from the previous year is printed. These are all available from ASME. The series commenced in 1961 and will probably continue into the indefinite future.

American Society for Testing and Materials (Philadelphia, Pennsylvania) This society has so far sponsored six symposia in the field of cavitation and impingement erosion (1961, 1966, 1969, 1973, 1975, 1977) and will very likely continue this series probably every few years. The proceedings in each case are published in book form after about one year. Previous such books are Refs. 85 to 88.

---

† Now *J. Fluids Engr.*
‡ Presently "Cavitation and Multiphase Flow Forum."

International Congress on Rain Erosion and Associated Phenomena Four such Congresses have so far been held, sponsored jointly by the British Royal Air Force and the Air Force of the German Federal Republic, on a three- or four-year schedule. It is possible that this series will be continued. These have been held either at Meersburg, GFR, or the Royal Aircraft Establishment, Farnborough, England. In all cases the proceedings have been published by the Royal Aircraft Establishment, Farnborough. References 80 and 98 to 100, for example, are from that source, but numerous other interesting and pertinent articles exist which have not been cited here.

Royal Society (London), Symposium on Impact Phenomena (1965) A symposium was held by the Royal Society (London) in 1965 on the subject of droplet-impact phenomena, and the papers later published in a special volume of the *Transactions of the Royal Society* (e.g., [117]). Numerous papers of classical interest on the subject exist in this volume.

Other sources Numerous other technical societies and journals have published many papers and articles of interest. These include (but not exclusively) International Association for Hydraulic Research (IAHR), American Society for Civil Engineers (ASCE), American Society for Chemical Engineers (ASChE), *La Houille Blanche*, British Institute for Mechanical Engineers (I.Mech.E)—to name some of the most important.

REFERENCES

1. Hammitt, F. G.: Effects of Gas Content upon Cavitation Inception, Performance, and Damage, *J. Hyd. Research (IAHR)*, vol. 10, no. 3, pp. 259–290, 1972.
2. Knapp, R. T., J. W. Daily, and F. G. Hammitt: "Cavitation," McGraw-Hill, New York, 1970.
3. Holl, J. W.: Nuclei and Cavitation, paper no. 70-FE-23, *Proc. Symp. on Role of Nucleation in Boiling and Cavitation*, May, 1970, ASME.
4. Rohsenow, W. M.: Nucleation with Boiling Heat Transfer, *Proc. Symp. on Role of Nucleation in Boiling and Cavitation*, May, 1970, ASME.
5. Hammitt, F. G., M. J. Robinson, and J. F. Lafferty: Choked-Flow Analogy for Very Low Quality Two-Phase Flows, *Nucl. Sci. and Engr.*, vol. 29, no. 1, pp. 131–142, 1967.
6. Jakobsen, J. K.: On the Mechanism of Head Breakdown in Cavitating Inducers, *Trans. ASME*, ser. D, vol. 86, no. 2, p. 291, 1964.
7. Spraker, W. A.: Two-Phase Compressibility Effects on Pump Cavitation, *Cavitation in Fluid Machinery*, ASME, pp. 162–171, 1965.
8. Wood, G. M., R. S. Kulp, and J. V. Altieri: Cavitation Damage Investigation in Mixed-Flow Liquid Metal Pumps, *Cavitation in Fluid Machinery*, ASME, pp. 196–214, 1965.
9. Johnson, V. E., and T. Hsieh: The Influence of Entrained Gas Nuclei Trajectories on Cavitation Inception, paper 7, *Proc. Sixth Naval Hydrodynamics Symp.*, 1966, Washington, D.C.
10. Yeh, H. C. and W. J. Yang: Dynamics of Bubbles Moving in Liquids with Pressure Gradient, *J. Appl. Physics*, vol. 19, no. 7, pp. 3156–3165, June, 1968.
11. Chincholle, L.: L'Effet Fusée et l'Erosion Mechanique de Cavitation, *Bull. Tech. de la Suisse Romande*, vol. 94, no. 19, pp. 269–279, 1968.

12. Gallant, H.: Research on Cavitation Bubbles, trans. no. 1, *Oesterreichische Ingenieur Zeitschrift*, no. 3, pp. 74–83, 1962; see also trans. no. 1190, Electricité de France, Chatou, France.
13. Baker, O.: Simultaneous Flow of Oil and Gas, *Oil Gas J.*, vol. 53, pp. 185–190, 1954.
14. Tong, L. S.: "Boiling Heat Transfer and Two-Phase Flow," John Wiley, New York, 1965.
15. Wallis, G. B.: "One-Dimensional Two-Phase Flow," McGraw-Hill, New York, 1969.
16. Wu, T. Y.: Cavity Flow Analysis—A Review of the State of Knowledge, *Cavitation State of Knowledge, ASME*, pp. 106–137, 1969.
17. Wu, T. Y.: Inviscid Cavity and Wake Flows, article in M. Holt (ed.), "Basic Developments in Fluid Dynamics," Academic Press, New York, 1968.
18. Knapp, R. T.: Recent Investigations of Cavitation and Cavitation Damage, *Trans. ASME*, vol. 77, pp. 1045–1054, 1955.
19. Ruggeri, R. S., and R. D. Moore: "Method for Prediction of Pump Cavitation Performance for Various Liquids, Liquid Temperatures, and Rotative Speeds," NASA TN D-5292, 1969.
20. Hammitt, F. G.: Collapsing Bubble Damage to Solids, *Cavitation State of Knowledge, ASME*, pp. 87–102, 1969.
21. Lord Rayleigh (John William Strutt): On the Pressure Developed in a Liquid During the Collapse of a Spherical Cavity, *Phil. Mag.*, vol. 34, pp. 94–98, August, 1917.
22. Besant, W. H.: "Hydrostatics and Hydrodynamics," art. 158, Cambridge University Press, London, 1859.
23. Jarman, P. D.: Sonoluminescence: A Discussion, *J. Acoust. Soc. Amer.*, vol. 11, no. 11, pp. 1459–1462, November, 1960.
24. Hickling, R.: Effects of Thermal Conduction in Sonoluminescence, *J. Acoust. Soc. Amer.*, vol. 35, no. 7, pp. 967–974, 1974.
25. Hickling, R.: Some Physical Effects of Cavity Collapse in Liquids, *Trans. ASME, J. Basic Engr.*, vol. 88, ser. D, pp. 229–235, 1966.
26. Ivany, R. D.: "Collapse of a Cavitation Bubble in Viscous, Compressible Liquid—Numerical and Experimental Analyses," Ph.D. thesis, Nuclear Engr. Dept., University of Michigan, Ann Arbor, Mich., 1965.
27. Ivany, R. D., and F. G. Hammitt: Cavitation Bubble Collapse in Viscous, Compressible Liquids—Numerical Analysis, *Trans. ASME, J. Basic Engr.*, vol. 87, ser. D, pp. 977–985, 1965.
28. Ivany, R. D., F. G. Hammitt, and T. M. Mitchell: Cavitation Bubble Collapse Observations in a Venturi, *Trans. ASME, J. Basic Engr.*, vol. 88, no. 3, pp. 649–657, 1966.
29. Canavelis, R.: "Contribution à l'Étude de l'Érosion de Cavitation dans les Turbomachines Hydrauliques," Ph.D. thèse, Faculté des Sciences, Université de Paris, 1966.
30. Shima, A., and K. Nakaj: The Collapse of a Non-Hemispherical Bubble Attached to a Solid Wall, *J. Fluid Mech.*, vol. 80, pt. 2, pp. 369–391, 1977.
31. Poritsky, H.: The Collapse or Growth of a Spherical Bubble or Cavity in a Viscous Fluid, *Proc. First Nat'l Congress Appl. Mech.*, 1952, pp. 823–825, ASME.
32. Gilmore, F. R.: "The Growth or Collapse of a Spherical Bubble in a Viscous Compressible Liquid," Calif. Inst. Tech. Rept. 26-4, Pasadena, Calif., 1952; see also *Proc. 1952 Heat Transfer and Fluid Mechanics Inst.*
33. Kirkwood, J. G., and H. A. Bethe: "The Pressure Wave Produced by an Underwater Explosion, I," OSRD no. 588, 1942.
34. Hickling, R., and M. S. Plesset: Collapse and Rebound of a Spherical Cavity in Water, *Physics of Fluids*, vol. 7, no. 1, pp. 7–14, 1964.
35. Hickling, R.: "I. Acoustic Radiation and Reflection from Spheres. II. Some Effects of Thermal Conduction and Compressibility in the Collapse of a Spherical Bubble in a Liquid," Ph.D. thesis, Calif. Inst. Tech. Div. of Engr., 1962.
36. Stahl, H. A., and A. J. Stepanoff: Thermodynamic Aspects of Cavitation in Centrifugal Pumps, *Trans. ASME*, vol. 78, pp. 1691–1693, 1956.
37. Jacobs, R. B., K. B. Martin, G. J. Van Wylen, and B. W. Birmingham, "Pumping Cryogenic Liquids," Nat'l Bur. Std. (U.S.) Rept. 3569, Boulder Labs. Tech. Memo 36, 1956.
38. Mitchell, T. M.: "Numerical Studies of Asymmetrical and Thermodynamic Effects on Cavitation Bubble Collapse," Ph.D. thesis, Nuclear Engr. Dept., University of Michigan, 1970; see also

T. M. Mitchell and F. G. Hammitt: Asymmetrical Cavitation Bubble Collapse, *Trans. ASME, J. Fluids Engr.*, vol. 96, no. 1, pp. 29–37, March, 1973.

39. Gavranek, V. V., D. N. Bol'shutkin, and V. I. Zel'dovich: Thermal and Mechanical Action of a Cavitation Zone on the Surface of a Metal, *Fiz. Metal. i Metalloved*, vol. 10, no. 2, pp. 262–268, 1960.
40. Florchuetz, L. W., and B. T. Chao: On the Mechanics of Vapor Bubble Collapse—A Theoretical and Experimental Investigation, *Trans. ASME, J. Heat Transfer*, vol. 87, ser. C, pp. 209–220, 1965.
41. Bonnin, J.: Début de Cavitation dans des Liquides Differents, *Bull. de Direction des Etudes et Recherches, Nucleaire, Hydraulique, Thermique*, ser. A, no. 4, pp. 53–72, 1970, Elec. de France; see also Incipient Cavitation in Liquids Other than Water, *1971 ASME Cavitation Forum*, pp. 14–16.
42. Devine, R., and M. S. Plesset: "Temperature Effects in Cavitation Damage," Calif. Inst. Tech. Div. of Engr. and Appl. Sci. Rept. 85-28, 1964.
43. Garcia, R., and F. G. Hammitt: Cavitation Damage and Correlations with Material and Fluid Properties, *Trans. ASME, J. Basic Engr.*, vol. 89, ser. D, pp. 753–763, 1967.
44. Hobbs, J. M. and A. Laird: Pressure, Temperature and Gas Content Effects in the Vibratory Cavitation Erosion Test, *1969 ASME Cavitation Flow Forum*, pp. 3–4; see also J. M. Hobbs, A. Laird, and W. C. Brunton: "Laboratory Evaluation of Cavitation Erosion Test," NEL Rept. 271, E. Kilbride, Scotland, 1967.
45. Hammitt, F. G., and D. O. Rogers: Effects of Pressure and Temperature Variation in Vibratory Cavitation Damage Test, *J. Mech. Engr. Sci.*, vol. 12, no. 6, pp. 432–439, 1970.
46. Kling, C. L.: "A High-Speed Photographic Study of Cavitation Bubble Collapse," Ph.D. thesis, Nuclear Engr. Dept., University of Michigan, 1970.
47. Kling, C. L., F. G. Hammitt, T. M. Mitchell, and E. E. Timm: Bubble Collapse near Wall in Flowing System, *1970 ASME Cavitation Forum*, pp. 41–43; also available as a motion picture film of the same title in ASME Film Library, ASME, New York.
48. Kling, C. L., and F. G. Hammitt: A Photographic Study of Spark-Induced Cavitation Bubble Collapse, *Trans. ASME, J. Basic Engr.*, vol. 94, ser. D, no. 4, pp. 825–833, 1972.
49. Gibson, D. C.: The Kinetic and Thermal Expansion of Vapor Bubbles, *Trans. ASME, J. Basic Engr.*, vol. 93, ser. D, no. 1, pp. 89–96, March, 1972.
50. Mitchell, T. M., and F. G. Hammitt: Collapse of a Spherical Bubble in a Pressure Gradient, *1970 ASME Cavitation Forum*, pp. 44–46, 1970.
51. Plesset, M. S., and R. B. Chapman: Collapse of a Vapor Cavity in the Neighborhood of a Solid Wall, *J. Fluid Mech.*, vol. 2, no. 47, p. 238, May, 1971.
52. Chapman, R. B.: "Non-Spherical Vapor Bubble Collapse," Ph.D. thesis, Calif. Inst. Tech., Pasadena, Calif., 1970.
53. Brunton, J. H.: The Deformation of Solids by Cavitation and Drop Impingement, *IUTAM Symp. on Un-Steady Flow of Water at High Speeds*, 1971, pp. 22–26, Leningrad, USSR.
54. Heymann, F. J.: On the Shock Wave Velocity and Impact Pressure in High-Speed Liquid-Solid Impact, *Trans. ASME, J. Basic Engr.*, vol. 90, ser. D, pp. 400–402, 1968.
55. Heymann, F. J.: High-Speed Impact between a Liquid Drop and a Solid Surface, *J. Appl. Phys.*, vol. 40, pp. 5113–5122, December, 1969.
56. Huang, Y. C.: "Numerical Studies of Unsteady Two-Dimensional Liquid Impact Phenomena," Ph.D. thesis, Mech. Engr. Dept., University of Michigan, 1971.
57. Petracchi, G.: Investigation of Cavitation Corrosion (in Italian), *Metallurgica Italiana*, vol. 41, pp. 1–6, 1944; English Summary, *Engr's Digest*, vol. 10, pp. 314–316, 1949.
58. Plesset, M. S.: Pulsing Technique for Studying Cavitation Erosion of Metals, *Corrosion*, vol. 18, no. 5, pp. 181–188, 1962.
59. Plesset, M. S.: The Pulsation Method for Generating Cavitation Damage, *Trans. ASME, J. Basic Engr.*, vol. 85, ser. D, pp. 360–364, 1963.
60. Shima, A.: The Natural Frequencies of Two-Spherical Bubbles Oscillating in Water, *Trans. ASME, J. Basic Engr.*, vol. 93, pp. 426–432, 1971.
61. Thiruvengadam, A., et al.: Experimental and Analytical Investigations on Liquid Impact, Erosion, *ASTM STP 474*, pp. 239–287, 1970.

62. Thiruvengadam, A.: A Unified Theory of Cavitation Damage, *Trans. ASME, J. Basic Engr.*, vol. 85, pp. 365–377, September, 1963.
63. Hammitt, F. G., et al.: A Statistically Verified Model for Correlating Volume Loss Due to Cavitation or Liquid Impingement, *ASTM STP* 474, pp. 288–311, 1970.
64. Heymann, F. J.: Discussion of Ref. 3, *ASTM STP* 474, pp. 312–322, 1970; and Toward Quantitative Prediction Liquid Impact Erosion, *ASTM STP* 474, pp. 212–248, 1970.
65. Hobbs, J. M.: Experience with a 20 kc Cavitation Erosion Test, *ASTM STP* 408, pp. 159–185, 1967.
66. Kemppainen, D. J., and F. G. Hammitt: Some Effects of Applied Stress on Early Stages of Cavitation Damage, *1970 Symp. on Fluid Mech. and Design of Turbomachinery*, September, 1970, Pennsylvania State University.
67. Young, S. G., and J. R. Johnston: "Effect of Cover Gas Pressure on Accelerated Cavitation Damage," NASA Rept. TMX-52414, 1968; also *Proc. ASTM*, 1969; see also "Effect of Cover Gas Pressures on Accelerated Damage Sodium," NASA Rept. TM D-4235, 1967.
68. Hobbs, J. M. and A. Laird: Pressure, Temperature and Gas Content Effect in the Vibratory Cavitation Erosion Test, *1969 ASME Cavitation Forum*, June, 1969, pp. 3–4.
69. Cooley, W. C., F. L. Beck, and D. L. Jaffe: "Design of a Water Cannon for Rock Tunneling Experiments," Terraspace Rept. FRA-RT-71-70, Bethesda, Md., February, 1971, prepared for Office of High Speed Ground Transportation.
70. Honnegger, E.: Tests on Erosion Caused by Jets, *Brown Boveri Review*, vol. 14, no. 4, pp. 95–104, 1927.
71. Canavelis, R., and G. Domergue: "Comparison of the Resistance of Different Materials with a Jet Impact Test Rig," Direction des Etudes et Recherches Rept. HC/061-230-9, Electricité de France, November, 1967.
72. Canavelis, R.: Jet Impact and Cavitation Damage, *Trans. ASME, J. Basic Engr.*, vol. 90, ser. D, pp. 355–367, 1968.
73. Ripken, J. F.: A Test Rig for Studying Impingement and Cavitation Damage, *ASTM STP* 408, pp. 1–21, 1957.
74. Hobbs, J. M.: Discussion of Ref. 73, 1967.
75. Bogachev, I. N. and R. I. Mints: Cavitation Erosion and Means for its Prevention (English trans. A. Wald), *Izdatel'stvo Mashinostroenie*, Moscow, 1964.
76. Filippov, G. I.: Personal communication, Moscow Engr. Inst., 1975.
77. Krzyzanowski, J.: "On Erosion Rate Droplet Size Relation for an Experiment with a Rotating Sample," University of Michigan ORA Rept. UMICH 03371-17-T.
78. Koutny, A.: Cavitation Erosion Resistance of Materials in Laboratory and Operating Conditions, *Proc. Fifth Conf. on Fluid Mech.*, 1975, pp. 491–512, Hungary Acad. of Science, Budapest.
79. Huang, Y. C., and F. G. Hammitt: Cavitation Within an Impinging Liquid Droplet, *ASME 1972 Cavitation Forum*, pp. 9–10.
80. Brunton, J. H., and J. J. Camus: The How of a Liquid Drop During Impact, *Proc. Third Int'l Congr. on Rain Erosion*, August, 1970, Royal Aircraft Est., Farnborough, England.
81. Brunton, J. H.: The Deformation of Solids by Cavitation and Drop Impingement, *IUTAM Symp. on Non-Steady Flow of Water at High Speeds*, June, 1971, pp. 24–25, Leningrad, USSR.
82. Engel, O. G.: Water Drop Collision with Solid Surface, *J. Res. Nat'l Bur. Stds.*, vol. 54, no. 5, p. 281, May, 1955.
83. Thomson, R. T. and R. J. Haydruk: "An Improved Analytical Treatment of the Denting to Thin Sheets by Hail," NASA Rept. TN D-6102, January, 1971.
84. Krzyzanowski, J.: Semi-Emperical Criteria of Erosion Threat in Modern Steam Turbine, *Trans. ASME, J. Eng. for Power*, vol. 93, pp. 1–12, January, 1971.
85. "Erosion by Cavitation or Impingement," *ASTM STP* 408, 1967, ASTM, Philadelphia, Pa.
86. "Characterization and Determination of Erosion Resistance," *ASTM STP* 474, 1970, ASTM, Philadelphia, Pa.
87. "Erosion, Wear, and Interfaces with Corrosion," *ASTM STP* 567, 1973, ASTM, Philadelphia, Pa.
88. "Selection and Use of Wear Tests for Metals," *ASTM STP* 615, November, 1975, New Orleans.

89. Wahl, N.: "Supersonic Rain and Sand Erosion Research, Part I," Bell Supersonic Aircraft Co. Tech. Rept. AFML TR-69-287, 1970.

90. Schmitt Jr., G. F., W. G. Reinecke, and G. D. Waldman, Influence of Velocity, Impingement Angle, Heating, and Aerodynamic Shock Layers on Erosion of Materials at Velocities of 5500 ft/sec (1700 m/s), *ASTM STP* 567, pp. 219–238, 1974.

91. Fyall, A. A.: Practical Aspects of Rain Erosion of Aircraft and Missiles, *Phil. Trans. Roy. Soc. (London)*, ser. A, vol. 260, pp. 131–167, 1966.

92. Fyall, A. A.: Rain Erosion Single Impact Studies, *IUTAM Symp. on High Speed Flows*, June, 1971, pp. 38–39, Leningrad, U.S.S.R.

93. Smith, A., R. P. Kent, and R. L. Armstrong: Erosion of Steam Turbine Blade Shield Materials, in G. B. Wallis (ed.), "One-Dimensional Two-Phase Flow," pp. 125–158, McGraw-Hill, New York, 1969.

94. Bowden, F. P., and J. H. Brunton: The Deformation of Solids by Liquid Impact at Supersonic Speeds, *Proc. Roy. Soc. (London)*, ser. A, vol. 264, pp. 433–450, 1961.

95. Hoff, G., G. Langbein, and H. Reiger: Material Destruction Due to Liquid Impact, *ASTM STP* 408, pp. 42–69, 1966.

96. Kenyon, H. F.: "Erosion by Water Jet Impacts, Parts I and II," Associated Electrical Industries Ltd. Rept. T.P. IR 5587, England, January, 1967.

97. Timm, E. E., and F. G. Hammitt: "A Repeating Water Gun Device for Studying Erosion by Water Jet Impacts," ORA Rept. 02643-1-PR, University of Michigan, April, 1969.

98. Hammitt, F. G., et al.: Laboratory Scale Devices for Rain Erosion Simulation, *Proc. Second Meersburg Conf. on Rain Erosion and Associated Phenomena*, August, 1967, Meersburg, FGR; available from Royal Aircraft Est., Farnborough, England, pp. 87–124 (edited by A. A. Fyall).

99. Hammitt, F. G.: Impact and Cavitation Erosion and Material Mechanical Properties, in A. A. Fyall (ed.), *Proc. Third Int'l Cong. on Rain Erosion and Associated Phenomena*, August, 1970, Royal Aircraft Est., Farnborough, England.

100. Fasso, G.: Paper in *Proc. Second Meersburg Conf. on Rain Erosion and Associated Phenomena*, August, 1967, Meersburg, FGR.

101. Huang, Y. C.: "Numerical Studies of Unsteady, Two-Dimensional Liquid Impact Phenomena," Ph.D. thesis, Mech. Engr. Dept., University of Michigan, July, 1971; obtainable from University Microfilms, Ann Arbor, Mich.

102. Huang, Y. C., F. G. Hammitt, and W. J. Yang, "Hydrodynamic Phenomena During High-Speed Collision Between Liquid Droplet and Rigid Plane," *Trans. ASME, J. Fluids Engr.*, vol. 95, no. 12, pp. 276–294, 1973.

103. Hwang, J. B.: "The Impact Between a Liquid Drop and an Elastic Half-Space," Ph.D. thesis, Mech. Engr. Dept., University of Michigan, 1975.

104. Elliot, D. E., J. B. Marriott, and A. Smith: Comparison of Erosion Resistance of Standard Steam Turbine Blade and Shield Materials on Four Test Rigs, *ASTM STP* 474, pp. 127–161, 1969.

105. Timm, E. E.: "An Experimental Photographic Investigation of Vapor Bubble Collapse and Liquid Jet Impingement," Ph.D. thesis, Chem. Engr. Dept., University of Michigan, 1974.

106. Timm, E. E., and F. G. Hammitt: Bubble Collapse Adjacent to a Rigid Wall, A Flexible Wall, and a Second Bubble, *1971 ASME Cavitation Forum*, pp. 18–20.

107. Gibson, D. C.: "The Collapse of Vapour Cavities," Ph.D. thesis, Cambridge University, Cambridge, England, 1967.

108. Gibson, D. C.: Cavitation Adjacent to Plane Boundaries, *Proc. Conf. on Hyd. and Fluid Mech.*, Inst. Engrs., *Australia*, 1968, pp. 210–214.

109. Heymann, F. J.: On the Shock Wave Velocity and Impact Pressure in High-Speed Liquid-Solid Impact, *Trans. ASME, J. Basic Engr.*, vol. 90, p. 400, July, 1968.

110. Heymann, F. J.: High-Speed Impact Between a Liquid Drop and a Solid Surface, *J. Appl. Physics*, vol. 40, pp. 5113–5122, December, 1969.

111. Hwang, J. B., and F. G. Hammitt: High Speed Impact Between Curved Liquid Surface and Rigid Flat Surface, *Trans. ASME, J. Fluids Engr.*, vol. 99, ser I, no. 2, pp. 396–404, June, 1977; see also ORA Rept. UMICH 012449-10-T, University of Michigan, October, 1975.

112. Hwang, J. B., and F. G. Hammitt: Transient Distribution of the Stress During the Impact Between a Liquid Drop and an Aluminum Body, *Proc. BHRA Third Int'l. Symp. on Jet Cutting Tech.*, May 11–13, 1976, Chicago; see also ORA Rept. UMICH 012449-8-T, University of Michigan, 1976.
113. Fyall, A. A.: Single Impact Studies of Rain Erosion, *Shell Aviation News*, p. 374, 1969.
114. Brunton, J. H., and J. J. Camus: The How of a Liquid Drop During Impact, in A. A. Fyall (ed.), *Proc. Third Int'l Cong. on Rain Erosion*, August, 1970, Royal Aircraft Est., Farnborough, England.
115. Brunton, J. H.: The Deformation of Solids by Cavitation and Drop Impingement, in A. A. Fyall (ed.), *Proc. Third Int'l Cong. on Rain Erosion*, August, 1970, p. 24, Royal Aircraft Est., Farnborough, England.
116. Field, J. E.: High Velocity Liquid Impact and Cavitation Studies, in A. A. Fyall (ed.), *Proc. Third Int'l Cong. on Rain Erosion*, August, 1970, pp. 15–16, Royal Aircraft Est., Farnborough, England.
117. Brunton, J. H.: Physics of Impact and Deformation; Single Impact, *Phil. Trans. Roy. Soc. (London)*, ser. A, vol. 260, pp. 79–85, 1966.
118. Reinhard, K. G.: "Turbine Damage by Solid Particle Erosion" (preprint), ASME paper no. 76-JPGC-Pwr-15.
119. Sheldon, G. L., J. Maji, and C. T. Crowe: Erosion of a Tube by Gas-Particle Flow, ASME paper no. 76-WA/Mat-9, *Trans. ASME, J. Engr. Mat'l and Tech.*, pp. 138–142, 1977.
120. McMillan, L. D., and D. A. Eitman: "Heat Shield Materials Erosion at Hypersonic Velocities—vol. III, Particle-Type Experimental Results," AFML-TR-76-68, vol. III, WPAFB, Ohio, November, 1976.
121. Adler, W. F.: Personal Communication, May, 1977.
122. Huber, M. T.: Contribution to Theory of Contact of Elastic Solids, *Ann. Phys. (Leipzig)*, vol. 14, p. 153, 1904.
123. Love, A. E. H.: The Stress in a Semi-Infinite Solid by Pressure on Part of the Boundary, *Trans. Roy. Soc. (London)*, ser. A, vol. 228, p. 377, 1929.
124. Tait, P. G.: Report on Some of the Physical Properties of Fresh Water and Sea Water, Report on Scientific Results of Voyage, H.M.S. "Challenger," *Phys. Chem.*, vol. 2, pp. 1–71, 1888.
125. Hammitt, F. G., J. B. Hwang, A. Mancuso, D. Krause, and S. Blome: Liquid Film Thickness Tests—Wet Steam Tunnel, *Proc. Inst. Mech. Engr., Durham Conf. on Steam Turbines*, April, 1976, Durham, England.
126. Hammitt, F. G., and W. Kim: Liquid Film Thickness Measurements in University of Michigan Wet Steam Tunnel, *1976 ASME Cavitation and Multiphase Flow Forum*, pp. 24–27.
127. Hammitt, F. G., W. Kim, and S. Krzeczkowski: Investigation of Behavior of Thin Liquid Film with Co-Current Steam Flow, *Proc. Two-Phase Flow and Heat Transfer Symp. Workshop*, Oct. 18–20, 1976, CERI, University of Miami.
128. Mikielewicz, J., and J. R. Moszynski: Breakdown of a Shear Driven Liquid Film, *Trans. IFFM, Polish Acad. Sci.*, vol. 66, pp. 3–9, 1975, Gdansk, Poland.
129. Mikielewicz, J., and J. R. Moszynski: Minimum Thickness of a Liquid Film Flowing Down a Solid Surface, *Int. J. Heat and Mass Transfer*, vol. 19, pp. 771–776, 1975.
130. Bankoff, S. G.: Stability of Liquid Flow Down Heated Inclined Plate, *Int. J. Heat and Mass Transfer*, vol. 14, p. 377, 1971.
131. Krantz, W. B. and S. L. Goren: Stability of Thin Liquid Films Flowing Down a Plane, *Ind. Eng. Chem. Fundam.*, vol. 10, no. 1, p. 91, 1971.
132. Hallett, V. A.: Surface Phenomena Causing Breakdown of Liquid Films During Heat Transfer, *Int. J. Heat and Mass Transfer*, vol. 9, p. 283, 1966.
133. Hartley, D. E., and W. Murgatroyd: Criteria for Break-Up of Thin Liquid Layers Flowing Isothermally over a Solid Surface, *Int. J. Heat and Mass Transfer*, vol. 7, p. 1003, 1964.
134. Anshus, B.: On the Asymptotic Solution to the Falling Film Stability Problem, *Ind. Engr. Chem. Fundam.*, vol. 11, no. 4, 502, 1972.
135. Nayfeh, A. H., and W. S. Saric: Stability of a Liquid Film, *AIAA Journal*, vol. 9, p. 750, 1971.
136. Hewitt, G. F., and P. M. C. Lacey: The Breakdown of the Liquid Film in Annular Two-Phase Flow, *Int. J. Heat and Mass Transfer*, vol. 8, p. 781, 1965.

137. Plate, E. J., and G. M. Hidy: Laboratory Study of Air Flowing over a Smooth Surface onto Small Water Waves, *J. Geophys. Res.*, vol. 72, pp. 4627–4641, September, 1967.

138. Norman, W. S., and V. McIntyre: Heat Transfer to a Liquid Film on a Vertical Surface, *Trans. Inst. Chem. Engr.*, vol. 38, pp. 301–307, 1960.

139. Murgatroyd, W.: The Role of Shear and Form Forces in the Stability of a Dry Patch in Two-Phase Film Flow, *Int. J. Heat and Mass Transfer*, vol. 8, p. 297, 1965.

140. Zuber, N., and E. W. Staub: Stability of Dry Patches Forming in Liquid Films Flowing over Heated Surfaces, *Int. J. Heat and Mass Transfer*, vol. 9, p. 897, 1966.

141. Orell, A., and S. G. Bankoff: Formation of a Dry Spot in a Horizontal Liquid Film Heated from Below, *Int. J. Heat and Mass Transfer*, vol. 14, p. 1835, 1971.

142. Puzyrewski, R., and T. Krol: Numerical Analysis of Hertz–Knudsen Model of Condensation Upon Small Droplets in Water Vapor, *IFFM, Polish Acad. Sci.*, 1970, Gdansk, Poland.

143. Gyarmathy, G., H. P. Burknard, F. Lesch, and A. Siegenthaler: Spontaneous Condensation of Steam at High Pressures: First Experimental Results, *Inst. Mech. Engrs. Conf., Heat and Fluid Flow in Steam and Gas Turbines*, April, 1973, paper C 66/73, p. 182, University of Warwick.

144. Gyarmathy, G., and H. Meyer: Spontaneous Condensation—Parts I and II, *Forsch Ver. dt. Ing.*, CEGB trans. no. 4160, p. 508, 1965.

145. Kim, W.: "Study of Liquid Films, Fingers and Droplet Motion for Steam Turbine Blading Erosion Problem", Ph.D. Thesis, Mech. Engr. Dept., University of Michigan, Ann Arbor, Michigan, 1978 (also available as DRDA Rept. UMICH 014571-4-T, Univ. Mich., Ann Arbor, Mich.); see also W. Kim, F. G. Hammitt, Investigation of Liquid Finger Behaviour at Trailing Edge of Simulated Steam Turbine Stator Blade, *Proc. 2nd Multiphase Flow and Heat Transfer Symposium-Workshop*, CERI, Univ.-Miami, Miami Beach, Florida, 16–18 April, 1979; also available as DRDA Rept. UMICH 014571-5-T, Univ. Mich., Ann Arbor, Mich.

146. Yilmaz, E., F. G. Hammitt, and A. Keller: Cavitation Inception Thresholds in Water and Nuclei Spectra by Light Scattering Technique, *J. Acoust. Soc. Amer.*, pp. 329–338, February, 1976; see also ORA Rept. UMICH 01357-38-T, University of Michigan, November, 1974.

147. Keller, A.: Influence of Cavitation Nucleus Spectrum on Cavitation Inception, Investigated with a Light Counting Method, *Trans. ASME, J. Basic Engr.*, vol. 94, ser. D, no. 4, pp. 917–925, December, 1972.

148. Puzyrewski, R., and S. Krzeczkowski: Some Experiments on Droplet Generation and Its Motion in the Aerodynamic Wake (in Polish), *Trans. IFFM, Polish Acad. Sci.*, no. 29–31, 1966, Gdansk, Poland.

149. Walters, P. T.: Optical Measurement of Water Droplets in Wet Steam Flows, *Inst. Mech. Engrs.*, Conf. pub. 3, pp. 66–74, 1973.

150. Ryley, D. J., and K. A. Tubman: Spontaneous Condensation in High-Pressure Expanding Steam, *Proc. Conf. on Large Steam Turbines, IFFM, Polish Acad. Sci.*, 1974, Gdansk, Poland.

151. Puzyrewski, R.: High and Low Pressure Effect on Gas Dynamics of Condensing Vapors, *Polish Acad. Sci.*, vol. XXI, no. 10, pp. 83–865 to 88–865, 1973.

152. Puzyrewski, R.: On Measuring the Thickness of Thin Water Films (in Polish), *Trans. IFFM, Polish Acad. Sci.*, no. 29–31, pp. 343–347, 1966, Gdansk, Poland.

153. Wilson, C. T. R.: *Phil. Trans. (London)*, ser. A, vol. 189, pp. 265–307, 1897.

154. Deich, M. E., G. V. Tsiklauri, V. K. Shanin, and V. S. Danilin: Investigation of Flows of Wet Steam in Nozzles, *High Temperatures*, vol. 10, no. 1, January–February, 1972 (translated from Russian by Consultants Bureau, Plenum Pub. Corp., New York, 1972).

155. Deich, M. E., Yu. I. Abramov, and V. I. Glushkov, Regarding the Mechanism of Moisture Flow in Turbine Nozzle Blade Channels, *Teploenergetika*, vol. 17, no. 11, pp. 34–38, 1970.

156. Deich, M. E., and G. V. Tsiklauri: Supercooling and Structure of a Stream of Wet Steam Flowing Out of a Convergent Nozzle, *High Temperature*, vol. 2, no. 3, pp. 454–463, May–June, 1964.

157. Deich, M. E., V. F. Stepanchuk, and G. A. Saltanov: Calculating Compression Shocks in Wet Steam Region, *Teploenergetica*, vol. 12, no. 4, pp. 81–84, 1965.

158. Comfort, W. J., T. W. Alger, W. H. Giedt, and C. T. Crowe: "Calculation of Two Phase Dispersed Droplet-in-Vapor Flows Including Normal Shock Waves," ASME paper no. 76-WA/FE-31, 1976.

THREE

BUBBLE GROWTH AND NUCLEATION

3-1 GENERAL GROWTH AND NUCLEATION

3-1-1 General Background

The understanding and predictability of bubble nucleation is a major problem for both cavitation and boiling heat transfer. There is a copious literature concerning both subjects. Much of that applying to cavitation is summarized in Refs. 1 and 2 and that for boiling in Refs. 3 and 4. The relatively standard and well-known part of that literature will not be covered in full detail here; rather it is summarized fairly briefly. In addition, some relatively new and specialized information will be discussed more fully.

3-1-2 Models and Theories

General There are two major nucleation models, both of which no doubt apply to some extent in all cases, but their significance differs substantially between boiling and cavitation applications. These are the stationary crevice model, pertinent primarily to boiling cases, and the entrained nuclei model, applied primarily to cavitation but no doubt also operative for boiling, i.e., there is the possibility of both stationary and mobile inception "nuclei." "Nucleus" in this sense refers to agglomerations of gas or vapor molecules of sufficient size to allow later growth, upon the imposition of additional heat energy in the case of boiling or reduced liquid pressure in the case of cavitation, into conventional "bubbles." These then, by definition, constitute the presence of "boiling" or "cavitation."

Stationary nuclei are generally assumed to be harbored in microcrevices of various shapes [3, 4] in an adjacent wall, while traveling nuclei are assumed entrained with the main stream. Entrained nuclei are primarily pertinent for cavitation. Of course, cavitation can also be initiated in some cases from stationary nuclei [3, 5] in crevices in the guiding wall in the minimum-pressure region, and boiling can, of course, be influenced in some cases by the presence of traveling nuclei. In either case, there is the fact that unless nuclei of appreciable size exist, boiling and cavitation, under the minimum superheat or underpressure conditions under which they are commonly observed, could not exist. In most simple terms, this conclusion results from the static force balance at the bubble wall between the surface tension and pressure differential:

$$P_v - P_L = \frac{2\sigma}{R} \tag{3-1}$$

where the subscript v signifies vapor (and/or gas) and L liquid, P signifies pressure at the bubble wall, and σ is surface tension. Of course, Eq. (3-1) is not pertinent when R approaches molecular dimensions.

From a slightly different viewpoint, all liquids possess a substantial tensile strength which greatly exceeds the observed underpressures required to provoke bubble nucleation [1–4]. Thus the generally observed existence of nucleation under many engineering conditions indicates the presence of "nuclei," i.e., small "microbubbles." Then generally small underpressures of the magnitude predicted by Eq. (3-1) generally cause these microbubbles to "nucleate" into the normally visible size range (~ 0.1 mm) associated with normal boiling or cavitation.

A model which predicts stabilized nuclei, both stationary and entrained in the flow, is thus required. However, a simple, uncoated, static vapor microbubble is not possible, since its existence according to Eq. (3-1) requires an internal pressure greater than the vapor pressure of the surrounding liquid to overcome the effects of surface tension, which otherwise would cause its immediate collapse. The persistence of a simple gas microbubble is not possible either. According to Eq. (3-1), the internal gas pressure would be greater than the saturated gas pressure in the surrounding liquid, so that the gas would escape relatively quickly into the liquid by diffusion effects if the bubble were in the "microbubble" size range. If it were larger, it would quickly escape by buoyancy effects, at least in a static liquid. This situation gives rise to the "microbubble paradox," the present resolution of which is described next.

Microbubble paradox The "microbubble paradox" refers to the fact that experimental values of underpressure or superheat necessary to provoke bubble nucleation are generally at least an order of magnitude less than the theoretically predicted values of liquid tensile strength would indicate should be the case, even though extreme care is taken to degassify and purify the liquid. Thus the existence of stabilized gas or vapor microbubbles or, conceivably, unwetted solid particles must be postulated to explain the observed nucleation threshold values. Historically, from the viewpoint of cavitation, this situation was first documented in the papers of Harvey [6–9], who proposed (1944–1947) the mechanism still generally believed to be pertinent in most cases. This model is discussed in detail in Ref. 1, but will also be covered here. The same mechanism in stationary crevices is also believed to be generally predominent in boiling. While Harvey's work involved fluid-dynamic cavitation, many later and more precise experiments have utilized ultrasonic radiation of the test liquid.

For present illustrative purposes, consider a glass water-filled beaker, irradiated by an ultrasonic sound field of known and adjustable strength, so that the imposed pressure oscillation necessary to provoke visible bubble nucleation can be determined. This then provides a method for measuring bubble-nucleation thresholds for various liquid conditions. If the water is carefully filtered to remove solid particles down to an arbitrarily small size, and is also well deaerated so that any entrained gas bubbles will either quickly go into solution or rise to the surface, there is then no apparent mechanism for the persistence of nucleation centers. Nevertheless, in such an experiment it is invariably found that the pressure-oscillation magnitude is much less than would be expected if full pure-liquid tensile-strength values were realized. This is the case even if long settling times are provided to allow very small gas bubbles to dissolve and larger bubbles to rise to the surface under easily calculable buoyancy effects. Also, as previously mentioned, pure vapor bubbles could not exist under such conditions due to surface-tension effects. Hence the "microbubble paradox" and the necessity for the postulation of a "stabilized nucleus" mechanism exist.

Many explanations of this "paradox" have been suggested, including the continued generation of new "nuclei" by the action of cosmic-ray bombardment [10, 11], similar to the conventional "bubble-chamber" mechanism used in high-energy physics for the study of nuclear radiations. While this and other specialized mechanisms may in fact sometimes contribute, it is generally assumed today that the mechanism originally proposed by Harvey and his colleagues [6–9] is probably the most important.

Harvey microparticle and crevice mechanism   Harvey and others [6–9] proposed entrained microparticles in the liquid, containing in themselves unwetted acute-angle microcrevices. The proposed mechanism is illustrated in Fig. 3-1 (reproduced from Ref. 1, for convenience). Essentially, an "unwetted," i.e., "hydrophobic," acute-angle crevice in an entrained microparticle is required. In this case, as shown in Fig. 3-1, the meniscus curvature is such that if the crevice walls are unwetted the pressure in the gas pocket at the apex of the crevice is less than that in the adjacent liquid. According to Eq. (3-1), this liquid-gas pressure differential will increase as the radius of curvature decreases. This will occur if the interface advances more deeply into the acute-angle crevice. Thus the interface will advance into the crevice, with gas solution and diffusion across the interface continuing, until the gas pressure becomes sufficiently low, following Eq. (3-1), that it is equal to the partial pressure of the gas in the liquid. Further solution and diffusion of gas into the liquid will not occur, since no further solution-concentration gradient will then exist. Thus the gas cavity would become stabilized over indefinite

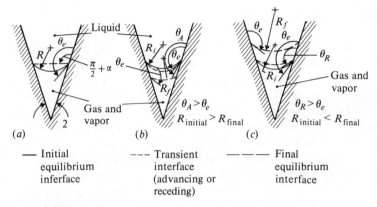

— Initial   --- Transient   ——— Final
  equilibrium   interface   equilibrium
  inferface   (advancing or   interface
                        receding)

R_i = initial equilibrium interface
R_f = final equilibrium interface

Figure 3-1 Stabilization of a gas-vapor pocket in a hydrophobic crevice. (*a*) Liquid saturated with gas. Interface in an equilibrium position with radius R and contact angle $\theta_e > \pi/2 + \alpha$. (*b*) Liquid undersaturated with gas. Liquid advances with reduced R and $\theta_A > \theta_e$. As the gas solution proceeds, θ_A decreases until at equilibrium the contact angle is again θ. (*c*) Liquid supersaturated with gas. Liquid recedes with increased R and $\theta_R < \theta_e$. As the gas release proceeds, θ_R increases. For equilibrium within the crevice, the contact angle is again θ_e (Knapp, Daily, and Hammitt [1]).

time periods by this mechanism, whereas the same geometry with wetted crevice walls would lead to complete extinction of the cavity by gas-solution effects.

The degree of gas-saturation content in the liquid would not affect the eventual fate of the gas pocket, only the time required for it to be achieved, according to this model. Thus any degree of possible deaeration would not eliminate nucleation "nuclei" completely, since zero gas content is not completely attainable. Also, the type of gas and liquid involved would not affect the eventual outcome, assuming that less-than-infinite solubility for the gas in the liquid exists. Thus the mechanism appears pertinent to all liquid-gas combinations. Of course, increased gas pressure for fixed total gas content would result in reduction in nuclei size, as originally demonstrated by Harvey [6–9], and corresponding increase in nucleation threshold, i.e., increase in required "underpressure," or reduction in actual liquid pressure. This was later demonstrated by Knapp [1, 12] in experiments with cavitating venturis. The original Harvey work (1944 and 1945), as a matter of interest, was motivated by the effects of bullet impact upon flesh, and the resultant cavitation induced in the wake of a bullet passing through a human body, for example. This cavity is presumably responsible in many cases for most of the resulting damage.

The effect of vapor within such an acute-angle crevice is the same as that of gas, since the vapor can exist in perpetuity at a pressure sufficiently less than that of the surrounding liquid. Thus it would not be extinguished by any degree of "subcooling," or suppression pressure in the case of cavitation. The effects of gas and vapor together are obviously simply additive.

Alternative nuclei-stabilization mechanisms

Fox and Herzfeld organic skin model [13]   The most prominent models for micro-nuclei stabilization, other than that of Harvey discussed above, are probably those assuming a bubble-wall interface consisting of some impurity material capable of withstanding compressive forces, and preventing or reducing gas diffusion between liquid and microbubble contents. If such a bubble wall were presumed, it would be capable of allowing an internal gas or vapor pressure less than that of the surrounding liquid, in spite of the effects of surface tension or subcooling, so that a nucleus-stabilizing mechanism would be provided. One of the earliest and best-known proposals of this sort was that of Fox and Herzfeld [13], who postulated the existence of an organic skin formed of impurities in the liquid, particularly pertinent to cases of relatively impure natural water or sea water. This skin would be capable of withstanding some compressive stress and also prevent or reduce the outward diffusion of gas or vapor. This model is discussed in more detail in Ref. 1. Later experiments by Bernd [14] seemed to confirm its existence, at least in some cases, although the original authors later disclaimed its probable importance. In the opinion of the present author, it seems likely that the Fox-Herzfeld model may apply to some extent in some cases but that it is probably less important than that of Harvey.

Unwetted-mote model   It is apparent that small particles, or "motes," which are entirely or partially unwetted by the liquid, can provide bubble-nucleation "nuclei," even without the presence of gas, other than the inevitably present vapor of the liquid. The idea was advanced earlier by Plesset [15] that such unwetted motes would provide "weak spots" in the liquid, i.e., "stress raisers" from the viewpoint of solid mechanics, about which tensile failure of the liquid will occur at underpressures much less than the theoretical strength of the pure liquid.

Regeneration models

1. *Cosmic and nuclear radiation.* An alternative to a microbubble-stabilization model is one assuming a constant balance between microbubble destruction and regeneration. Since some continuous elimination of nuclei must occur in almost all real situations, due to such effects as solution, disentrainment, entrapment in static regions, centrifugal disentrainment effects, etc., it must also be true that some countering mechanisms, which create nuclei, must exist in most real cases. A relatively exotic, but probably operative in some cases, mechanism was suggested by Sette and others [10, 11], i.e., the action of cosmic radiation. They verified the existence of this effect in a case involving ultrasonic cavitation by comparing nucleation sonic-pressure threshold variation in their test beaker, with and without external lead shielding. Of course, it is well known [16] that high-energy radiation of this sort can, under some conditions, e.g., bubble chamber, cause bubble nucleation. While the cosmic-ray effect is probably not significant in most engineering cavitation cases, it is certainly quite

likely that nuclear radiation effects can importantly affect nucleation thresholds in liquid-cooled nuclear reactor situations [17–20]. Of course, the effect must always be in the direction of promoting nucleation and thus reducing required underpressures or superheats. This has especially practical importance in sodium-cooled reactors because of the appearance in some cases of large "sodium superheats" [21].

2. *Gas entrainment.* A less exotic, and no doubt generally much more important, mechanism for the generation of "nuclei" is that of gas entrainment, which may occur wherever a free surface is present. The high level of entrained microbubbles in normal tap water presumably results from the effect of such free surfaces as found in storage reservoirs, etc. Presumably, there is also a high level of gas entrainment in most engineering equipment such as power-plant circuits and components, nuclear reactor flow circuits, the open ocean or other large bodies of water, and in most other important engineering applications. Thus a substantial lack of nuclei for bubble nucleation is to be expected only in very special cases where extraordinary degrees of gas removal and liquid purity are maintained.

Another practically important mechanism for the generation of entrained nuclei is that associated with "rectified diffusion" of gas or heat.

Gas and heat diffusion effects

1. *Conventional diffusion.* Many operative mechanisms pertinent to either generation or regeneration of nuclei involve conventional diffusion of both gas and heat. Such mechanisms are predominant, particularly in the case of bubble nucleation from stationary crevices as in most boiling cases [3, 4], and will be discussed later in further detail in that connection. However, they are obviously also of importance for stabilized entrained nuclei, such as have been discussed in the foregoing sections. This is the case where either (*a*) the bubble motion carries it into regions where the partial pressure of the gas dissolved in the liquid is greater than that within the entrained nucleus or (*b*) the liquid temperature in some region of the circuit, through which the entrained nucleus passes before reaching the cavitation region, is greater than the saturation temperature corresponding to the vapor pressure existing within the bubble. However, the effects of such conventional diffusion are relatively slow so that, in many cases involving entrained nuclei, adequate time is not available to make such conventional diffusion important. This is obviously not a limitation for cases of nuclei in stationary crevices, in general, for either boiling or cavitation.

2. *Rectified diffusion effects.* The effects of diffusion of either gas or heat can be greatly enhanced by the existence of "rectified diffusion." This is particularly important for cases of ultrasonic cavitation, involving thousands of cycles of pressure oscillation per second, as perhaps originally investigated by Plesset [22]. Under conditions of ultrasonic cavitation, gas nuclei are "inflated" rapidly and substantially by the effects of "rectified diffusion," i.e., the gas-diffusion

effects upon a bubble in an oscillating pressure field are nonsymmetrical, so that there is a net inward flow of gas. In simplest terms, the bubble surface area is augmented during the low-pressure part of the cycle and diminished during the high-pressure portion. Thus more inward gas diffusion occurs during the low-pressure portion of the cycle than outward diffusion during the high-pressure portion, when the surface area is relatively reduced. An analogous rectified diffusion effect can also be postulated for heat or any other quantity, the transfer of which depends essentially upon the linear diffusion law, i.e., "Fick's law."

Rectified diffusion effects, particularly for gas, were first noted and studied [22] for ultrasonic cavitation, where at least the gas effect is now generally accepted as of major importance. Since these rectified transfer effects depend upon large and rapid pressure oscillations, they are obviously not pertinent to ideally steady-state flowing systems. However, the existence of substantial turbulent-pressure fluctuations in most engineering flowing systems provides the necessary pressure oscillations for at least the existence of such rectified bubble-wall transfer mechanisms. Some computerized studies of such effects for flowing loops are reported, but no definitive results have yet been reported to my knowledge. Aside from the possibility of rectified turbulent diffusion mechanisms for potential nucleation nuclei in flowing systems, there is also the possibility of relative velocities between nuclei and mean liquid velocity in such systems, i.e., "slip," particularly predominant for gas or vapor (rather than solid) particles, with or without turbulent effects. Such relative velocities would add to the apparent diffusion effects, since actual transport phenomena rather than simple static diffusion would then be involved. Obviously, a quantitative solution of this overall problem would be highly complex.

Relative velocity effects   An important practical effect of relative (slip) velocity in cavitating systems is the "windshield" effect, i.e., gas or other low-density nuclei entrained in a liquid flow will tend to avoid regions of increasing pressure and will be "attracted" in the direction of decreasing pressure by the action of slip velocity effects. This problem was investigated in some detail by Daily and Johnson [23] concerning cavitation-inception effects. In general, it is clear that a homogeneous distribution of entrained gas or vapor nuclei in a liquid flow approaching a submerged object will become nonhomogeneous in the vicinity of the object. There will then develop a relative scarcity of nuclei in regions of rising pressure, such as near stagnation regions on the leading edge of an object or on the outside of a pipe bend. On the other hand, there will be relatively increased nuclei population density in the low-pressure region along a submerged body where cavitation inception would be expected to occur or on the inside of a bend. Thus this "windshield effect" will in general increase the likelihood of cavitation inception for a given spectrum of upstream nuclei.

Of course, the nuclei trajectories relative to the mean liquid flow will depend upon nuclei size, shape, and density relative to liquid density. Their trajectory will depend upon the balance between fluid-dynamic drag forces, and inertial and pressure forces. Thus very small nuclei having a relatively large ratio of surface area to volume will exhibit less slip relative to the liquid than larger bubbles, the shape of which in their growth or collapse will also influence the problem.

Yeh and Yang [24] investigated these effects for larger bubbles in the author's laboratory, considering the effects of bubble-profile asymmetries in a cavitating venturi and showing that either negative or positive slip was possible, depending upon the relative velocity and other parameters as the bubbles entered the diffuser section of the venturi. These calculations were confirmed experimentally in the same laboratory [25]. A very recent study of this same venturi flow geometry, but neglecting bubble asymmetries, is reported by Chincholle and Sevastianov [26], confirming the Yeh-Yang results with a somewhat simplified calculating model.

Special shear flow effects   Separated flows, such as those downstream of an orifice, nozzle, or valve discharging into a submerged region, differ substantially in terms of cavitation-inception sigma† from nonseparated flows, such as, for example, in a well-designed venturi or along a hydrofoil section of relatively gradual curvature. The inception sigmas for such separated flow geometries are in fact much greater than are normally encountered for "streamlined flows," as discussed in more detail in later sections of this book as well as Ref. 1 and elsewhere. The mechanism motivating this greatly increased cavitation sigma is presumably the presence of microvortices in the separated region, the centers of which present very substantial and local underpressures. Such vortex cavitation is also common in the trailing vortices from propellor blades, unshrouded pump or turbine blades, or hydrofoils. In many cases, this type of cavitation does not lead to substantial damage effects, since the vortices in such cases as orifices and nozzles, propellor blades, etc., collapse far from material objects. This, however, may not be the case for unshrouded pump or turbine blades. The strength of the microvortices would, of course, depend strongly on viscosity and other fluid parameters, discussed later in the section on viscous effects.

Normal boundary-layer flow, of course, also presents a shear flow region where microvortices and other turbulent manifestations would no doubt influence cavitation sigma. This situation was investigated particularly by Daily and Johnson [23] and later by Arndt and Ippen [27, 28], with results discussed in Ref. 1, where the probable effects of boundary-layer turbulence upon inception sigma are discussed and tabulated. Holl, in related studies [29–31], investigated the effects of finite and regular forms of wall roughness upon inception sigma.

Another very interesting effect of wall proximity is related to the fact that for potential flow assumptions [32] it can be shown that the pressures within the fluid cannot be less than some wall pressure, i.e., the locations of minimum and maximum pressure for a potential flow regime must be along a boundary, assuming other than simple one-dimensional translatory flow. Thus some streamline curvature must give actual minimum pressure along a wall, i.e., wall pressure must

† "Inception sigma" = $\sigma_{INC} = (p - p_v)/(\rho v^2/2)$, where p and v are reference pressure and velocity.

be somewhere less than minimum mainstream pressure, thus indicating the probability of bubble inception within a boundary layer, where the above discussed effects of boundary-layer turbulence and vortices would be operative.

Liquid cavitatability   In principle, it should be possible to compute cavitation thresholds, or the boiling superheat requirement, from essentially "basic principles;" i.e., if the detailed characteristics of the operative entrained or stationary "nuclei" were known, as well as the details of the flow regime, then it should be possible with modern-day computer techniques to accurately compute nucleation thresholds for any given case. While attainment of this capability is no doubt the desired objective of research in this field, it has not yet been attained to a useful degree, because of the very large complexities of the problem, as discussed in the foregoing sections. Hence, another alternative, at least from the viewpoint of nuclei content, would be the use of a "standard cavitator" to measure the cavitatability of the liquid in use as it exists under the particular application. For example, such a device could be used to measure the "nucleation resistance" of molten sodium in a nuclear reactor circuit under the particular conditions pertinent to that circuit. A simple small-scale venturi in a bypass loop might be used for this purpose. The use of such a venturi for water tests was suggested recently by Oldenziel [33]. Work toward the development of an ultrasonic device for testing the cavitatability of sodium has also been reported [34]. In principle, the adoption of this technique would avoid the detailed measurement of nucleation "nuclei" spectra and the subsequent calculation of their effect upon bubble nucleation. They would be replaced by a simple measurement of cavitation sigma (or superheat) under a simple standard condition, and then its calculation from this measured point in the realistic geometry.

Pertinence of entrained nuclei models to stationary crevice nucleation   Many of the previous sections obviously apply particularly to "traveling nuclei" which, as is generally believed, are of primary importance in cavitation but less so in boiling, where stationary crevices in the heated walls are presumably of predominant importance. Of course, if in a particular case these are absent, the possibility of the attainment of large superheats would then be obviated by the presence of entrained nuclei. This factor could be of considerable importance in such applications as the sodium-cooled nuclear reactor, where the possibility of substantial "sodium superheat" has been an important problem related to reactor safety. In many cases of conventional cavitation, there is also the possibility of an important contribution from nuclei entrapped in stationary crevices in the region of minimum pressure.

   While some of the mechanisms discussed in the foregoing obviously do not apply to stationary crevices, or to boiling applications in general, some obviously do apply almost directly, particularly those concerned with liquid tensile strength and the necessity for stabilization mechanisms for nuclei. The role of gas-diffusion effects in this case differs somewhat between stationary and traveling nuclei. For example, in many cases involving traveling nuclei, sufficient time is not afforded to allow gas-diffusion effects to become important. The growth (or collapse) of a small bubble passing through a region of strong pressure gradients in a flowing system is not sensitive primarily to gas-diffusion effects, but rather to the effects of surface tension, pressure, and inertia. As previously indicated, these effects may become important in cases of very rapid and repeated pressure variation through "rectified diffusion" effects, such as encountered in "ultrasonic cavitation" or conceivably in flowing cavitation through the effects of turbulent pressure and velocity variations. However, the situation is very different for nuclei entrapped in a stationary crevice in a flow field. In this case, since the nucleus is stationary, it cannot be assumed that sufficient time for substantial conventional diffusion effects will not be available. Thus there is no difficulty imposed upon the continued growth of a gas or vapor nucleus entrapped in the low-pressure region of a cavitating body, provided the dissolved gas content of the liquid is sufficient. Of course, in this case also there is the possibility of added rectified diffusion effects due to pressure variations from turbulence or pressure pulsations in the main flow, as, for example, from the action of a centrifugal pump, trailing vortices from a submerged object, or similar cause.

3-1-3 Bubble-Growth Dynamics

General From a relatively general viewpoint, the subject of bubble dynamics is discussed in detail in Chap. 4, primarily from the viewpoint of cavitation (Sec. 4-1), but boiling bubble dynamics are also included and the differences between boiling and cavitation analyses are discussed. Both growth and collapse are considered, but again the emphasis is upon collapse, since that is of more pressing interest for the cavitation case. In general, the emphasis is on the intermediate portion and termination of collapse, rather than initial growth, i.e., "nucleation," which is the subject of the present chapter. It is the purpose of the present section to consider the differences between bubble-dynamics analyses pertinent especially to the nucleation phase as opposed to more general bubble dynamics, including the collapse portion.

Nucleation versus general bubble dynamics As discussed in detail in Chap. 4, the problem of single-bubble dynamics in the general case is controlled by the interaction of the effects of pressure differential between bubble contents and external liquid, inertial effects, viscous effects, and heat-transfer restraints. The latter are primarily important in boiling cases, but are also important in the final phases of symmetrical collapse. Surface-tension effects are not important during these phases of bubble dynamics, but asymmetrical effects induced by external asymmetries are of major importance in collapse, controlling the type of collapse which occurs.

For nucleation, on the other hand, surface tension is of major importance [Eq. (3-1)] in controlling the size of "nucleus" which will be available for future growth into actual bubbles, generally of visible size.

Initial growth is then most closely controlled by the balance between internal and external pressure and surface tension. The initial rate of growth may also depend upon either inertial or thermal restraints, or in some cases upon gaseous diffusion effects, as already discussed. The latter are usually not of importance for cases of cavitation bubble collapse, since insufficient time is afforded. Gaseous diffusion, however, is of major importance in nucleation, since this depends upon the stabilization and gradual growth of small gaseous or vaporous "nuclei." Whether these are located in stationary wall crevices or in entrained microparticles, adequate time for gaseous or thermal diffusion effects is often available. For entrained nuclei, sufficient time for important diffusion effects during the passage through the minimum pressure region, where cavitation occurs, may not exist. However, prior to their arrival at the minimum pressure point, relatively long exposure to the surrounding liquid may exist. This is especially the case for closed-loop geometries. To prevent the gradual accumulation of nuclei of this type, some large cavitation research tunnels include special "resorbers" for entrained nuclei [1]. The behavior of entrained nuclei in such circuits is complicated, from the viewpoint of analysis, by the effects of turbulent pressure and velocity oscillations to which they are exposed, providing the possibility of "rectified diffusion" effects, as discussed earlier.

Stability and asymmetrical effects As previously indicated, asymmetrical effects are very important in bubble collapse and cavitation damage, since both photographic and theoretical evidence show that the basic form of the collapse is entirely controlled by these effects, so that the basic Rayleigh spherical-collapse model [1] becomes essentially inapplicable in many cases. However, analogous asymmetrical effects are not of comparable importance in bubble nucleation. Of course, spherical symmetry would not normally be assumed in any case for nucleation, except for "homogeneous nucleation" from the main stream where it is in fact applicable, since nucleation originates otherwise from acute-angle crevices which obviously control the bubble geometry.

On both theoretical and experimental grounds, spherical bubble growth is inherently much more stable than collapse. As initially shown by Plesset [35, 36], bubble collapse from an initially spherical shape is inherently unstable while growth from an initially spherical shape is stable; i.e., a small asymmetric perturbation imposed upon a spherically collapsing bubble will grow in many cases, but this is not true to the same extent for a growing bubble. This situation is somewhat analogous to the well-known Taylor instability [37, 38] involving the acceleration of a low-density fluid toward one of relatively higher density in a planar geometry, as in a gravity or centrifugal field. The situation, however, is reversed for the spherical geometry, where the acceleration of the heavier fluid toward the lighter fluid, as in the bubble-collapse case, is the more unstable.

The question of these bubble-collapse and growth instabilities is discussed in considerable detail in Ref. 1, along with the related work of Plesset concerning particularly the relation between the Taylor instability for a planar case and the instabilities applicable to the spherical geometry. These instabilities presumably originate from miscellaneous small perturbations of an initially spherical shape. However, in most cases much more substantial asymmetries result from virtual mass effects related to such asymmetrical factors as wall proximity, external liquid translatory velocity, external pressure or velocity gradients, or imposed accelerations such as gravity or centrifugal effects in many common flow cases. These questions are discussed in further detail in Chap. 4.

3-1-4 Magnetic and Electrostatic Field Effects

Various recent experiments from both the USSR and Japan indicate the existence of significant effects of magnetic fields and electrostatic fields on cavitation damage [39, 40] and bubble growth [41] or nucleation sigma [42–44, 48, 49]. While some of these tests were in liquid metals [39, 42–49] others were in ordinary tap water [40, 41], and most involved magnetic rather than electrostatic fields. If these effects do in fact exist in water, they may indicate a strong ionized layer along the liquid-vapor interface. At the moment, this seems the only credible mechanism. The effects in liquid metals [42] may be also operative through rearrangement of upstream velocity profiles.

These effects are of interest not only from the viewpoint of basic science but also in connection with various fusion power-reactor concepts involving the pumping of liquid-metal coolants through strong magnetic fields. Another less exotic application involving water is that of the cooling of certain computer components.

Recent experiments in the author's laboratory have been conducted for both tap water and mercury using "ultrasonic cavitation" with a magnetic-field strength in the cavitating region of ~ 6 kG [19]. This is only about one-tenth of the magnetic-field strength pertinent to the fusion reactor concepts. Our tests [19] indicated no measurable effect on either nucleation thresholds or nuclei spectra for water, measured using laser light scattering (discussed in Sec. 3-2). From these results, any such effect must have been less than ~ 3 percent for cavitation threshold and $\gtrsim 1$ percent for nucleus spectrum [19]. These negative results are not, however, in accord with those of the other previous investigators [39–49]. However, our experiment was the only one involving nucleation thresholds with ultrasonic rather than flowing cavitation. Hence, it may be assumed that perhaps this type of cavitation is less sensitive to magnetic-field nucleation effects than are flowing cavitation regimes. Intuitively, this does not seem unreasonable, since much greater underpressure values are involved with the ultrasonic cavitation experiment.

The general subject of electrohydrodynamic (EHD) and magnetohydrodynamic (MHD) effects upon heat transfer in liquids is well covered in a recent report by Jones [44].

3-2 EFFECTS OF AIR AND GAS CONTENT

3-2-1 General Background

The major past and on-going studies of the effects of gas content upon cavitation are reviewed in this section. Those studies are considered first which provide information directly applicable to the estimation of air-content effects upon cavitation-inception sigma or other performance parameters. Next, studies are considered which are of a more basic nature. Finally, those studies pertinent to a prediction of the effects of gas content upon cavitation damage are discussed. Conclusions, where possible, and recommendations for future directions of research are included.

Nucleation thresholds of liquids, whether in cavitation or boiling, depend very strongly on the population and size spectrum of microbubbles (gas "nuclei") in the liquid. In addition, the violence of collapse of cavitation bubbles appears also to be strongly influenced by the quantity of gas (free or dissolved) in the liquid. Cavitation damage is thus usually found to be reduced for "gassy" liquids. While the existence of substantial effects of gas (or air) content upon both the inception of cavitation (or boiling) and cavitation damage has long been recognized, no full understanding of these effects, or any simple and reliable method for their prediction, is available. A large quantity of pertinent, sometimes conflicting, technical literature exists. Reasonably complete survey reports have been published [50].

Air-content effects The effects of air content (or gas content in general) upon cavitation can be considered under three main headings, of which the first is probably the most important, i.e.:

1. Cavitation-inception sigma
2. Flow regime, torque, power, head, efficiency, noise, vibration, etc., for well-developed cavitation
3. Cavitation erosion

There is evidence of important air-content effects under each of these three headings, and these will be discussed where appropriate. However, a major portion of the research has been concentrated under the first category for which theoretical treatments are more possible. This category is also the subject of the present section.

The available information can also be divided as follows:

1. That which is directly applicable to the prediction of field and laboratory performance
2. That which is primarily of the nature of basic research which can then be used to clarify the observed trends or make meaningful predictions from observed trends

3-2-2 Data Directly Applicable to Sigma and Performance Effects of Air Content

There have been numerous fairly systematic and comprehensive studies of the effects of air content upon cavitation-inception sigma (or boiling nucleation), and also upon its effects, after initiation, upon head, efficiency, power, etc. However, it is difficult to apply this information in general because of the large number of independent parameters having an apparently very important influence upon the results. The importance of some of these has not been recognized until recently, and it seems probable that insufficient understanding of the overall phenomena exists to define and construct a basic test which would provide data entirely applicable to the various prototype or model devices at this time. There are two primary reasons for this:

1. *Experimental difficulty.* No readily practical and usable method exists as yet for the measurement of the number, size, and location distribution of the very small entrained gas "nuclei" in the flow from which audible and visible cavitation develop, the effects of which can be measurable upon machine performance, etc. These "microbubbles" probably cover the diameter range of 10^{-5} to 10^{-3} cm, thus being invisible to the unaided eye. Knowledge of the total gas content is insufficient in itself to determine sigma effects.

2. *Theoretical difficulty.* It is not possible to describe in sufficient detail actual flow patterns in order to delineate the pressure and velocity history, or the trajectory, of a given gaseous "nucleus," assuming that its position and condition at a given instant of time were known by such a measurement as that discussed above. The realistic problem is two or three dimensional (depending upon the type of device) and essentially biphasic in nature (even if only a question of cavitation inception), since the trajectory of the low-density entrained nuclei may not be even approximately that of the liquid if important pressure or velocity gradients exist. Finally, turbulence must be considered, since turbulent fluctuations importantly influence gas-diffusion effects into and from the nuclei, as well as affecting the likelihood of cavitation inception through the application to the gas nucleus of instantaneous pressures, which may be considerably below the time-mean pressure. These questions were discussed in the last section.

Thus a complete solution of the air-effects problem seems unlikely at this time. However, an improving theoretical understanding of the phenomena involved, the increasing availability of large computers, and the on-going development of instrumentation techniques continually reduce the gap between the possibilities of basic investigations and their direct application to model and prototype devices. Pertinent studies date from the 1930s. These are considered under the continuing efforts of various institutions, universities, companies, etc., rather than as isolated papers by individuals. More complete detail is provided elsewhere [1].

Numachi and Kurokawa, Institute of High-Speed Mechanics, Tohoku University, Sendai, Japan [51–56]   This group working in the late 1930s apparently conducted the earliest comprehensive investigations of the effects of total air content upon cavitation sigma. Both a venturi test section and an isolated profile were used. Tests were in distilled water, tap water, and salt water [51–56]. Water temperature covered the range 10 to 40°C and air content 0.3 to 1.3 times saturation as STP, measured by Van Slyke.

In general, a fairly linear rise of inception sigma with air content was observed with greater effect at lower temperatures. The change in sigma was considerable, ranging from up to tenfold in the venturi to 20 to 50 percent for profile tests. Figure 3-2 is typical of their results.

Edstrand, Lindgren, and Johnsson, Swedish State Shipbuilding Tank, Göteborg, Sweden [57–59]   This group has reported a very comprehensive series of experiments on the effects of air content on inception sigma and other performance parameters for marine propellers in both tap water and sea water (1946–1950, Refs. 57, 58). More recent data are found in Ref. 50. A substantial effect of relative air content ranging between 0.23 and 0.73, i.e., relative to saturation at STP, was noted with efficiency, as relative air content dropped from 0.62 to 0.48 for a given advance coefficient and sigma. Figures 3-3 and 3-4 are typical. Their results differ considerably from those of Numachi and others [51–56], and the geometries tested

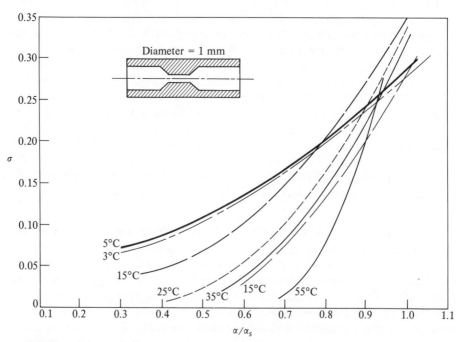

Figure 3-2 Cavitation inception sigma in venturi as function of relative air content compared to saturation at STP (Numachi data [51–55]).

are very different. Some difference between tap and sea water was found, though not so great as that found by Numachi. They also noted a considerable effect upon inception sigma of the rate of lowering of pressure to obtain cavitation [59]. They thus confirmed that inception sigma is not determined solely by the flow

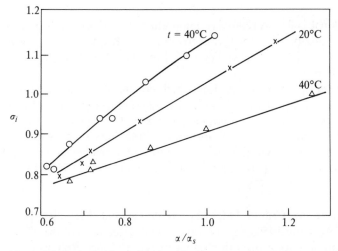

Figure 3-3 Cavitation inception sigma for flow past circular section as function of relative air content compared to saturation at STP (Edstrand [57]).

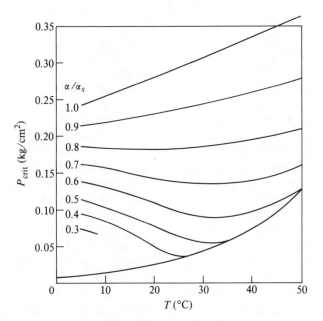

Figure 3-4 Critical pressure as function of water temperature in Numachi venturi for different relative air contents (compared to STP) (Gutsche [49] and Edstrand [57]).

parameters and total air content. They found also that the type of cavitation influenced the magnitude of the air-content effect. For inception, they found large effects for "laminar cavitation" (steady-state cavities). The variation in performance parameters such as torque, thrust, and efficiency was little affected, however, by large differences in total air.

Escher-Wyss and Vuskovic [60] A relatively comprehensive series of tests on air-content effects upon both performance and erosion in a Kaplan turbine was reported by Vuskovic working at Escher-Wyss [60] in 1940. The damage effects will be discussed in Chap. 5.

Vuskovic observed a tip-vortex cavitation (which he considered to be simply "gaseous" rather than true "vaporous" cavitation, to use present-day terminology), the inception of which was sensitive to air content. However, this type of cavitation in his tests had little effect upon turbine-performance parameters such as torque, etc. Vuskovic observed "true" cavitation, which comprised a heavier and whiter cloud, which does affect turbine performance, but which is not sensitive to air content.

Miscellaneous venturi tests: Crump [61, 62], Williams and McNulty [63], Ziegler [64] In the 1940s and 1950s, various relatively isolated tests of the effects of total air content upon cavitation-inception sigma are reported, e.g., Refs. 61 to 64. The tests are rather similar to the earlier tests of Numachi and others [51–56]. However, the results of the various investigators do not agree quantitatively, and there is scatter in the individual data sets. In general, inception sigma is reduced for reduced air content and, for the lowest air contents, liquid tensions sometimes appear. A "hysteresis effect" is noted in one of these studies [63], showing again the impossibility of representing air-content effects in terms of total air content only, i.e., presumably the hysteresis effect is due to changes in nuclei content between inception and desinence, though total air content remains constant.

Large water-tunnel investigations in the United States: Pennsylvania State University (Penn State), University of Minnesota (U-Minn), and California Institute of Technology (CIT) [65–71] Work in the United States on gas-content effects upon inception sigma for submerged objects has been largely conducted since World War II, especially in the large water tunnels of Penn State, CIT, and U-Minn. While the work at CIT has not involved air-content effects specifically, it has served as a good basis for comparison with other investigators.

The work of these various institutions (pertinent to the present discussion) has centered upon the observed difference in sigma between incidence and "desinence," so named by Holl [1, 67]. It was generally found that desinent sigma exceeds inception sigma, so that a "hysteresis" exists. This difference appears to decrease with increased velocity or size, so that it is actually a "scale effect." It was further found that desinent sigma data exhibit much less scatter than incident. Inception sigma depends upon the rate of lowering of pressure, as also reported by Lindgren and Johnsson [59].

All the above effects appear closely connected with the details of the nucleation process from entrained or stationary "nuclei." Hence, knowledge of the distribution and size spectra of these nuclei is needed. Recent work at U-Minn has been aimed in this direction. Riplen and Killen [68] found an equilibrium of entrained gas nuclei to be attained in a closed tunnel for given tunnel conditions. They also found no hysteresis if the free gas conditions were maintained, so that hysteresis ceases to be a scale effect under these conditions. The U-Minn investigators also developed a method for the continuous monitoring of the free gas distribution, using its effects upon velocity of propagation of a pressure pulse and the attenuation of pressure pulses with resonant bubbles [1, 68–71].

Bassin d'Essais des Carènes, Paris [72–74]   A comprehensive series of tests has been conducted by Bindel and others on the effects of the variation of total air content upon inception sigma for various ogives, hydrofoils, and propellors [72–74]. In agreement with Vuskovic at Escher-Wyss [60] and Edstrand and others [57–59] in Sweden, they found that the effects of velocity and total air content upon sigma depend strongly on the type of cavitation. They observed bubble (or "burbling") cavitation, cavitation by lamina (steady-cavity), and vortex cavitation, although all types did not appear in all tests. Figures 3-5 is typical. In general, an increase of air content or velocity reduces sigma. Their results are

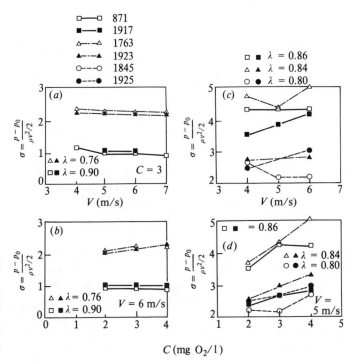

Figure 3-5 Types of cavitation observed on propellors and effects of oxygen content and velocity on inception sigma (Bindel and Riou [72]).

qualitatively similar to those of both Edstrand and Vuskovic, in that they found large effects of air for vortex cavitation and little effect with laminar cavitation. The data for "burbling cavitation" are less consistent between Bindel and his colleagues and the others.

A further result of Bindel and others is that velocity and air-content effects are not independent. In all cases where there is an effect of air content, an increase therein favors the appearance of cavitation. However, the effect of velocity increase upon sigma is not consistent in direction, and depends upon other conditions.

National Physical Laboratory water-tunnel tests, Teddington, England [75] Propellor-performance tests are reported by Silverleaf and Berry [75] in which total air content was varied. An ultrasonic transmission method was tried unsuccessfully to distinguish entrained from dissolved gas. Again it was found that the development of cavitation is favored by an increase of air content.

Colorado State University water tunnel [76, 77] Tests upon cavitating valves and orifices ranging in diameter from 1 to 40 in are reported by Tullis and Govindarajan [76, 77] for a once-through pipe system in which air content was not measured. Size scale effects were found, as expected, but there was little evidence of the effect upon inception sigma of probable large variations in total air and entrained air content in these tests.

SOGREAH, Grenoble, France [78–84] A series of tests have been reported [78–84] from SOGREAH upon cavitation in regions of strong shear such as the wake region downstream of an orifice wherein air content, velocity, and temperature were varied (to increase the Reynolds number variation). A substantial effect upon sigma of total air content was found (Fig. 3-6), particularly in the range of moderate air contents (30 to 60 percent saturation at STP). Also, a strong effect of the previous pressure history of the water was found, again indicating the necessity of a more detailed specification of entrained air content than that provided by total air content. A "bubble microscope" [83, 84] for the direct observation and photography of the entrained gas nuclei has been studied in this laboratory. A difficulty with this approach is that only a very small field can be sampled, and counting and classification of particles is very tedious.

University of Michigan venturi studies [85–89] A relatively long and comprehensive study of both damage and performance effects in venturi test sections has been made in the writer's laboratory at the University of Michigan [85–89]. Considerable work on gas effects on cavitation sigma for both water and mercury has been done. The inclusion of liquid metals is of basic as well as practical interest because gas solubility is usually much reduced for such fluids. The cavitation-sigma studies have been in two parts:

1. Cavitation venturi tests for geometrically similar venturis over a range of throat

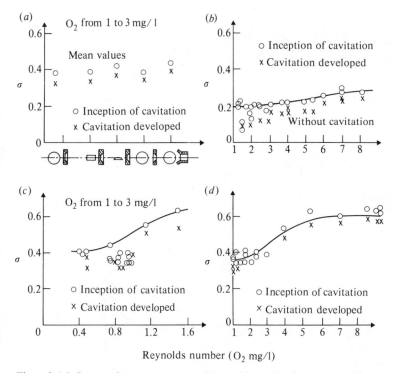

Figure 3-6 Influence of oxygen content and Reynolds number for various orifice shapes on inception sigma (Duport [82]). (*a*) Influence of form jet and of the nozzle. (*b*) Influence of O_2 (ref. test section). (*c*) Influence of Reynolds number (circular orifice). (*d*) Influence of O_2 (circular orifice).

diameter from 3 to 18 mm with 6° included angle for water and mercury with considerable variation of temperature and velocity [85–89]. Figure 3-7*a* and *b* shows the flow path and typical water results.

2. Development of a modified Coulter-counter [85, 89] system and light scattering, for example, for measurement of gas nuclei size and population distribution, and correlation with sigma. Typically for this small high-speed tunnel (to 70 m/s) the entrained gas volume is only about 10^{-6} of the total. The effects of gas content and velocity are substantial (Fig. 3-8). For high gas contents, sigma decreases strongly with velocity, passes through a minimum, and then increases. This behavior is similar to that found by Jekat [90] in an axial-flow "hubless" inducer for air-saturated water.

For low gas contents in these venturi tests (Fig. 3-8) sigma increases monotonically with velocity. The separation between sigma curves is much greater at low than at high velocity, consistent with work by Holl [67] and Bindel [72–74]. Since this separation is approximately inversely proportional to dynamic head, it appears that the gas pressure in the bubbles is constant over the range tested. Following the approach suggested by Holl [67]:

$$\sigma = \sigma_0 + \sigma_{\text{gas}} \tag{3-2}$$

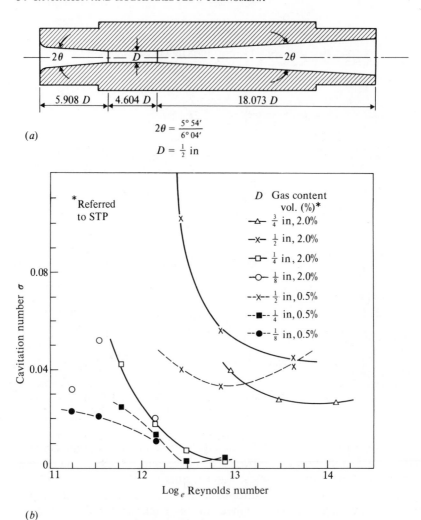

Figure 3-7 (a) University of Michigan venturi flow path ($\frac{1}{2}$ inch throat), (Hammitt, et. al. [86–88]). (b) Inception sigma vs. Reynolds number, University of Michigan venturis ($\frac{1}{8}$ to $\frac{3}{4}$ inch throats) (Hammitt et. al. [86–88]). σ is defined in Fig. 3-8.

where

$$\sigma_{gas} = (kP_{gas_{sat}})/(\rho V_t^2/2)$$

$P_{gas_{sat}}$ = gas pressure to which water is exposed

k = proportion of saturation pressure actually in bubble

It can be computed from the data, but with considerable scatter which is presumably largely due to the differing pressure histories of the water. Figure 3-9

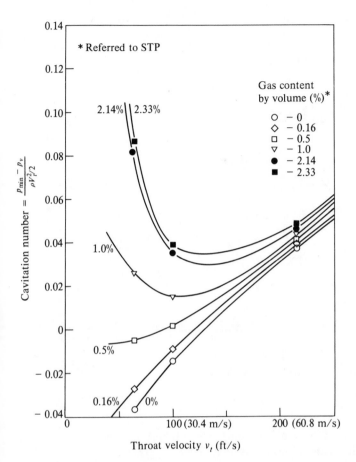

Figure 3-8 Cavitation inception sigma vs. throat velocity $-\frac{1}{2}$ inch (12.7 mm) venturi (Hammitt et al. [86–88]).

shows the effect of pre-pressurizing of water. In our present tests, k averages 0.009 for water and 0.058 for mercury.

Swedish State Power Administration [91]   Figure 3-10 shows data by Fallström [91] from tests upon a Kaplan turbine. The results are consistent with those of Vuskovic [60], also upon a Kaplan turbine. It is Fallström's opinion [91] that air content within the range tested affects only the initial appearance of bubbles in this type of machine.

Technischen Hochschule, Darmstadt, West Germany   A recent doctoral dissertation by Gast [92] from Darmstadt reports on experiments involving cavitation upon a submerged object in a water tunnel, in which the effects of air content upon inception sigma were studied. Consistent with other work, it was found

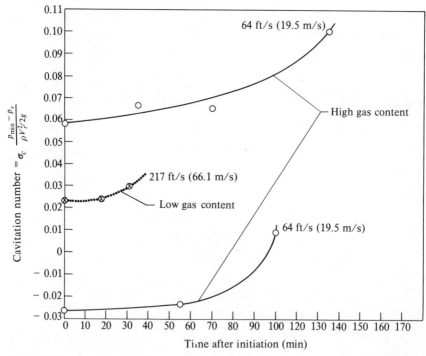

Figure 3-9 Prepressurization effects on cavitation sigma, $\frac{1}{2}$ inch (12.7 mm) venturi (Hammitt et al. [80–82]).

that sigma increased for higher air contents. Some theoretical justification based on the dynamics of individual bubbles is included.

National Engineering Laboratory, East Kilbride, Scotland, and University of Durham [93–96] While no air-content work from the water tunnels at NEL have been reported, to the writer's knowledge, some relatively basic work relating air content and nucleation under nonflowing conditions has been supported by NEL at King's College, University of Durham. Nucleation thresholds in static samples (water and organic liquids) under ultrasonic irradiation as a function of total air content, pressurization history, and other forms of pretreatment, such as centrifuging, were reported by Richardson and others [93–96]. For fixed total air content it was found that nucleation thresholds were strongly influenced by the distribution between entrained and dissolved portions, and nuclei diameters. Generally only the entrained portion was of importance. They developed a technique to measure the entrained gas spectra [93–96] based upon the attenuation of an ultrasonic beam caused by bubbles of a size resonant with the imposed frequency. This approach is similar to that used at the University of Minnesota [68–71], previously discussed. Iyengar and Richardson also experimented with a light-scattering instrument for the same purpose [96].

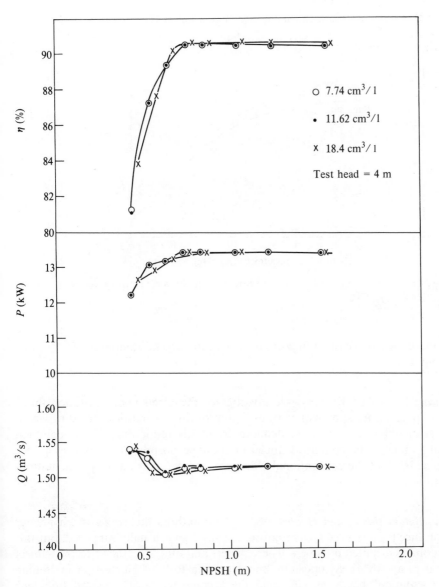

Figure 3-10 Air-content (STP) effects on Kaplan turbine (P. G. Fallström, Swedish State Power Administration, Stockholm, Sweden). [91].

Miscellaneous nucleation studies for nonflowing systems

Galloway [97] Figure 3-11 shows the strong increase in cavitation threshold (comparable to a decrease in sigma in the flowing test) observed by Galloway [97] for water and also for benzene from a static test in an ultrasonic field as air content

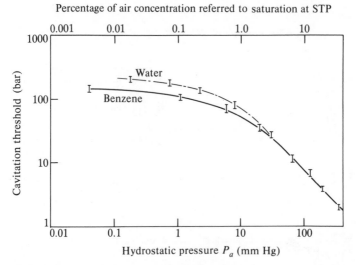

Percentage of air concentration referred to saturation at STP

Figure 3-11 Cavitation threshold of water and benzene as function of relative air content (compared to STP) (Galloway [97]).

is reduced. For this case of high-frequency excitation, substantial liquid tensions are found.

Hayward **[98]**   At NEL Hayward investigated the effects upon cavitation threshold of prepressurization with various liquids of various purities, including water, in a small flowing system. He deduced from this result that, of the common liquids, water alone contained nuclei of the type postulated by Harvey [6–9]. Thus he felt that the conventional Harvey model was not the major cavitation-nucleation mechanism.

Ward, Balakrishnan, and Hooper **[99]**   These authors theorize a much greater importance for dissolved (versus entrained) gas than is usually supposed, but this viewpoint is disputed by others. These papers and others pertinent to the subject appear in an ASME Symposium Booklet, "The Role of Nucleation in Boiling and Cavitation" [2, 4, 99, 100] and the accompanying "Discussion Booklet." Reference 2 by Holl in this Symposium Booklet is a particularly good survey of the nucleation state of the art pertaining to cavitation.

Nystrom and Hammitt, University of Michigan **[101, 102]**   Ultrasonic cavitation-threshold tests in molten sodium by these authors [101, 102] show the existence of large liquid tensions to nucleate cavitation under high-frequency irradiation (tension increasing with frequency), consistent with the previously discussed results of Galloway [97] for water and organic liquids.

3-2-3 "Nuclei" Measurement Techniques and Inception Sigma

General background Entrained "nuclei," which are generally too small to be visible to the unaided eye, can be measured and counted only through the use of sophisticated instrumentation. For transparent liquids, the first and most obvious technique is that of high-speed photography combined with a high-resolution microscope. An instrument for this purpose has in fact been developed at SOGREAH [83, 84]. Practical difficulties for its routine use lie in the very restricted field of view and the tedious nature of data reduction. A related and improved technique is the use of holographic photography. Many focal planes can be recorded with one holograph. Automatized data-reduction techniques

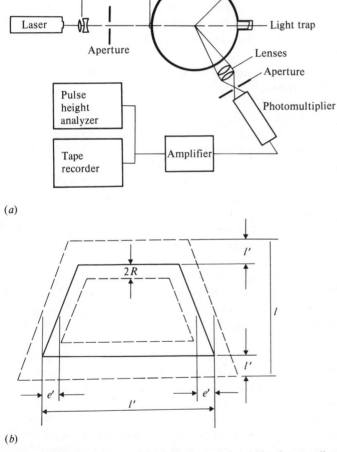

(a)

(b)

Figure 3-12 (a) Basic layout of the optical measuring device for recording gas nuclei. (b) Projection of the control volume on a plane perpendicular to the flow direction, and designation of the boundary regions (Yilmaz [118]).

have been developed [103, 104], but the equipment is costly and complex. The last report compares the results of several optical methods. A very useful optical technique which has recently been developed for this purpose is that of light scattering [103–109]. Instruments of this type using laser light have been developed at the University of Munich by Keller [105], and by Landa and his colleagues at Hydronautics, Inc. [103, 104]. Figure 3-12 is a schematic of this system. In practice laser light was used in these cases because of its well-controlled and collimated nature. Scattered light is measured at any convenient scattering angle by a photomultiplier. Light is scattered to the photomultiplier only when a particle is present in the field of view, the volume of which is defined by the optics of the system. The energy of the scattered-light pulse is theoretically proportional to the volume of the particle. However, calibration with particles of known volume is necessary. The particle size to be measured in the usual case (~ 10 to $1000\ \mu$m) is in the range convenient for the scattering of visible light. Such an instrument could thus not measure individual particle size if this were, for example, in the range of 0.01 to 1 μm, which is in fact the case for the primary droplet content of wet-steam flows (discussed in Chap. 7). A somewhat similar system was also studied at NEL [96].

One of the earliest tools used for the detection and measurement of entrained cavitation nuclei is that of imposed ultrasonic irradiation [68–71, 93–95, 110–114]. Microbubble resonances in the range 1 to 10 MHz can be exploited for this purpose. An input signal of a narrow frequency band will be attenuated by microbubbles of the resonant size for that particular frequency, and thus the number of such bubbles within the beam volume can be estimated. Use of other frequency bands can then produce a microbubble spectrum, similar to that from the light-scattering method already discussed. Figure 3-13 [111] shows the relation between bubble diameter and resonant frequency (inversely proportional).

Other acoustic systems which have been used for the purpose of microbubble-spectra measurement involve sound scattering as well as the effect of a very small "void fraction" on the speed of sound in the liquid. The latter approach can determine a total entrained gas volume, but the distribution into particle size can only be inferred [112, 113]. This approach may not be useful for relatively small gas volume, which may range from 10^{-6} to 10^{-9} in different cases. The acoustic techniques have the inherent advantage that they can be applied to nontransparent liquids, where they represent perhaps the only feasible approach, such as for sodium.

Still another effect which has been utilized to produce this type of measurement is that of gas "nuclei" upon the electrical conductivity of the liquid. If a sample of the test liquid (water, etc.) is considered as an electrolyte and is drawn through a microorifice in the wall of a glass test tube, for example, the passage of a gas microbubble through the orifice will provide a measurable pulse in the electronic circuitry measuring the resistance between points in the test liquid inside and outside the glass test tube. The strength of this pulse is proportional to the volume of the gas particle. This system, known as a "Coulter counter," is shown schematically in Fig. 3-14, as well as typical results. It has been developed [85, 89,

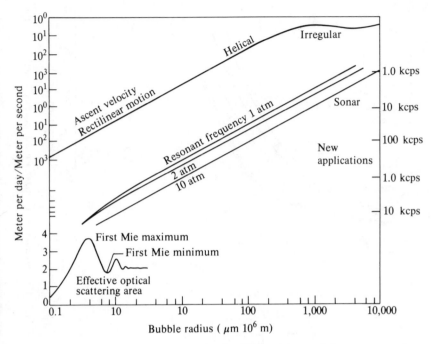

Figure 3-13 Classification of bubble size (from [111]) Vitro laboratories).

115–118] for use at the University of Michigan, along with the laser light-scattering system [105, 106, 118] for use in both venturi and vibratory cavitation systems. A comparison between the results obtained [106, 118] shows agreement within a factor of <2 (Coulter system provides the lower count), only if about 0.1 percent NaCl is added to the water. The necessity of such an additive, plus that of obtaining a currect water sample in the instrument which is outside the tunnel, tend to limit the utility of the Coulter-counter system for this purpose. However, it is a relatively simple and low-cost instrument, which can be easily adapted to some opaque liquids. The application to molten sodium, however, involves difficult material problems which may not be solvable.

International Towing Tank Conference (ITTC) studies [119, 120] A recent "round-robin" test program to measure cavitation-inception sigma for various selected "head forms" in various cavitation tunnels was sponsored by ITTC [119, 120]. The large resultant scatter in inception sigma between different installations for the same head forms and otherwise identical nominal tunnel conditions has motivated ITTC to encourage the development of readily usable and practical instruments for the measurement of entrained gas nuclei spectra in cavitating water tunnels and to organize a comparison between the results of some of the presently available techniques. For this purpose, a comparative test between microscopic photography, holographic photography, and laser light scattering was

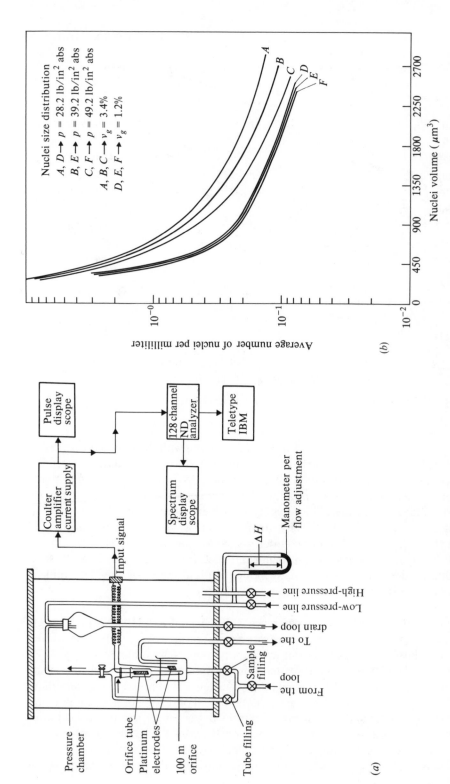

Figure 3-14 (a) Schematic of cavitation nuclei spectrum set-up (Ahmed and Hammitt [85]). (b) University of Michigan Coulter counter, schematic diagram and nuclei size spectrum (Ahmed and Hammitt [85]).

conducted at the SOGREAH cavitation tunnel in Grenoble [121]. The Coulter counter was not included, but the comparison between that instrument and the laser scattering system was made at the University of Michigan [19, 20, 106, 107, 118], as already discussed. The results of the SOGREAH experiment [120] were that the holographic and light-scattering systems agreed quite closely, but not the microscopic photography system. No similar direct comparison between the ultrasonic and other systems has yet been made. Hence, it appears that much remains to be done before a standardized system for gas nuclei spectrum measurement will be available, but that reasonable agreement has already been demonstrated between the laser scattering, holographic, and Coulter-counter systems. However, of these, only the latter can be conceivably applied to nontransparent liquids.

University of Michigan nuclei spectra and sigma tests [85–89, 122] Over the past few years at the University of Michigan a series of studies to measure microbubble spectra in water and correlate with cavitation thresholds in both venturi and vibratory cavitation facilities has been conducted involving five separate Ph.D. theses [87, 89, 116, 118, 123]. Nuclei spectra have been measured with both Coulter-counter and laser light-scattering systems, and cavitation thresholds have been measured in both vibratory and venturi systems. Water, distilled and with various additives, has been used. Cavitation thresholds have also been measured in mercury and sodium [86, 101, 102, 123], but no nuclei measurement was possible. Nuclei sizes can, however, be inferred from cavitation-threshold measurements for all the liquids tested [122].

The results of this work as they pertain to threshold correlations in general will be discussed in an appropriate later section. Here, the particular features pertaining to the entrained nuclei spectra measurements will be considered.

The earliest particle spectrum studies at the University of Michigan involved the application of the Coulter counter to a closed cavitating venturi loop with water [85, 89, 115, 122]. These tests showed that the entrained portion of the total gas, measured by "Van Slyke" [85, 89], has relative volume only of the order 10^{-6} in our facility and that the entrained volume decreased constantly for long periods of tunnel operation (1 to 2 weeks) after start-up. Thus the possibility of constantly changing entrained portions, even over very long periods in a closed-loop facility where the total gas content (~ 1 to 2 percent by volume at STP) is maintained constant, is demonstrated.

The above-described venturi program with Coulter counter was followed by two in-series Ph.D. investigations [116, 118] in a vibratory cavitation facility shown schematically in Fig. 3-15a to d. This is again essentially a closed-loop sealed facility where total gas content can be measured (Van Slyke), varied, or maintained constant. However, as opposed to the venturi tests conducted in our "high-speed water tunnel" [124], the vibratory test loop is of small-bore glass tubing with a small plastic circulating pump, so that high degrees of water purity can be maintained. The rate of circulation is sufficient only to allow measurement of entrained nuclei content, so that velocity effects are nil. This loop and

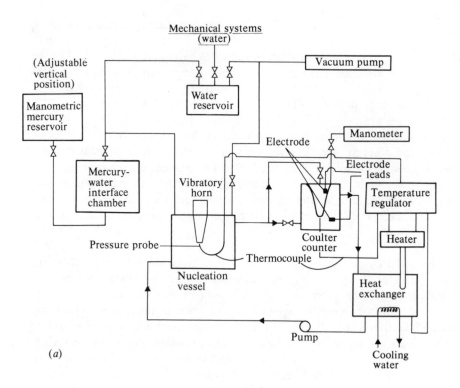

(a)

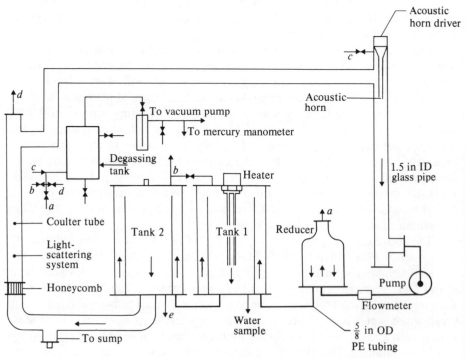

(b)

74

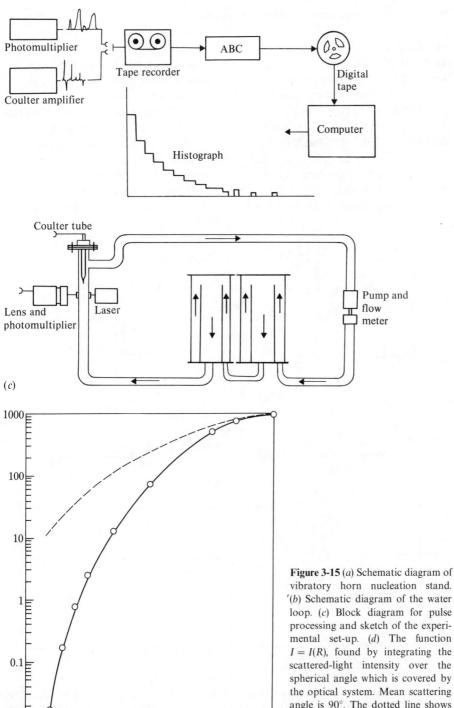

Figure 3-15 (*a*) Schematic diagram of vibratory horn nucleation stand. '(*b*) Schematic diagram of the water loop. (*c*) Block diagram for pulse processing and sketch of the experimental set-up. (*d*) The function $I = I(R)$, found by integrating the scattered-light intensity over the spherical angle which is covered by the optical system. Mean scattering angle is 90°. The dotted line shows the scattered intensity versus the nucleus radius as obtained from the geometrical approximation ($I \sim R^2$).

the vibratory cavitation facility may be considered as providing an essentially static test. Nevertheless, glass stilling tanks were included in the second investigation [116] to allow more complete detrainment of larger microbubbles.

Typical microparticle spectra from the two investigations are shown in Fig. 3-16a and b. Considering the anticipated differences between the Coulter-counter results of Pyun [116] with unsalted water and the later primarily laser scattering results of Yilmaz [118] in the facility with stilling tanks added, the overall microbubble spectra are reasonably similar, at least over the midportion of the range where the majority of the particles exist. For the very small particles, Pyun shows a decrease in number with diameter, so that his results show a "most probable size," whereas the results of Yilmaz show a continually increasing number

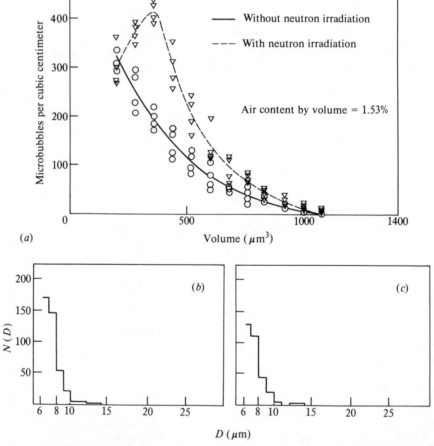

Figure 3-16 (a) Microbubble spectrum in a given sample of tap water at $P = 15.8$ lb/in^2, $T = 70°$F, and $t = 24$ hours. (b) Nucleus histogram of untreated tap water, obtained with the optical method (sample Nr. 1). Volume of tested water 0.36 cm^3, $N_{tl} = 406$; N (nuclei per cm^3) = 1130 (c) Nucleus size histogram of degased tap water obtained with the optical method (sample Nr. 2) $N_{tl} = 328$; N (nuclei per cm^3) = 911.

as size is decreased to the limits of the instrumentation (~ 5 μm). This discrepancy is not explained, but previous results from the water tunnel at the University of Munich by Keller [105] agree with Yilmaz in this respect.

While the particle spectra of Pyun and Yilmaz agreed reasonably well over the main part of the diameter range, no valid comparison can be made for the few larger particles, which no doubt exist in either case but may well escape measurement entirely. However, it is reasonable to assume that these would be reduced in number (by perhaps orders of magnitude) by the stilling tanks incorporated by Yilmaz but not used by Pyun (Fig. 3-15b). This hypothesis is confirmed by the fact that the thresholds measured by Pyun were of the order 1 to 2 lb/in^2 whereas those of Yilmaz were of the order 4 to 5 bar. This discrepancy is of major importance, because it shows that reasonably detailed microbubble spectra, such as obtained in the case at present under discussion, may not be sufficient to predict cavitation threshold. It is probably only the few very large microparticles which control, rather than the distribution in general and the "most probable size." These few large particles may be much more difficult to detect in practice than measuring the overall spectrum.

Another possible cause, aside from the possible existence of a few larger bubbles, of the low thresholds measured by Pyun (as compared to Yilmaz in almost the identical facility) may be the effect of "rectified diffusion" [125–127] in inflating originally small microbubbles to those of sufficient diameter to explain the low measured thresholds. However, this effect should have existed in the Yilmaz experiment to the same degree as in that of Pyun, since the dissolved gas contents were similar.

The first relatively comprehensive cavitation nucleation experiments at Michigan, conducted before the availability there of particle spectra-measuring techniques, were the Ph.D. theses of Ericson [87] and Nystrom [123]. Ericson's work was conducted in venturi systems (and has already been mentioned) and that of Nystrom in a vibratory facility. Ericson studied effects upon cavitation sigma, including overall gas content as measured by Van Slyke, in water and mercury, while Nystrom studied effects in high-temperature sodium in a sealed vibratory facility. The effects of imposed frequency (14.5 to 24.5 kHz) as well as temperature (500 to 1500°F) were investigated. While it was not possible to measure entrained particles in either of these investigations, their "effective" diameter can be inferred from the cavitation-threshold measurements. As a first approximation, the well-known static balance equation between surface tension and pressure differential of bubble contents to the surrounding liquid can be used, (Eq. 3-1), so that

$$\Delta p = \frac{2\sigma}{R} \qquad \text{and} \qquad R = \frac{2\sigma}{\Delta p} \qquad (3\text{-}3)$$

From this calculation it was found that, for the sodium tested, the "effective" nucleus radius ranged from ~ 10 to ~ 600 μm. This radius increased with increased temperature and decreased with imposed frequency [101, 102]. However, over most of the test range, the effective radius varied from 9 to 30 μm, which is about the

Table 3-1 Rectified diffusion parameters for sodium and water [102]

Temperature, °F	Fluid	Concentration C, gr/gr	Diffusion coefficient, D, cm^2/s	CD, cm^2/s
68	Water-air	2.4×10^{-5}	2.0×10^{-5}	5.0×10^{-10}
500	Sodium-argon	6.9×10^{-11}	7.0×10^{-5}	4.8×10^{-15}
800	Sodium-argon	4.4×10^{-8}	13.0×10^{-5}	5.7×10^{-13}
1000	Sodium-argon	3.5×10^{-8}	18.0×10^{-5}	6.3×10^{-12}
1500	Sodium-argon	1.10×10^{-6}	31.0×10^{-5}	3.4×10^{-11}
500	Sodium-helium	1.1×10^{-8}	11.5×10^{-5}	1.3×10^{-12}
800	Sodium-helium	1.1×10^{-7}	21.5×10^{-5}	2.4×10^{-11}
1000	Sodium-helium	3.5×10^{-7}	30.0×10^{-5}	1.0×10^{-10}
1500	Sodium-helium	2.2×10^{-6}	52.0×10^{-5}	1.1×10^{-9}

same range found in our vibratory cavitation water tests [116, 118]. The apparent much-larger size nucleus, i.e., smaller threshold, found in the maximum-temperature, minimum-frequency tests may stem from the increased importance of rectified gas diffusion (discussed earlier) at the high-temperature condition [101, 102, 122, 127]. As opposed to most liquids, the solubility for inert gas in sodium increases rapidly with temperature. These results are summarized in Table 3-1 from Ref. 102. The increased radius with reduced frequency is no doubt due to the reduced importance of inertial resistance to bubble growth at lower frequency. Conditions thus approach those predicted by the static balance [Eqs. (3-1) and (3-3)].

3-2-4 Major Basic Research Trends

Since the overall objective of predicting in advance air-content effects upon the cavitating behavior of various apparata has not yet been attained, many studies of an empirical nature such as those already reviewed have been necessary. To attain a predicting capability, more basic work is required. Three major issues are outstanding:

1. Measure (or compute) the upstream nuclei spectrum.
2. Compute nuclei trajectory and growth or collapse rates during their passage between the region where the spectrum is known and the region of cavitation. Since the flow is likely to be turbulent, three-dimensional, and biphasic, this is probably infeasible at present, even for large computers. The applicability of adequately simplified flow models is then the question.
3. Compute the effect of the cavitating flow on the machine. At present it is not feasible to fully implement any of the above. However, many studies, as already discussed, have sought to improve these capabilities. Such miscellaneous studies are discussed below.

Bubble nucleus and fluid-flow calculations

Pennsylvania State University (Penn State) and California Institute of Technology (CIT)   Theoretical and experimental studies in these laboratories have considered both the behavior of bubble nuclei attached to a wall (Refs. 65, 66, and 128 at CIT, for example) and those growing in the flowing stream (Refs. 5 and 67, for example, by Holl and others at Penn State). This latter theoretical work has been applied successfully to the venturi sigma measurements at the University of Michigan [86–88], previously discussed. However, important features such as turbulent effects upon gas diffusion have been largely neglected.

Individual bubble studies   Many individual bubble studies exist in the literature, but are too numerous to list here. Many are not pertinent to nucleation. One of the more comprehensive and applicable early studies, however, is that of Gallant [129]. Another of special interest, in that it involves overall bubble trajectories, is that by Johnson and Hsieh [130]. This is one of the few studies which considers the effects of pressure and velocity gradients upon bubble trajectories.

3-2-5 Air- or Gas-Content Effects upon Cavitation Damage

Studies of the effects of air content upon cavitation damage are far more limited [131, 132] than those upon cavitation inception and performance effects. However, air content should have a substantial effect, at least in some cases, through the action of the following opposing mechanisms:

1. Higher air contents favor cavitation, providing an increased number of bubbles, the collapse of which may be damaging.
2. Higher air contents within individual bubbles reduces collapsing-wall velocities and hence pressure radiation into the surrounding liquid [87, 88]. Field observations show that large quantities of free air in damage-prone regions reduces damage rates [1].

   More detailed consideration of these mechanisms indicates the strong probability that for very high gas contents (saturation and above), an increase in air will reduce damage through the "cushioning" effect on individual bubble collapse, and perhaps through more rapid attenuation of shock waves in the surrounding liquid. On the other hand, for very low gas contents, sigma is substantially increased if gas content is further reduced [87–89, 97], so that the reduction in the number of bubbles is more important than the increased collapse violence for individual bubbles. Thus cavitation damage is reduced if gas content is further reduced, and, in fact, in some cases cavitation may disappear entirely.

   Beyond the individual bubble-collapse studies already discussed there are very few theoretical studies of the effects of gas content upon cavitation damage. One such recent study is that by Smith and Mesler [133].

   The known experimental results will be discussed next.

Venturi tests at Holtwood Laboratory, Safe Harbor Water Power Corporation, by Mousson [134]   The earliest report of tests of air-content effects upon cavitation damage is that by Mousson [134] in 1937 (Fig. 3-17) for runs made in a special damage venturi. For substantial rates of air injection (range of 1 to 2 percent), damage was substantially reduced (air content itself was not measured). The proportionate air flow to reduce damage markedly increased with water velocity, so that injection power loss may become substantial at high water velocity.

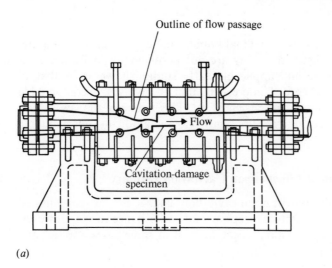

(a)

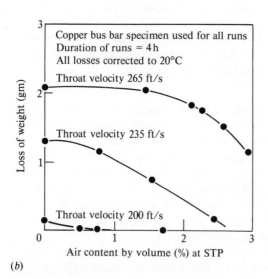

(b)

Figure 3-17 Effects of air injection upon cavitation damage in venturi (Mousson [134]). (a) Venturi test section. (b) Effects of air injection upon cavitation damage in venturi (Mousson [134]).

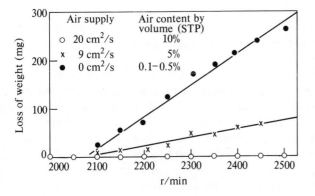

Figure 3-18 Effects of air content upon cavitation damage in rotating disc apparatus (Rasmussen [135, 136]).

Venturi tests at Escher-Wyss by Vuskovic [60] Vuskovic's air-content damage tests in 1940 [60] were made in a venturi similar to that of Mousson [134] at a velocity of 60 m/s (lowest velocity of Mousson). No actual weight-loss measurements were made, but air content was measured. Consistent generally with Mousson's results, there was a steady reduction of damage as air content was increased from 0.3 to 1.7 percent of saturation at STP.

Rotating disc and venturi test by Rasmussen [135, 136] In 1955 Rasmussen used both a rotating-disc apparatus and a special damage venturi (Shal'nev type). Air contents were measured. Figures 3-18 and 3-19 show typical results for the rotating disc and the venturi, respectively. Again, damage decreased continuously and substantially as air content was increased from near zero to about 10 percent by volume (i.e., about five times saturation at STP), thus covering a range similar to

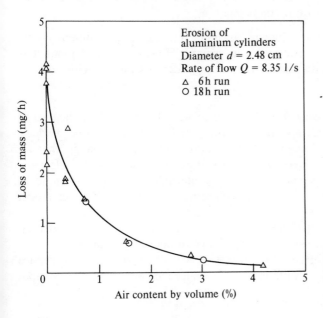

Figure 3-19 Effects of air content upon cavitation damage in venturi upon aluminum alloy (Rasmussen [135, 136]).

that of Mousson [134] and Vuskovic [60]. The proportionate decrease depends both on material and the type of test, being much greater for aluminum than for cast iron and greater for the venturi than for the rotating disc.

Nonflowing vibratory damage tests by Hobbs [137, 138] and Sirotyuk [139]   Air-content effects upon damage in a static vibratory-type test were measured by Hobbs and others [137, 138] in 1967 and 1969, and Sirotyuk [139] in 1966. It is well known that damage rates maximize in a test of this sort for an intermediate temperature [1, 137–141]. It has been suggested that the reduction at the low-temperature end is due to increased gas solubility at low temperature and hence increased gas content, since tests are normally conducted in an open beaker with free surface period Hobbs' results show little effect on damage over the gas-content range 0.1 to 1.0 times saturation at STP, tending to disprove this hypothesis. Hobbs does show a reduction in damage near the upper end of his gas-content range, which is much lower than that used in the flowing tests, so that his results are not inconsistent with these tests. He also shows a reduction in damage at low gas content, which the flowing tests did not, presumably due to the lack of "nuclei" under this condition. The results of Sirotyuk [139] are relatively similar.

Cathodic protection and gas content [142, 143]   Cathodic protection to suppress electrochemical effects in cavitation was apparently first suggested by Petracchi [142] in 1944. Later work by Plesset [143] suggested that the damage reduction might be partially due to the gas cushioning effects of the electrolytic hydrogen released at the wall. Thus the success of cathodic protection may be partially a gas-content effect.

3-2-6 Conclusion and Recommendations

Much remains to be done in achieving an understanding of gas-content effects upon cavitation and bubble nucleation in general. However, at this point it is possible to formulate certain important recommendations and conclusions, for example:

1. In general, variation of gas content has little observable effect upon the overall performance of machines operating well into the cavitating regime. It does, however, often have an important effect upon inception sigma, in that an increase in air causes an increase in sigma. It can thus importantly affect the prediction of inception sigma for prototype machines from tests on models, if there are differences in water quality with respect to gas nuclei between model and prototype conditions.
2. Air content can importantly affect cavitation-damage rates if, as in some cases, it establishes the existence and quantity of cavitation itself. Also, large amounts of air, usually well in excess of saturation, will often substantially reduce cavitation damage, probably because of the reduced violence of bubble collapse in a gassy liquid.

3. The importance of air content upon inception sigma depends upon the type of cavitation, i.e., bubble, laminar (steady cavity), vortex cavitation, etc.; bubble cavitation is most sensitive. The type of cavitation found depends on geometry and other flow parameters.
4. Predictions of gas-content effects upon cavitation are not possible if only total gas content is known. It is necessary to assure water of similar population and size spectra of "nuclei," as well as total gas content, if gas-content "scale effects" are to be avoided.

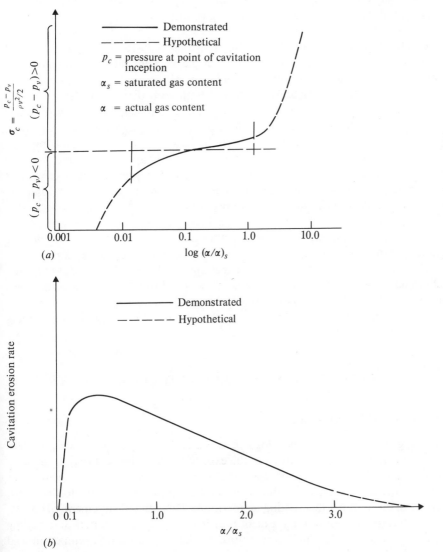

(a)

(b)

Figure 3-20 (a) Inception sigma vs. relative air content (hypothetical example). (b) Erosion rate vs. relative air content (hypothetical example).

5. A general capability for a fully theoretical prediction of gas-content effects upon cavitation is not within the present state of the art, since the flow is biphasic, multidimensional, and turbulent. None of these three factors can be feasibly handled in general, even alone. However, mathematical models of limited applicability are helpful to indicate at least the trends to be expected. Much has been done in this direction and much remains.
6. Another essential capability for the prediction of gas-content effects is the easy and practical measurement of the gas nuclei spectra in a flowing system. An alternative and complementary capability is the "calibration" of the liquid for "cavitability" using a standard cavitating device.

Finally, it appears from all the foregoing that only rather vague guidelines can be drawn for the quantitative effects of gas content either upon inception sigma or damage rate for an untested condition. However, fairly firm qualitative results can be utilized, which are consistent both with the existing experimental and theoretical studies, and with the pertinent physical laws. Figure 3-20a and b shows such hypothetical curves for both inception sigma and damage rate. The curve for inception sigma (Fig. 3-20a) is based upon the fact that for very small gas contents the tensile strength of the liquid is appreciable, and for very large gas contents a large liquid pressure is required to prevent rapid growth of gas bubbles, i.e., "gaseous cavitation." The same concept leads to the damage curve (Fig. 3-20b). No cavitation should occur for extremely low gas content due to insufficient nuclei. For somewhat higher gas content, the nuclei population would approach sufficiency, so that a further increase in gas content would not significantly increase the number of cavitation bubbles and hence damage. For very large gas contents, the cushioning effect upon bubble collapse would become predominant and damage would decrease. The typical values shown in both curves (Fig. 3-20) are estimates from experimental results already discussed.

Some quite new reports [45–47] provide good new information on the status of gas-content effects in flow machinery.

3-3 SCALE EFFECTS IN MACHINES AND COMPONENTS

3-3-1 General Background

Following the accepted terminology in the literature, cavitation "scale effects" cover any important deviations from the expectations of classical cavitation laws. These "classical laws" are based upon the assumption that cavitation will occur whenever the static pressure in the region of minimum pressure falls to the vapor pressure of the liquid at the existing temperature. This general statement, of course, also applies to cases of boiling, except that it is usually then stated that boiling will occur whenever the fluid temperature reaches the saturation temperature at the existing pressure. In either case it is further presumed that the pressure in the noncavitating (or nonboiling) portions of the liquid, i.e., the entire flow regime for preinception conditions, will follow conventional scaling laws for single-phase flow of an incompressible liquid, i.e., for conditions of geometric and dynamic "similarity," pressure (or "head," if elevation differences are involved) differentials are proportional to differentials of velocity squared.

Important deviations from such classical scaling conditions have commonly been found to exist for cavitating flows under conditions involving differences of size, i.e., "scale," and hence the name "scale effects." However, it is found that substantial deviations also exist in most cases also for differentials in velocity and pressure. Of course, velocity and pressure effects are related to their own interdependence in most cases. However, if geometric and dynamic similarity are maintained between model and prototype, the origin of important size scale effects is not obvious. Also, there are secondary effects which must be involved, such as changes in turbulence level, Reynolds number, etc. Unfortunately, it has not yet been possible to correlate such size or velocity effects in terms of simple parameters such as Reynolds number. These factors are discussed in more detail in the following sections.

In addition to size, velocity, and pressure scale effects, there are, most importantly, "thermodynamic" and gas-content effects. The former are primarily due to the increased vapor density and pressure within cavitation bubbles (or other cavities) for some liquids, or at higher temperatures for any. This leads to an increased restraint upon either bubble collapse or growth, due to the necessary condensation or evaporization of the vapor and the concomitant requirement for transport of the resultant latent heat to or from the bubble wall. Otherwise, the vapor pressure within the bubble will either rise or decrease, thus retarding collapse or growth. "Thermodynamic" effects upon collapse and damage are discussed in detail in Chap. 4, and the parallel effects upon nucleation and growth will be further discussed in this section.

Effects related to gas content within the liquid, while not being perhaps truly "scale effects," produce results very similar to those of the more conventional scale effects. Hence, these cannot in general be understood without first "sorting out" the effects due to differences in gas content. As explained in considerable detail in Sec. 3-2, these effects are primarily related to the population spectra of "micro-bubbles" from which the macrobubbles comprising a normal cavitation or boiling regime originate. In addition, there are also effects of gas content upon damage as explained in Chap. 4, i.e., increased gas within the bubbles restrains collapse due to strongly increasing internal pressure as this gas is compressed. Also, the presence of even a very small quantity of gas within bubbles (see Chap. 4 for more quantitative information) promotes bubble "rebound," which itself appears to contribute importantly to the resultant damage. Hence gas content can create important variations from classical expectations for both inception and damage, and can thus be legitimately considered as a "scale effect."

As explained in the foregoing, cavitation scale effects apply to all important aspects of cavitation, i.e., inception and damage, as well as to performance. Inception and damage effects are probably in general of greater, but not exclusive, importance. By "performance effects" here is meant such items as changes in efficiency, power, head, thrust (from a propellor), and shape of characteristic curves such as, for example, "head" versus "NPSH." For instance, the "sharpness" of fall-off of head with reducing NPSH often differs between the model and prototype as a typical "scale effect." This could be important in the determination of "inception," since inception is usually defined in terms of a certain percentage of fall-off of head or power from a pump or turbine (or thrust or power for a propellor) from the noncavitating head at the same pump (or turbine or propellor) speed and flow, rather than in terms of the first acoustic (or visible) manifestation of cavitation. These latter indications would perhaps be less sensitive to possible scale effects as compared to the overall externally measured performance curves such as the head-NPSH curve, since no measurable change in head will occur until quite considerable cavitation exists, and often the first externally observable change is in the reverse direction, i.e., a small amount of cavitation may slightly improve performance. However, insufficient information is as yet available to determine the validity of the foregoing supposition concerning scale effects and "first" inception.

The scale effect due to the "thermodynamic-effect" mechanism may be more dependent upon performance scale effects than those due to changes in velocity, pressure, or size, since it could be argued that no thermodynamic-effect changes will exist until cavitation has become relatively well developed beyond the "minimum" (acoustic or visible) inception condition. This is especially true for thermodynamic damage effects, since damage occurs only when cavitation is relatively extensive. The characteristics of the entire cavitation field are certainly strongly affected by thermodynamic effects, i.e., the size and number of bubbles existing, since their development from the nucleation state is relatively restrained when thermodynamic effects are important. Thus the cavitation-performance effects upon the machine are also reduced in these cases, as is well known in the commercial pump field. Of course, in addition, the damage is further drastically reduced in such cases by the reduced collapse violence of individual bubbles, if initial size and number of bubbles were the same as in the absence of the thermodynamic-effect mechanism. This strong damage fall-off with increased temperature for a given liquid at a given suppression pressure is well documented and discussed in Chaps. 4 and 5.

In addition to the various "scale effects" discussed in the foregoing, there are also the effects upon inception, performance, and damage which are incurred through the change of liquids or of liquid temperatures for the same liquid. Assuming that classical cavitation scaling laws are met (i.e., cavitation "sigma," suction specific speed,† Thoma sigma, or other similar scaling parameters are held constant) and the performance of different liquids, or alternatively the same liquid at different temperatures, are compared, significant differences in all phases of cavitation performance are often observed. Presumably these differences are due to differences in the many liquid properties which influence cavitation behavior

---

† $S = (\text{r/min})(\text{gal/min})^{1/2}\text{NPSH}^{3/4}$ in English units, for pumps. A dimensionless form is also possible, and no doubt preferable, but is not in wide use. A separate form for turbines also exists.

beyond those considered in the classical scaling laws already discussed. All these questions are discussed in some detail with respect to specific machines in the next section.

3-3-2 General Machinery Effects

Pumps and turbines—general   Fluid-flow components of major importance, and perhaps least subject to exact analyses because of the rotating and complex nature of the flow, are turbomachinery units such as pumps and turbines. It is conceivable that meaningful progress could be made at this time in the analyses necessary for cavitation prediction by highly sophisticated computer studies of the flow in the rotor to determine locations and magnitudes of minimum pressure. However, this does not seem yet to have been accomplished to the necessary degree. If this were accomplished to the necessary extent, presumably "scale effects" would become predictable, qualitatively and quantitatively. Since it is not possible at present, however, to predict these scale effects, it is necessary to draw upon all experience available, where cavitation-performance data exist, comparing geometrically similar units or the same unit at different speeds or with different liquids.

Effects of size, velocity, and pressure   There is much experimental information at present proving the existence of large cavitation scale effects due to size, velocity, and pressure changes for geometrically similar turbomachines. However, the causative mechanisms are at present only slightly (if at all) understood, and detailed numerical studies are not as yet sufficiently reliable to provide trustworthy predictions. It is thus impossible in general to predict either the magnitude or direction of these scale effects in the absence of specific model-to-prototype tests for specific designs.

Indicative of the above, sophisticated cavitation pump work is being carried on at the National Engineering Laboratory (NEL). The NEL group at Glasgow have long experience with, and are world leaders in, cavitation pump research and design [144–146]. To quote Pearsall [146] in a recent article (1974) concerning pump cavitation-inception scale effects (i.e., variation of inception sigma with model size, speed, etc.):

> On cavitation inception there is so little information that no reliable conclusions can be drawn. There is some evidence that air content alters the trends of scale effects. Most of the tests on pump performance breakdown have shown an improvement with higher speeds and larger sizes (..., i.e., higher suction specific speed). Some tests, however, show the reverse effects.

The present author fully agrees.

It thus appears that very careful model tests are necessary to achieve a given objective regarding cavitation performance and that the possible differences between model and prototype due to speed and size changes, etc., are much greater than the uncertainties between different liquids. In any case, acoustically instrumented prototype tests are clearly necessary to verify the full detailed cavitation performance of a new pump.

Carefully instrumented acoustic tests, summarized by Hammitt [147], do indicate that in many cases considerable bubble activity exists for NPSH values ~4 times that corresponding to the standard commercial 3 percent head drop-off point. Thus cavitation damage may well occur for NPSH values considerably greater than those corresponding to the commercially defined cavitation-inception point. Industrial experience has often indicated that damage is most severe near the start of head drop-off ("knee of curve").

Typical specific results

1. *University of Michigan (U-M) pump tests* [148, 149]. Typical specific data result from tests at U-M using a low specific speed centrifugal pump tested over a range of speed, flow, and temperature in both water and mercury [148, 149].

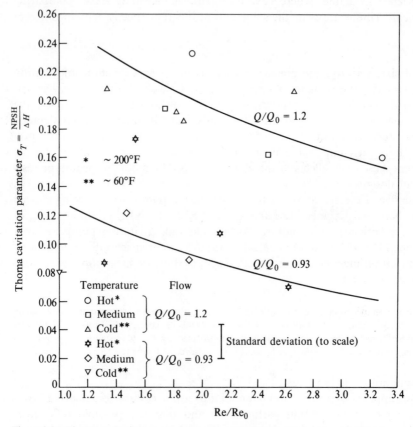

Figure 3-21 Thoma cavitation parameter vs. normalized Reynolds number. U-M tests.

This unit is a small single-stage sump-type centrifugal pump used to power the U-M mercury cavitation loop. Specific speed is 740 in gallons per minute, revolutions per minute, and foot units, and design flow at 1800 r/min is ~40 gal/min (~9.1 m³/h) for $\Delta H = 40$ ft (12.2 m).

In this loop, pump cavitation can only be obtained by reducing the sump pressure to near vacuum, balancing vacuum-pump capacity against stuffing-box leakage. The rather large scatter of the data is probably a result of this test procedure.

A long-radius elbow immediately upstream of the pump suction is partially responsible for the low suction-specific speed values. In addition, the pump was designed for high-temperature liquid-metal operation rather than good cavitation performance.

The data points (Fig. 3-21), with one exception, are the result of several repetitive runs (varying from two to six). In each case a standard deviation was calculated and from these an average value computed, which is shown in the figure. Data of sufficient precision to be statistically meaningful have been obtained (Figs. 3-21 to 3-23). The onset of cavitation is defined as sufficient

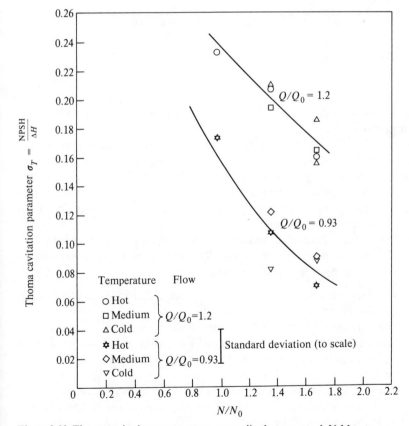

Figure 3-22 Thoma cavitation parameter vs. normalized pump speed. U-M tests.

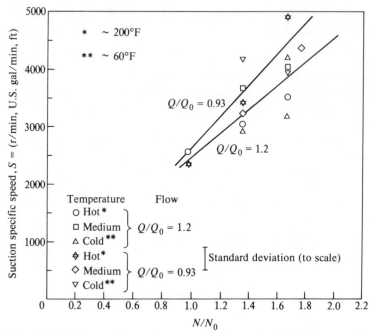

Figure 3-23 Suction specific speed vs. normalized pump speed. U-M tests.

cavitation to cause a decrease of 5 percent in the pump head at a fixed speed and system resistance.

(a) *Water results* [148, 149]. Pump speed was varied over a range of about 1.7, flow (i.e., ratio of actual to design flow) over a range of 1.3, and water temperature over the range from about 80 to 160°F (27 to 71°C). Tests were run at three distinct speeds, which are denoted in the figures as the ratios of actual to design speeds, two flow ratios, and three temperatures (including the extremes of the range above).

The resulting Thoma† cavitation parameters are presented in Figs. 3-21 and 3-22 plotted against normalized Reynolds number and normalized pump speed, respectively. The figures disclose:

(i) The data can be correlated reasonably well either in terms of Reynolds number or pump speed, i.e., velocity.

(ii) The data divide naturally according to the flow ratios, i.e., the Thoma parameter for the higher flow ratio (somewhat above design flow) is higher than for the lower (close to design flow), for the same N/N_0 or Re/Re_0.

(iii) There is a substantial decrease in the Thoma cavitation parameter as the Reynolds number or velocity is increased (30 to 50 percent for a speed increase of 75 percent).

† NPSH/ΔH_{pump}.

The data from Fig. 3-22 are replotted in Fig. 3-23 in terms of the suction specific speed, which varies over almost a 2 : 1 range from about 2500 to 4500. The direction of variation and the relation between high and low flow curves, of course, follows from the previous curves.

2. *Mercury and water results.* It was found for water and mercury, considered together, that the Thoma cavitation parameter decreased virtually on a single smooth curve as the normalized pump speed N/N_0 increased, for a fixed flow coefficient. Although the pump speeds with mercury and water did not overlap due to equipment limitations, it appears from these data that the Thoma cavitation parameter for a given flow coefficient is a function solely of pump speed, regardless of fluid (Fig. 3-24).

The Thoma cavitation parameter also decreased for increasing normalized Reynolds number for both water and mercury, when considered separately

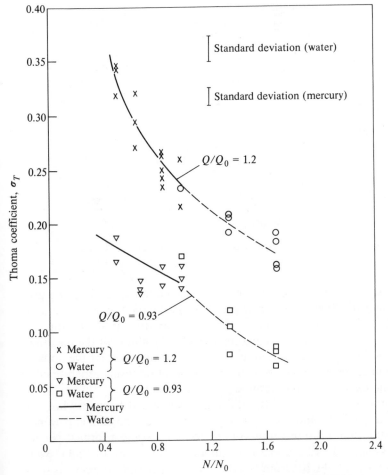

Figure 3-24 Thoma cavitation parameter vs. normalized pump speed. U-M tests.

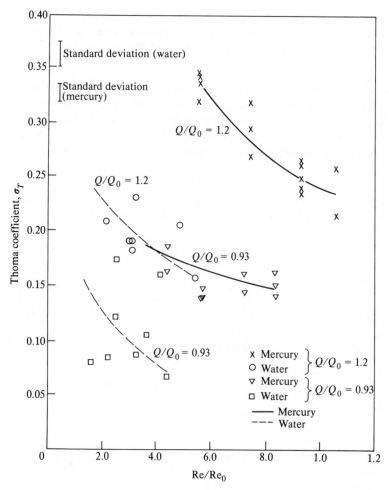

Figure 3-25 Thoma cavitation parameter vs. normalized Reynolds number. U-M tests.

(Fig. 3-25), although the curves for the two fluids did not coincide. For a given flow coefficient and Reynolds number, the Thoma cavitation parameter is about twice as large for mercury as for water. This variation is in the direction predicted by the thermodynamic parameters [1, 50], although the magnitude of the thermodynamic effect cannot be predicted. It may be that the apparent correlation in terms of velocity is actually a result of opposing separate effects due to Reynolds number and thermodynamic parameters.

As mentioned previously, little difference was noted between "hot" and "cold" water ($\sim 70°C$ and $27°C$). However, the thermodynamic-parameter [1, 150] equilibrium ratio of vapor volume to liquid volume formed per unit head depression, differs by a factor of about 5 from "hot" to "cold" water, but by a factor of about 10^7 from "cold" water to mercury. Mercury is thus a fluid for which very little thermodynamic effect would be expected, i.e., it is a "very cold water" in this respect.

Figure 3-26 is a plot of suction specific speed versus normalized pump speed. It shows, of course, simply the inverse trend from the Thoma parameter plots, ranging from about 2500 in gallon per minute units for low speed with mercury to about 4000 for high speed with water.

A hysteresis effect in the ΔH versus NPSH curves was noted for both water and mercury. The pump head tends to be higher for a given NPSH while NPSH is being increased, rather than decreased, through the pump cavitation region. A typical curve from the mercury data (Fig. 3-27) illustrates the effect. Since the average passage time for fluid around the loop is about 10 s (and the time between the readings and reversal of pressure variation for the runs is much longer), no explanation is readily apparent. However, cavitation-inception hysteresis is a common observation in large water tunnels giving rise to Holl's suggested terminology of "incidence" and "desinence" [151]. Obviously, only a

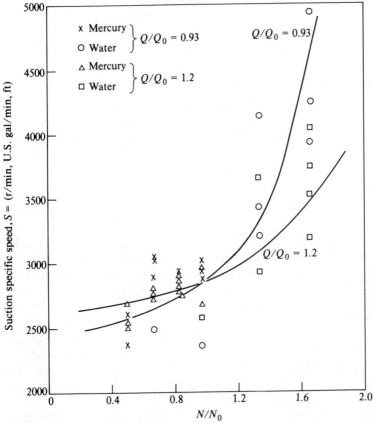

Figure 3-26 Suction specific speed vs. normalized pump speed for water and mercury at two different normalized flow conditions using Berkeley model $1\frac{1}{2}$ WSR centrifugal pump.

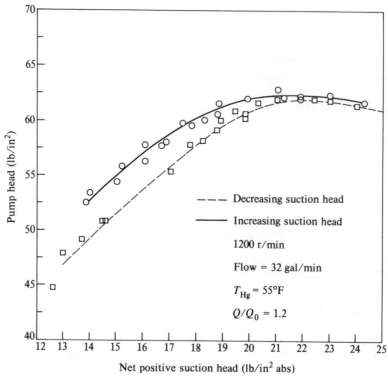

Figure 3-27 Net positive suction head vs. pump head with increasing and decreasing suction head (illustrating hysteresis effect)—Berkeley model $1\frac{1}{2}$ WSR centrifugal pump with mercury as the working fluid.

detailed study and visualization of the flow in the impeller could shed light on this and other phenomena, which complicate at present the evaluation of pump scale effects.

3. *General trends.* The specific results discussed in the foregoing indicated that an increase in speed (or Reynolds number) caused a substantial reduction in cavitation sigma (increase in suction specific speed) over the range tested. They also indicated that pump speed alone was a more successful correlating parameter than Reynolds number, at least for the comparison between mercury and water. Somewhat similar tests for venturis, to be discussed later, indicate the same in this regard. However, the trend of improved cavitation-inception performance for increased speed or Reynolds number is not always found, and in fact the opposite trend has been observed in other cases. In general, the direction and magnitude of this scale effect cannot be predicted in the present state of the art [144] for cases where specific test results are not available. No doubt they depend upon various parameters such as specific speed, detailed pump design, etc. There are numerous fragmentary results published in the literature, but since no definable trends yet emerge, these are not cited in detail

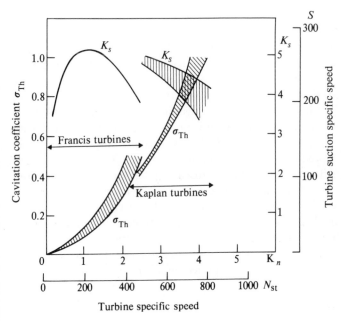

Figure 3-28 Cavitation limits for turbines.

here. For many flow components, including both venturis and pumps, cavitation performance can be expected to improve for an increase in velocity or Reynolds number for relatively low velocity and Reynolds number, but a reverse trend may occur for higher velocity or, perhaps, Reynolds number. This favorable trend has sometimes been accepted as valid, e.g., for some designs of rocket turbopumps. However, the unfavorable trends certainly also exist in some cases.

Test results including both a favorable trend at relatively low velocity and an unfavorable trend at increased velocity were reported, e.g., by Jekat [152] for tests on an axial-flow inducer pump, and were also observed in our venturi tests [148].

4. *Turbine scale effects.* Much of what has already been said applies to water turbines [1, 144, 153–155] as well as pumps, but there are essential differences, as discussed by Pearsall [144]. Since water turbines are usually very large machines, efficiency is more important than for most pumps. Second, a water turbine differs hydraulically from a pump in that the cascade of blades is an accelerating cascade, rather than a diffusing set as in a pump. Thus, minimum pressure occurs near the cascade exit rather than near the inlet as for pumps.

Analysis of test results on many turbines allows an empirical correlation of the critical Thoma coefficient† against specific speed [155], as shown in Fig. 3-28. An empirical correlation has been fitted to this data by Karelin [154] and is reported and discussed by Pearsall [144].

---

† $\sigma_{Thoma} = H_{sv}/H_{turbine}$, where H_{sv} is suppression head, i.e., head above vapor head.

Venturi geometry

General background As illustrated by Fig. 3-29, for example, important and substantial cavitation scale effects due to changes in size, velocity, pressure, thermodynamic parameters, gas content, etc., for geometrically similar venturi flow paths may be expected. Even though the geometry is much simpler than that of centrifugal pumps, it is still true that the theory is unfortunately lacking to allow the prediction of either the magnitude or direction of these effects, except with relatively rare exceptions, as, for example, the gas-content effect, discussed in Sec. 3-2, where some predicting ability appears to exist. Venturi flows differ basically from orifice flows, which will be discussed separately, since a non-cavitating venturi flow is a nonseparated flow, while that from an orifice definitely involves separation, even in the absence of cavitation. Hence, vortices downstream of an orifice are presumably the most important cavitation-inception mechanism for such "separated flows," while they are often of minor importance in venturi flows. They may, however, be of primary importance in some instances of pump cavitation, especially those involving unshrouded impellers.

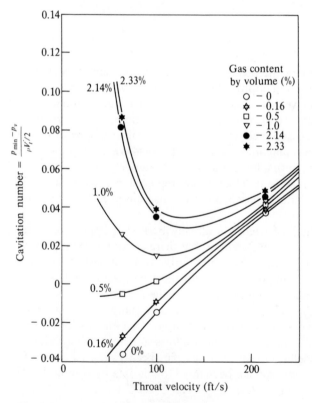

Figure 3-29 Cavitation-inception number vs. throat velocity for $\frac{1}{2}$ inch Plexiglas venturi at room temperature and several contents in water [159].

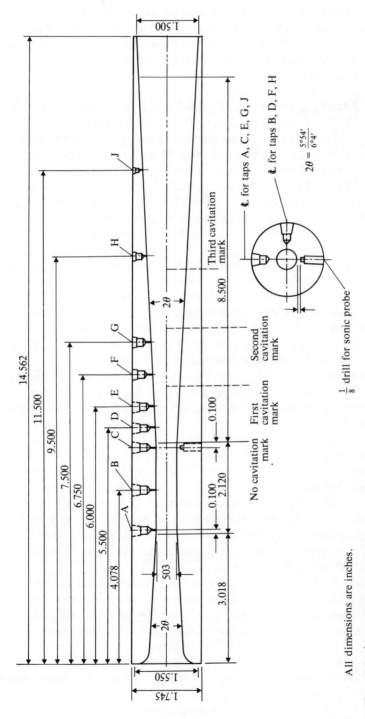

All dimensions are inches.

Figure 3-30 $\frac{1}{2}$ inch (12.7 mm) cavitating venturi test section.

One of the most comprehensive sets of cavitation tests involving venturis is that conducted over the past 10 to 15 years at the University of Michigan. Hence, these results will be reviewed here. Other venturi cavitation investigations have also been reported, but, to the author's knowledge, there is no conflict reported with the U-M data.

University of Michigan (U-M) and Electricité de France (EdF) tests   The U-M tests included both water and mercury. The basic venturi geometry (Fig. 3-30) for a nominal $\frac{1}{2}$-in throat diameter involves a 6-degree included-angle nozzle and diffuser section separated by a cylindrical throat (2.25 L/D). Geometrically similar units were tested for various throat diameters ranging from $\frac{1}{8}$ to 1 in (0.318 to 2.54 cm). Throat velocities ranged from ~ 12 to 70 m/s. Cavitation condition ranged from inception to fully developed, i.e., extending well into the diffuser section. Gas content in many cases was controlled and measured. Both water and mercury were used as test fluids, and a relatively broad temperature range was investigated. For low-temperature tests the venturis were of Plexiglas, and cavitation inception was detected both acoustically and visually. In some cases high-speed motion pictures of the cavitating regimes were made.

The investigations were extended also to sodium by tests at EdF, where stainless-steel venturis of identical flow-path geometry were used. EdF also performed water tests in the same venturi with results consistent with those from U-M. The sodium tests were at a single temperature and throat diameter, but velocity was varied to the extent possible, so that a sodium velocity scale effect was also measured for this geometry. These sodium results will be reviewed in a later section. Thus water, sodium, and mercury have been tested in identical geometry, involving tests in two different laboratories and over differing parameter ranges.

1. U-M water tests

(a) *Inception sigma.* Figure 3.30 shows the basic flow-path dimensions and indicates the various cavitation conditions ("degrees of cavitation") used. Figure 3-31 shows a typical normalized wall axial static-pressure profile for "sonic initiation," i.e., first detectable sound by stethoscope or hydrophone due to cavitation. "Visible initiation" differed from "sonic" in that it involved a continuous ring of cavitation at the exit to the cylindrical throat, the position at which cavitation first appeared.

Figure 3-32 shows typical cavitation sigma† versus throat Reynolds number curves for "sonic initiation" for both $\frac{1}{4}$- and $\frac{1}{2}$-in throat venturis with varying conditions of temperature and gas content. It is noted that there is a very substantial decrease in sigma with increased Reynolds number (or velocity) from ~ 0.2 to ~ 0.0. The data from all these tests for differing degrees of cavitation are combined in Fig. 3-33. It is noted that the velocity or Reynolds number effect upon sigma decreases strongly for increased degree of cavitation,

---

† $\sigma = (p_{min} - p_v)/\rho V^2$, where p_{min} is minimum throat pressure.

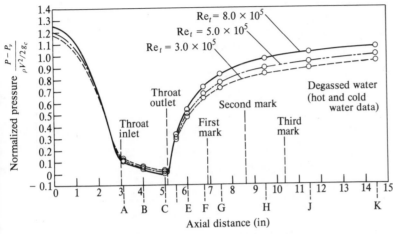

Figure 3-31 Normalized pressure vs. axial position, sonic initation. $\frac{1}{2}$ inch (12.7 mm) test section. U-M venturi.

i.e., it is less pronounced for more fully developed cavitation. However, to some extent, sigma always decreases for these tests for increased Reynolds number. However, later tests (e.g., Fig. 3-29) showed that a minimum sigma was obtained, after which further increase in Reynolds number corresponded to an increasing sigma, i.e., an "optimum" Reynolds number was observed. This

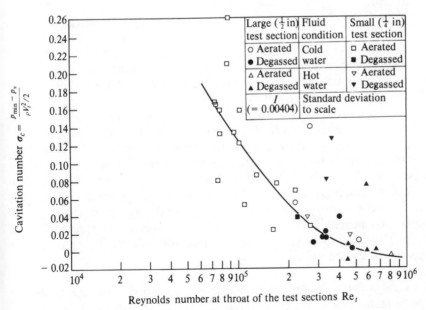

Figure 3-32 Cavitation number vs. throat Reynolds number in a cavitating venturi for sonic initiation (12.7 and 6.35 mm venturis).

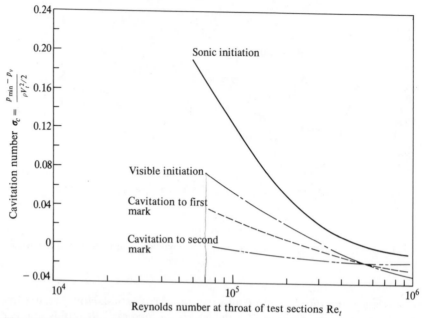

Figure 3-33 Comparison of cavitation number vs. throat Reynolds number for various degrees of cavitation (12.7 and 6.35 mm venturis). U-M venturi.

is, of course, consistent with the Jekat pump results [152] previously mentioned.

Figure 3-34 illustrates the U-M water and mercury sigma results in the same venturi geometry for various "degrees of cavitation" as a function of throat Reynolds number. While the trends are in all cases the same for water and mercury, the water and mercury data do not collapse to single curves when plotted against Reynolds number. Rather, they are separated by a factor of ~10. This is consistent with the previously discussed U-M pump results between water and mercury, where it was found that pump speed was a much better correlating parameter than Reynolds number. However, Reynolds number did correlate reasonably well the separate water (or mercury) sigma results (e.g., Fig. 3-34). It thus appears that the best correlating parameter at present would involve the product VD alone, rather than the complete Reynolds number. In any case, Reynolds number does not appear to be a successful correlating parameter between water and mercury, where a very large factor in kinematic viscosity (or density) is involved. As explained later, Reynolds number does seem reasonably successful as a correlating parameter between water and sodium.

(b) *Loss coefficient.*† As previously indicated, cavitation scale effects include "performance effects" as well as inception. The U-M tests included the measure-

† Loss coefficient = $(p_{in} - p_{out})/(\rho V^2/2) = 0$ for ideal flow.

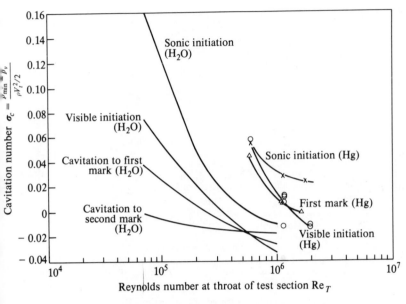

Figure 3-34 Cavitation number vs. Reynolds' number for several cavitation conditions with mercury and water ($\frac{1}{2}$ inch venturi—12.7 mm).

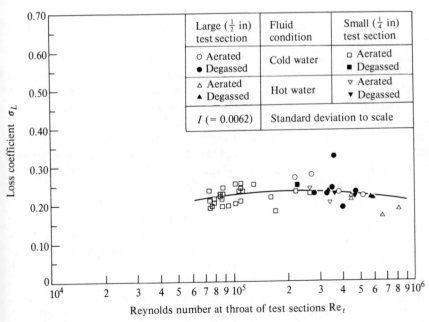

Figure 3-35 Loss coefficient vs. throat Reynolds number in a cavitating venturi for sonic initiation (12.7 and 6.35 mm venturis).

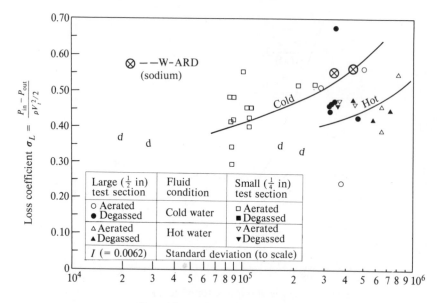

Figure 3-36 Loss coefficient vs. throat Reynolds number in a cavitating venturi for cavitation to first mark (12.7 and 6.35 mm venturis).

ment of "loss coefficient" as well as sigma. The venturi loss coefficient for noncavitating flow is, of course, a relatively insensitive function of Reynolds number, and so is that for cavitation inception. However, there is a large effect for more developed cavitation conditions (Figs. 3-35 to 3-37). Figure 3-35 shows the inception loss coefficient as a function of throat Reynolds number, indicating that it is ~0.2. Presumably it is essentially a function of the friction losses for the basic venturi geometry used. The friction losses for this venturi design are relatively large because of the small nozzle and diffuser angles and the relatively long cylindrical throat. Our later tests with much shorter and more abrupt venturi geometries [156] showed that loss coefficients were then less than 0.1, depending upon details of the geometry.

Figures 3-36 and 3-37 show loss coefficient as a function of throat Reynolds number for relatively well-developed cavitation conditions. In these cases it increases with Reynolds number for fixed degree of cavitation, reaching values as large as ~0.9. Two points from sodium results achieved in recent Westinghouse venturi tests [157] are included in Fig. 3-36. These agree closely with the U-M water results. Loss coefficient also depends strongly on the "thermodynamic effects," as expected [150, 158], i.e., the losses are less for "hot" than for "cold" water. Presumably this would also be the case for sodium, although no comprehensive sodium test results are yet available, to the author's knowledge.

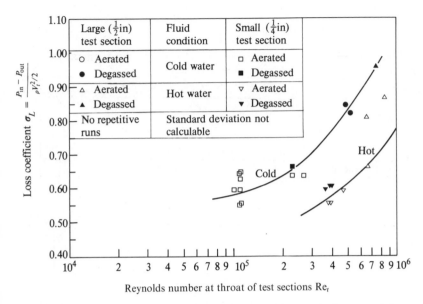

Figure 3-37 Loss coefficient vs. throat Reynolds number in a cavitating venturi for cavitation to second mark (12.7 and 6.35 mm venturis).

(c) *Gas-content effects.* Some discussion of gas-content effects, though they are beyond the basic scope of this report, is necessary for an understanding of the U-M and EdF venturi results. As shown in Fig. 3-29, the optimum velocity (or Reynolds number) feature of the sigma curves mentioned above appears to be true only for relatively high gas contents. The differential between these curves can be computed quite closely [159–161] following a model suggested by Holl. A permanent gas content within the bubbles is then assumed proportional to the total liquid-gas content (entrained and dissolved).

The generally expected effect of gas content is carefully reviewed in Sec. 3-2. In general it is expected that gas content will have little effect over the moderate gas-content range, causing only a gradually increasing sigma. For very low gas contents, so that there are insufficient "nuclei" for inception, sigma should increase strongly, i.e., the "sodium superheat" problem may be encountered. For very high gas content, sigma should increase strongly, so that the cavitation would become essentially "gaseous cavitation" as opposed to conventional vaporous cavitation.

2. *Electricité de France (EdF) and University of Michigan (U-M) sodium comparisons* [162, 163]. As already mentioned, cavitation-inception sigma tests in water and sodium were conducted by EdF at their Chatou Laboratory in a stainless-steel venturi of flow path identical to the venturis previously tested at U-M. Gas content was neither measured nor controlled, but was probably typical for reactor loops. Both helium and argon were used as cover gas, but no difference was attributed thereto. Two test temperatures and velocities were

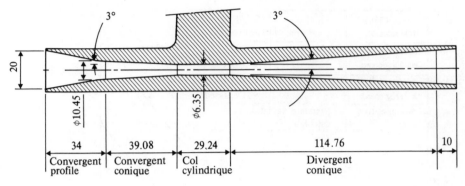

(a)

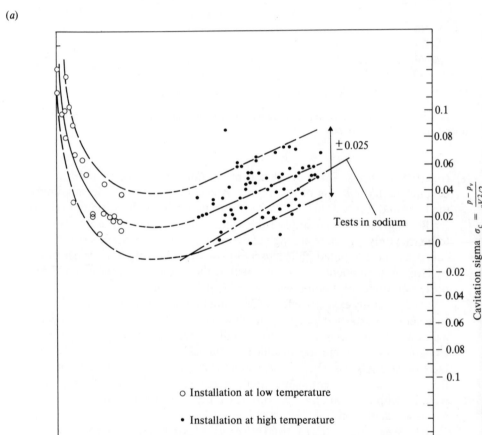

(b)

Figure 3-38 (a) Schematic of venturi tube (dimensions in mm). (b) Water cavitation-inception tests. Electricité de France.

used and the results plotted as a function of Reynolds number (Figs. 3-38 and 3-39). The water portion of the data is obviously very closely similar to the U-M water data (Fig. 3-29). The comparison between the EdF and U-M data sets is shown directly in Fig. 3-40 as a function of throat Reynolds number. It is noted that the EdF data fall within the envelope of the U-M curves, which differ because of differing gas contents. No more-exact direct comparison is possible, since the gas content of the EdF tests is not known. However, the major result from these tests is that inception sigma for water and sodium lay within the same scatter data band, which, even considering the substantial effects of velocity or Reynolds number, is still small enough to be considered negligible from an engineering viewpoint.

3. *Cadarache tests* (*French Atomic Energy Commission*). In later tests conducted by the French Atomic Energy Laboratory at Cadarache [164] in sodium

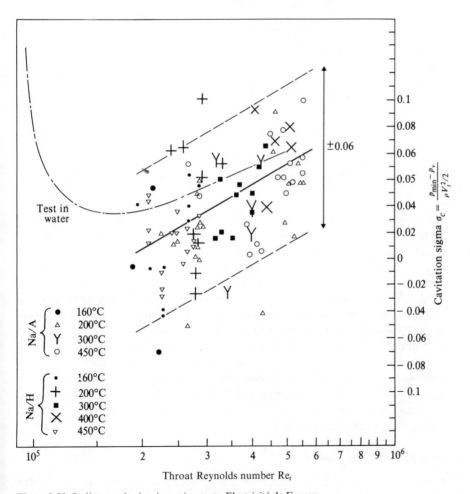

Figure 3-39 Sodium cavitation-inception tests. Electricité de France.

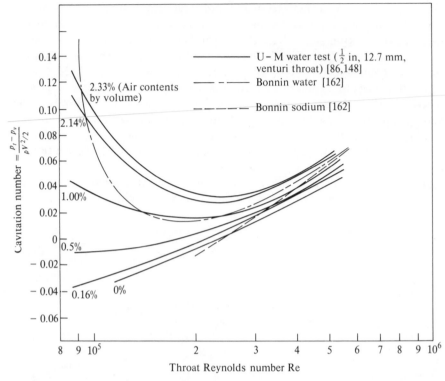

Figure 3-40 Cavitation sigma vs. Reynolds number in University of Michigan and Electricité de France.

for various orifices and nozzles, it was found that inception sigma depended significantly upon the cover gas used, i.e., helium or argon. This may be due to differing solubilities for these gases and hence differing quantities of entrained gas. The differences in NPSH were about 2 lb/in^2, and hence generally negligible from the engineering viewpoint.

4. *Westinghouse sodium tests.* It is possible to obtain two experimental points on the cavitating venturi loss coefficient in sodium from recent tests conducted by the Westinghouse Advanced Reactor Division (W-ARD). These agree closely with the U-M water venturi results (Fig. 3-36). This is probably partly fortuitous, since the venturi geometries used differed considerably.

5. *Kamiyama tests* [48]. A more recent study of the venturi geometry with entrained gas is that by Kamiyama and Yamasaki [48], where they explain the minimum sigma at given Reynolds number (U-M results) in terms of a "choking" analogy with compressible flow.

Orifices and Free Jet Flows

General background There is an important basic difference between venturi geometry, on the one hand, and orifices or nozzles, on the other, from the viewpoint of cavitation inception, performance, and damage. Venturis are essentially a streamlined or nonseparated flow, at least for only slight cavitation (or inception), while orifices and nozzles lack the diffuser geometry necessary to prevent such separation, with or without cavitation. Orifices and nozzles are, in fact, often called "free-shear" or "free-jet" flows for that reason, and the microvortices formed in the free-shear layer around the downstream jet are usually considered to be instrumental in the inception of cavitation, because of the strong underpressures generated at the center of such vortices. Such vortices are not of major importance for cavitation inception in venturis, at least for those with moderate diffuser angles. Of course, the venturi becomes a "free-jet" flow once developed cavitation is formed, but that does not influence the inception problem. Performance effects in terms of loss coefficient also differ, of course, between venturi and nozzle or orifice flows because of differing effects of cavitation upon downstream pressure recovery. Damage effects are also entirely different because much of the bubble-collapse activity is usually not adjacent to the pipe wall in orifices or nozzles, as opposed to venturis, unless the cavitation regime is very extensive, which is not usually the case for engineering applications of nozzles or orifices.

The published scale-effects data for orifices and nozzles [164–169] are apparently much less extensive than those for venturis, which are also not sufficient, as previously indicated, to allow prediction of inception sigma for cases for which actual experimental data are not available. However, the orifice or nozzle geometry is perhaps the simplest possible flow geometry for the theoretical prediction of inception sigma. Obviously it depends strongly on entrained gas microbubble spectra, which unfortunately are not usually measured for most test results. It also depends upon upstream turbulence levels which control instantaneous under-pressures, flow distribution approaching the restriction, and thermodynamic effects as well as the usual flow parameters.

Typical specific results

1. *Arndt and Keller* [165]. Figures 3-41 and 3-42 show typical "desinent"† sigma versus velocity and Reynolds number results from Arndt and Keller [165]. These show differing trends in that Fig. 3-41 indicates a decreasing sigma for increasing Reynolds number, while Fig. 3-42 indicates a strongly increasing sigma for increasing velocity. As for the U-M and EdF venturi tests (e.g., Figs. 3-29, 3-38 to 3-40), there is an optimum Reynolds number or velocity for high and low gas content (Fig. 3-41), and a more strongly decreasing sigma with increasing velocity for high gas content. In any case, the results are not sufficiently consistent to allow a confident prediction of inception sigma as a function of conventional flow parameters unless entrained gas contents are known. In this particular case these were in fact measured.
2. *Tullis* [166]. The most comprehensive cavitation scale-effects data available for both orifices [166] and valves [170–173]‡ have been published by Tullis and

---

† Obtained by raising rather than lowering pressure as in most inception tests.
‡ Discussed later.

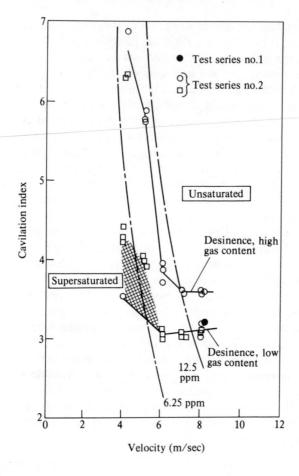

Figure 3-41 Cavitation desinence data at high and low gas content, 16 cm disk.

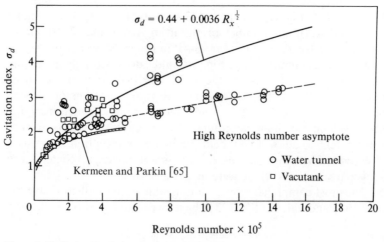

Figure 3-42 Cavitation desinence data, disk geometry.

colleagues. Again these are primarily experimental studies presenting empirical results and possible curve-fitting relations. Nevertheless, insufficient results from this and other sources are available to allow confident prediction of results for as-yet untested situations. In the case of cavitating orifices, there are two questions of interest, i.e., cavitation-inception sigma and loss coefficient. Since cavitating venturis, nozzles, and orifices are "choked-flow" devices, the changes in loss coefficient ("discharge coefficient") can be dramatic. However, presumably heavily cavitating orifices (or nozzles) are not likely to be useful flow components because of eventual damage considerations. Nevertheless, typical results from Tullis and Govindarajan [166] will be reviewed here.

Figure 3-43 shows the orifice discharge flow as a function of pressure differential, indicating the "choked-flow" nature of this type of flow component. Figure 3-44 shows inception sigma as a function of orifice-to-pipe diameter ratio, β. Sigma is here defined with respect to upstream pipe pressure rather than throat pressure, as used here for venturi results. A very strong effect of β is indicated for the various sizes of pipes tested. It is possible that at least the shape of these curves could be predicted by Bernoulli-type calculations, assuming some friction loss, although no such calculations have as yet been made. Such results would be useful in achieving predicting capability.

Figure 3-45, for the onset of "choking cavitation," again shows a strong scale effect with β, and also the effect of different pipe sizes. It appears probable that a relatively simple theoretical analysis could predict at least a significant portion of these trends, but such an analysis is not yet available. Figure 3-46 summarizes the Tullis results for various "degrees of cavitation," including inception and choking as extremes and also various intermediate conditions. Gas contents were not measured for any of these tests, though they were

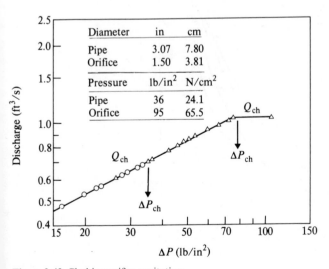

Figure 3-43 Choking orifice cavitation.

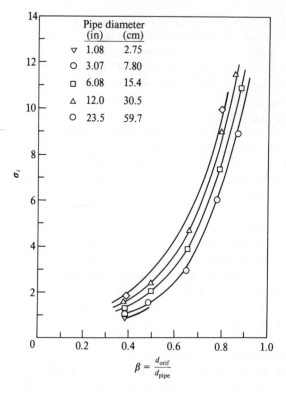

$$\beta = \frac{d_{\text{orif}}}{d_{\text{pipe}}}$$

Figure 3-44 Incipient cavitation—orifice geometry.

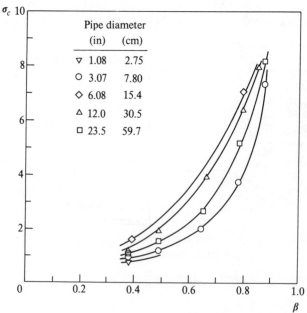

Figure 3-45 Critical cavitation—orifice geometry.

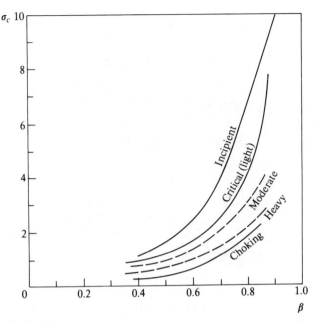

Figure 3-46 Cavitation limits for orifice in 3.07 inch (7.62 cm).

probably approximately constant and relatively low because of the nature of the experimental facility.

3. *SOGREAH.* A considerable test program to evaluate cavitating behavior of orifices and nozzles in water was conducted during the 1960s at SOGREAH, Grenoble, France [167, 168]. Typical curves from Ref. 167 are shown in Fig. 3-6, repeated here for convenience as Fig. 3-47. Figure 3-47a shows the effect, for fixed total oxygen content (measured in lieu of total air) of various nozzle and orifice forms upon cavitation-inception sigma, referred to upstream pressure and jet velocity. The effect upon inception sigma was relatively small for these various configurations.

Figure 3-47b shows the effect of dissolved air upon cavitation-inception sigma for air contents ranging up to ~ 1.6 times the saturated content at STP for a standard test section. Sigma almost doubles over this range. This is roughly consistent with Fig. 3-20a showing general expectations of air-content effects upon cavitation sigma. Figure 3-47d shows analogous results for an orifice, with even greater increase of inception sigma for a total gas-content increase over the same range.

Figure 3-47c shows the increase of inception sigma for an orifice for two gas contents (within the saturated range at STP) as a function of Reynolds number referred to jet velocity and diameter. These results are at least approximately consistent with the U-M and EdF results previously discussed for the higher Reynolds number range (Figs. 3-29 and 3-38 to 3-40).

From the SOGREAH tests it can be concluded that inception sigma for

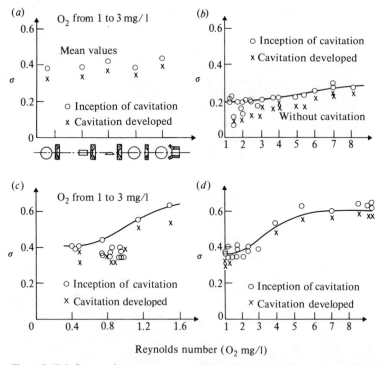

Figure 3-47 Influence of oxygen content and Reynolds number for various orifice shapes on inception sigma (Duport [82]). (*a*) Influence of form jet and of the nozzle. (*b*) Influence of O_2 (ref. test section). (*c*) Influence of Reynolds number (circular orifice). (*d*) Influence of O_2 (circular orifice).

both orifices and venturis is a strong function of Reynolds number as well as total gas content, even for moderate variation within the saturated range, but that Reynolds number alone is not a sufficient correlating parameter in that considerable data scatter remains.

4. *French Atomic Energy Commission—Cadarache Tests* [164, 169]. Sodium cavitation tests in nozzles and orifices is reported from the French Atomic Commission Laboratory at Cadarache [164, 169]. This work, as previously mentioned, indicated a substantial difference in inception sigma depending upon whether the cover gas was argon or helium. Also, as previously mentioned, no difference according to cover gas was found in the EdF venturi tests. The Cadarache results [164] also indicate a substantial ($\sim$20 percent) drop of inception sigma for increasing Reynolds number, over a Reynolds number variation range of only $\sim$1.5. This trend was found for either argon or helium cover gas. Over the relatively very-limited Reynolds number range tested, the difference in suppression pressure (NPSH $\cdot \rho$) is only $\sim$2 lb/in^2 ($\sim$0.14 bar), so that it could be argued that this difference is negligible for most engineering considerations. However, the differential in NPSH could become substantial over larger ranges of Reynolds number which would be encountered in most model-to-prototype tests. Thus, these Cadarache results

are consistent with the other scale-effect data reported, but again allow little predicting ability without further understanding of the basic mechanisms than is presently available.

Figure 3-48 from a published Cadarache report [169] does indicate a

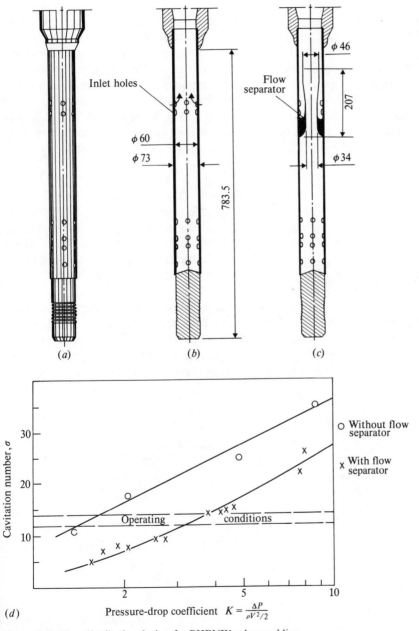

(a)   (b)   (c)

Inlet holes

Flow separator

ϕ 46

207

ϕ 60

ϕ 73

783.5

ϕ 34

Cavitation number, σ

30

20

10

○ Without flow separator

x With flow separator

Operating   conditions

2   5   10

(d)   Pressure-drop coefficient $K = \dfrac{\Delta P}{\rho V^2/2}$

Figure 3-48 Flow distribution devices for PHENIX subassemblies.

substantial change in inception sigma for sodium reactor components as a function of pressure drop, i.e., velocity. This figure is included only to show the existence of substantial scale effects in prototype sodium hardware.

Valves

General background   It is generally accepted, from whatever evidence is presently available, that very large cavitation-inception and performance scale effects exist for flow-control valves of all types. Since the basic flow patterns for valves are considerably more complex than those for nozzles, orifices, or venturis, even less basic understanding exists than for these components, so that little a priori predicting capability exists, as is also the case for the other components. Also, there is considerably less data available so far for valves. Furthermore, it is generally reported in a format quite different from that used for the other components, so that direct comparisons are difficult at present. Typical results will be discussed below.

Colorado State University (CSU) results [170–173]   Cavitation scale effects for valves are important from the viewpoint of inception, of course, and also for performance effects including "choking" characteristics. In addition, the problem is complicated by the fact that a given valve must be investigated for various valve openings, so that the problem on this basis alone is an order of magnitude more complex than that for orifices, nozzles, and venturis. The CSU results include all these factors for various types of valves.

There was no control or measurement of air content for these tests, though it was probably relatively low and constant, considering the nature of the experimental facility. The test fluid was in all cases cold water, so that vapor pressure was very low and essentially negligible in the calculation of sigma.

Figure 3-49 shows sigma for "moderate cavitation," i.e., about midway between

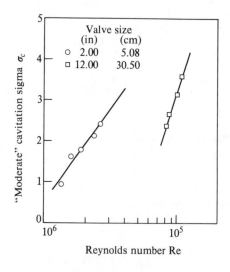

Figure 3-49 Size and pressure scale effects for ball valves (fully open).

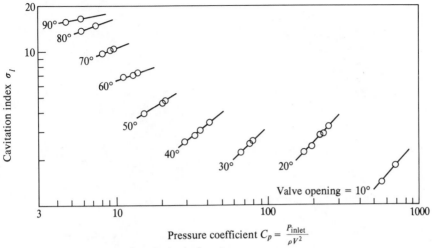

Figure 3-50 Pressure scale effects for butterfly valves.

inception and "critical" cavitation, beyond which major flow instabilities and vibrations occur, so that valve operation from an engineering viewpoint no longer appears feasible for "ball valves" ranging between 2- and 12-in pipe sizes for fully open conditions. A very strong scale effect in sigma, referred to upstream pressure,† is noted in terms of Reynolds number, which varies for a given valve size by a factor of ~ 3. Sigma over this increasing Reynolds number range increases by a factor of ~ 3 also. Thus the effect upon sigma is proportionately much greater than previously discussed for any of the other flow components considered. Reynolds number and VD do not correlate these results for the different valve sizes (Fig. 3-49).

Figure 3-50 shows inception sigma for "butterfly" valves for different valve openings, as a function of C_p, i.e., inlet static pressure divided by inlet kinetic pressure. This parameter is closely similar to inception sigma in this case, since the vapor pressure here is essentially negligible. Again strong scale effects are noted for each valve opening.

Figure 3-51 shows upstream velocity as a function of suppression pressure for "critical cavitation," i.e., maximum limiting cavitation, for different valve openings. Although the curves differ for the different openings, their slope is relatively constant on such a log-log plot.

Figure 3-52 shows the size effect for ball valves as corresponding to critical cavitation, V_c versus C_d, which is defined as inlet pressure divided by pressure drop across the valve. Figure 3-53 is a similar plot for orifices, showing that, in these terms, orifices and valves are indeed quite similar. The conversion of these curves into terms of cavitation sigma versus Reynolds number or velocity would obviously be highly useful, and is perhaps the logical next step.

† $\sigma = (P_{out} - P_v)/(P_{in} - P_{out})$, which differs considerably from the usual definition in the cavitation literature for other components.

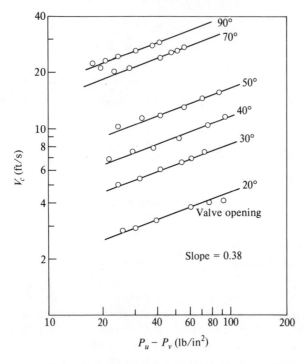

Figure 3-51 Variation of V_c with $P_u - P_v$.

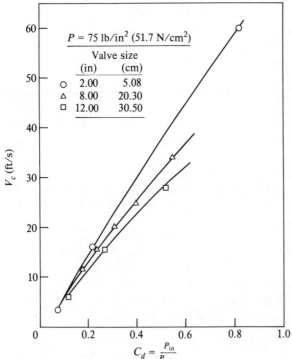

Figure 3-52 Effect of size on V_c for ball valves.

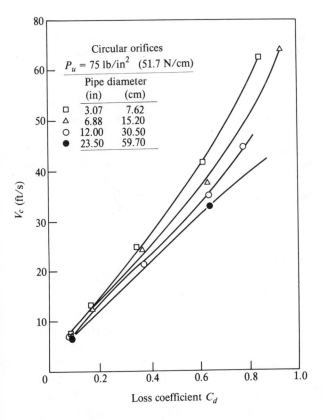

Figure 3-53 Size scale effects for circular orifices.

Recommended curve-fitting relations from the CSU data to allow prediction of prototype performance from model tests are presented by Tullis [173].

3-3-3 Miscellaneous Fluid Property Effects

Various possible effects of liquid viscosity upon cavitation-inception sigma† exist. These will be discussed in the order of their greatest probable importance, in the author's opinion. The extent of actual quantitative knowledge existing in each area is considered also.

Shear flow field and cavitation inception It is probable that the influence of viscosity upon cavitation-inception sigma for shear flows, such as those downstream of an orifice, or other sharp-edged apertures producing separated flow would be much more substantial than for nonseparated flows, such as in a cavitating venturi. This hypothesis is based upon the following reasoning. It is generally supposed [1, 166–168] that cavitation in such separated flows is a result

† Inception sigma $= (p_{ref} - p_v)/(\rho V^2/2)$, where p_{ref} is usually jet or throat pressure.

of the local underpressures existing in small turbulent vortices created by the high-shear region around the submerged jet discharging from a sharp-edged aperture, for example. The controlling importance of this mechanism is verified by the fact that inception sigma for orifices, etc., is of the order 0.4 [1, 166–168], whereas that for a venturi is of the order 0.0 to 0.1 [50, 148, 162, 167, 168, 174]. This difference can only be justified if the vortices in the separated region are quite powerful, so that their center pressure is indeed well below that of the mean flow in that region. However, the local pressure difference generated by such vortices depends strongly upon liquid viscosity. Few quantitative analyses of such vortex flows, where losses are considered, exist. Experiments for the purpose of investigating the hypothesized different magnitudes of sigma dependence upon liquid viscosity appear to be quite simple. Different mixtures of glycerine and water, for example, could be used to vary viscosity as desired in any small water-tunnel facility.

In conclusion, it might be supposed that inception sigma for separated flows would be somewhat decreased for liquids of increased viscosity, because the underpressures produced by the turbulent vortices would be reduced. Recent simplified numerical calculations by the writer do, in fact, show that losses due to viscous effects are substantial.

Development of vortices   The effect of viscosity upon the development of turbulent vortices, and the effect thereon on cavitation-inception sigma, is to a large extent discussed in the foregoing. However, the effect of increased liquid viscosity upon the strength and number of such vortices generated in the shear flow layer must also be considered. This appears to be a somewhat more complex analytical problem than that of the effect of viscosity on individual vortices, once they are in fact produced.

The effect of increased viscosity in producing more and stronger turbulent vortices by virtue of increased shear in the liquid layer surrounding the jet would obviously produce a trend opposite to that previously discussed, i.e., a decreased pressure differential between centers of vortices and the surrounding liquid due to increased viscosity. It is at this point impossible to predict which of these trends is the more important.

From one viewpoint, the importance of increased viscous shear stress around the jet due to increased viscosity may not be of major importance, since the contribution of conventional viscosity to shear in highly turbulent flows is usually small or negligible. It is rather a question of "eddy viscosity," which depends upon velocity gradients, etc., rather than viscosity itself. The effect of increased viscosity in decreasing the pressure differential across small traveling vortices may, on the other hand, be of major importance, since the local Reynolds number pertaining to the vortex flow itself is probably small and perhaps not even in the turbulent range. Hence, the effect of conventional (not "eddy") viscosity may be appreciable for this case.

Nondimensional fluid-dynamic parameters   The conventional nondimensional fluid-dynamic parameter involving viscosity and having the most obvious pertinence to cavitation-inception sigma is certainly Reynolds number. In fact, inception sigma has very often been correlated against Reynolds number [1, 50, 148, 162, 166–168, 174]. However, other parameters of possible importance are Prandtl number, Weber number (based on the presumed diameter of "nuclei"), Jacob number, Strouhal number, and some perhaps as-yet-undefined parameter involving velocity and distance (related to the time of exposure of "nuclei" to underpressure). Reynolds number is discussed first, and then those others which seem to be of possible pertinence. These latter will only be discussed very briefly, since they do not in fact involve viscosity, which is the subject of this section.

Boundary-layer thickness and Reynolds number   Reynolds number can be expected to be of importance, since it affects both boundary-layer thickness and level of turbulence. The boundary-layer thickness is probably of more importance for nominally nonseparated flows such as in a venturi, where it will directly affect the possibility of flow separation, leading perhaps directly to the existence of a region of cavitation. A heavily cavitating venturi is, in fact, a separated flow [148]. However, this is presumably not the case (at least for a well-designed venturi) for inception, where the quantity of cavitation is too limited to affect the overall flow. The boundary-layer thickness will have an influence on inception, since it will affect the entrainment of stationary "nuclei" growing from wall crevices. Presumably reduced boundary-layer thickness, i.e., increased Reynolds number and reduced viscosity, will thus add to the population of traveling "nuclei" and increase inception sigma.

Turbulence and Reynolds number   The other apparent, possibly major, influence of Reynolds number on cavitation-inception sigma is through its effect upon the level of turbulence and the resultant instantaneous underpressures resulting therefrom. The turbulence level, of course, depends upon both Reynolds number and the possible effect of upstream obstacles, elbows, roughnesses, etc. In the absence of such upstream obstacles, a fully developed turbulence level will exist which depends upon Reynolds number alone. The turbulent-pressure coefficient for well-developed turbulence is generally in the range 0.01 to 0.05, and hence would not greatly affect inception sigma for flow regimes involving separated flow where the sigma values are of the order 0.4. However, for streamlined nonseparated flows such as in a well-designed venturi, the underpressure effect of turbulence originating upstream would be of greater relative importance, since the measured sigma values [50, 148, 162] are of the same order as the pressure coefficient for turbulent eddies. Nevertheless, it is likely that these effects would be overwhelmed by the effects of upstream obstacles such as elbows, wall obstructions, etc., if these in fact exist in the case under consideration.

In brief summary, it appears that the effect of local underpressures due to well-developed turbulence, as characteristic of a particular Reynolds number, are likely to be relatively unimportant for separated flows (as orifices, or the flow downstream of blunt obstacles), but could be of significance for inception sigma for well-streamlined nonseparated flows.

Size and velocity scale effects Many experiments attempting to show the effect of Reynolds number may rather show direct velocity or size effects, which are in addition to the normal Reynolds number effects discussed above. Velocity and size are the parameters usually varied in tunnel tests from which most of these data are produced. These possibilities are discussed next.

The growth of "nuclei" into bubbles large enough to create a cavitating flow regime requires the existence of a positive pressure differential between the vapor or gas within the nucleus and the liquid pressure at a distance, for a time long enough for the growth to take place. Hence both magnitude of underpressure and time of exposure are involved. Since the basic bubble-growth (and collapse) mechanisms are highly complex and nonlinear [1], no tractable mathematical model (other than large computerized studies) exists to allow the estimation of the required interrelations between bubble underpressure and time of exposure. Due to the complex nature of these relationships in general, it seems highly probable that there are large influences of liquid velocity and model size upon cavitation-inception sigma beyond those included within the obvious Reynolds number effects already discussed. These presently unexplained effects are usually considered as "cavitation scale effects" [50, 148, 162, 166, 167, 174]. Present information shows that the direction of the velocity effect upon inception sigma can be either positive or negative. It is often negative for high entrained gas contents and streamlined objects, as venturis [50, 174], but positive for the same devices with low gas contents. It also appears to be usually positive for blunt bodies, including orifices [1, 166].

Other nonviscous flow-parameter effects Numerous other liquid and flow parameters, which do not involve viscosity, can affect inception sigma. Experiments to determine the effects of viscosity may also inadvertently involve changes in some of these. Those which seem most likely to affect bubble nucleation, and hence cavitation-inception sigma, are listed below.

1. *Prandtl and Jacob numbers.* Cavitation bubble growth (after the initial stage), as well as collapse, is usually assumed to be primarily governed by the interrelation of pressure and inertial effects. However, in the case of conventional boiling, "thermodynamic" restraints (discussed in detail in Chaps. 4 and 5) become more important than the effects of inertia. For cases of highly subcooled boiling, or cavitation in "hot" liquids (where vapor density is substantial), both inertial and thermal restraints become significant, as recently described by Bonnin, Bonnafoux, and Gicquel [162]. Damage tests in any liquid in a vibratory facility, including sodium [175], show that "thermodynamic effects" strongly reduce damage rates at high temperatures, as discussed in detail in Chap. 5. It is thus probable that these same effects influence nucleation in the same temperature range, though apparently no pertinent experimental data nor precise analyses are available for the nucleation case. These effects involve both heat conductivity of the liquid (Prandtl number) and heat capacity, including latent-heat effects of the vapor contents of the bubble (Jacob number),

or as expressed by various forms of a thermodynamics parameter B [150, 162].

2. *Weber number.* Initial bubble growth from a "nucleus" is presumably controlled by the balance between surface tension which restrains growth and under-pressure which provides the motivating force. A "Weber number" based on nucleus diameter is thus the conventional nondimensional flow parameter involved.

3. *Strouhal number.* For certain types of flow regimes, such as for orifices or the flow behind blunt obstacles in general, the traveling vortices, presumably giving rise to the underpressures necessary to provoke cavitation, are shed from the obstacle in "Karman vortex streets," whose spacing and shedding frequency is governed by a "nondimensional frequency," i.e., Strouhal number. For flows where cavitation inception depends upon these vortices, presumably the Strouhal number may be important.

Viscosity effects: individual bubble dynamic relations There is no doubt that in-creased viscosity reduces rates of bubble collapse or growth, assuming the other governing parameters of the problem to remain constant. However, pertinent experimental measurements are virtually nonexistent, and theoretical studies also, particularly those concerning bubble growth rather than collapse. Numerical studies of spherical bubble collapse, wherein the effects of viscosity were considered, were made at the author's laboratory by Ivany [131]. Figure 4-5 summarizes the effects for various values of viscosity for water. Canavelis [176] also investigated the effects of viscosity, among numerous other parameters, upon spherical bubble collapse. His results generally agree with those of Ivany, discussed below. While bubble collapse is discussed in much more detail in Chap. 4, it is introduced here only to indicate the possible effects of viscosity on cavitation inception.

Figure 4-5 shows bubble-wall velocities and Mach numbers existing at various stages of bubble collapse plotted against radius ratio based upon initial bubble radius. The classical Rayleigh solution, assuming neither liquid viscosity nor com-pressibility, is shown as an upper limit. Bubble-wall velocity and Mach number tend to infinity as radius tends to zero for this solution. The introduction of liquid compressibility brings both to large but finite limits. If spherical symmetry could be maintained through a large-enough radius ratio, these results [131, 176] would be quite precise for bubble collapse in cold water. However, high-speed photographs show that such symmetry generally cannot be maintained, as dis-cussed in detail in Chap. 4. The effect of internal gas in restraining collapse (and producing eventual rebound) is also included in Fig. 4-5.

The addition of terms to consider viscosity at a level similar to that of cold water does not significantly affect the collapse rates (Fig. 4-5). A "critical viscosity" for which collapse would be virtually stopped was first computed by Poritsky [177]. His theoretical value, derived under certain simplifying assumptions, since the study was conducted before the advent of large computers, indicated a value ~ 1470 times that of cold ($70°F$) water to produce this result. Thus the liquid viscosity necessary to significantly affect bubble collapse is that of liquids such as heavy oil. The Poritsky "critical viscosity" was confirmed by Ivany (Fig. 4-5). These results, and those of Canavelis, which apparently is the only other pertinent study, indicate that viscosity has little effect upon bubble collapse unless very large changes in viscosity for quite viscous liquids are involved.

It seems even less likely that there would be a significant influence of viscosity upon bubble growth, since the wall velocities for growth are typically much less than for collapse, and the term involving viscosity is proportional to velocity gradient whereas those involving other terms, such as surface tension and pressure differential, are not velocity dependent.

The term involving viscosity, as derived originally by Poritsky and discussed in detail in Chap. 4, appears in the force balance at the bubble wall, along with terms representing pressure differential and surface tension, as shown by

$$p_{\text{liq}} = p_i - \frac{2\sigma}{R} + 2\mu_{\text{liq}}\left(\frac{\partial u}{\partial r}\right)_R \tag{3-4}$$

where p_{liq} is liquid pressure near the bubble, p_i the pressure within the bubble, μ_{liq} the liquid viscosity, σ the surface tension, u the velocity, r the radial velocity, and R is the bubble-wall radius.

Asymmetric bubble effects and viscosity   Photographic data as well as numerical studies discussed in detail in Chap. 4 show that bubble collapse in flowing systems such as venturis, for example, is in fact highly asymmetric, resulting in the eventual formation of a liquid microjet to which the damage is partially attributed, rather than to shock waves, as in the classical Rayleigh model of spherical collapse, on which earlier calculations such as those by Ivany and Canavelis [131, 176], previously discussed, were based. Although viscosity was included in these numerical studies of asymmetric collapse, no comprehensive studies have yet been made of the effects of viscosity upon the microjet type of collapse as compared to the Rayleigh model. However, since, as previously discussed, viscosity for most ordinary liquids such as water does not appear to have a major effect, it seems equally unlikely that it will strongly influence the microjet model.

While bubble collapse has been shown to be strongly asymmetrical [1, 25, 26], as discussed here and in Chap. 4, the growth process is much more stable, also discussed here and in Chap. 4, so that spherical symmetry is a much better model for growth than for collapse. Hence, it appears most unlikely that the interrelation between viscous effect and possible lack of spheroidicity represents an important effect in the determination of cavitation-inception sigma.

Velocity of bubble relative to flow ("slip velocity")   The possible "slip velocity" of microbubbles, i.e., "nuclei,"† with respect to the main flow could certainly be affected significantly by changes of viscosity. Even in highly turbulent flows the relative motion of very small bubbles, motivated primarily by pressure differentials, would presumably be restrained primarily by viscous drag. For larger bubbles, such as might be characteristic of fully developed cavitation rather than inception, the effects of changes in viscosity would presumably be less than for "nuclei," since

---

† Vapor and/or gas filled.

inertial rather than viscous restraints would be controlling. However, even though "slip velocity" of nuclei might be significantly affected by changes in viscosity, it does not follow necessarily that this would represent an important influence on cavitation-inception sigma. While there appears to be no experimental confirmation of such an effect, the probable importance of an accelerating "slip" on bubble collapse in pump impellers and flowing devices has been investigated in France by Chincholle [26, 178, 179].

The generally recognized mechanism of bubble "slip velocity" in a liquid flow results in a positive slip in regions of decreasing pressure and negative slip in positive pressure gradients.† Thus, differences in resistance to motion between the vapor and liquid phases result in a relative motion between these. The slip phenomenon would thus reduce the penetration of "nuclei" into high-pressure regions because of negative slip and increase their rate of penetration into low-pressure regions. Thus, conceivably, some reduction of "superheat" or increase in cavitation-inception sigma due to increased slip might result. Increased slip would correspond to reduced viscosity, so that a mechanism relating viscosity to cavitation inception through changes in slip velocity is possible. However, there is generally no difficulty involved in the penetration into low-pressure regions of gas or vapor nuclei. Hence, it seems unlikely to the author that this effect would be of more than relatively minor significance for cavitation (or boiling) inception. It seems more likely that it would be of importance for cavitation damage, where the convection of larger bubbles to higher-pressure regions adjacent to submerged bodies, etc., would be reduced. The trajectory of larger bubbles is less affected by viscous effects. These trajectories were investigated by Chincholle [26, 178, 179] and also Johnson and Hsieh [130]. However, "ideal-fluid" assumptions were made so that viscous effects were neglected.

The overall effects of slip velocity over the entire flow loop may, of course, be important in determining the number and size of "nuclei," in that it may affect their solution or dissolution in different parts of the overall loop, since it affects "dwell time" in the different pressure regimes. However, only a very complex computer study, none of which has yet been reported, could shed light on this problem.

Growth of bubbles trapped in wall crevices   This subject has already been mentioned under the subject of Reynolds number effects, where the effects of Reynolds number on boundary-layer thickness and hence the entrainment into the main flow of such "trapped" bubbles were considered. Briefly, increased Reynolds number (reduced viscosity) would lead to higher shear stresses along the wall and hence earlier entrainment, i.e., entrainment of smaller bubbles. On the other hand, increased viscosity would increase viscous shear in the "sub-boundary layer," assuming a fixed Reynolds number and hence velocity profile. Thus, if viscosity of the liquid were increased but the Reynolds number maintained constant either by increased velocity or diameter, then an increased liquid velocity would result in increased entrainment capability for bubbles trapped in wall crevices. On the other

† Due to interactions of pressure, inertia, and shear.

hand, if viscosity were increased with constant VD so that Reynolds numbers were reduced proportionately, conflicting mechanisms would exist so that the direction of any trend on wall-bubble entrainment could only be determined by more detailed study of the particular situation. Apparently, no such detailed studies have yet been made.

Viscosity experiments using water-soluble polymers [180–186] In recent years there has been some experimentation upon the effects of the addition of small quantities of high molecular weight, water-soluble polymers to water upon its cavitation performance. The special high resistance to cross-velocity flow created by the viscoelasticity of such aqueous polymer solutions was expected to affect their cavitation performance to a much greater extent than would the concomitant increase in ordinary viscosity achieved otherwise. The results obtained so far have been somewhat inconsistent, so that application to cavitation-inception sigma differences attributed to viscosity changes for ordinary liquids is not clear.

Much of the cavitation work with high-polymer aqueous solutions has so far been applied to jet cavitation (or impact [185]) applications. It has been hoped that the jet-stabilizing effect of the polymer, in tending to suppress cross-flow turbulence, would increase the damaging capability of a given jet impact. For devices designed for liquid jet cutting of materials this would be advantageous, assuming that the special effect of the polymer was greater than the general damage-reducing effect of increased viscosity. The ratio of viscosity increase, caused by the addition of 100 ppm of polymer solutions of various types to distilled water, is ~4. This is about the same ratio of increase of viscosity when compared to cold water as would be obtained by substitution of 40 percent by volume of glycerol in water. The effects upon erosion rates from liquid jet impact might be assumed to be similar to those to be expected with cavitation. Hence, cavitation-damage tests of this kind would be useful in helping to verify the importance of the microjet impact mechanism in cavitation damage. In fact, recent cavitation-damage tests in a conventional† vibratory facility did show an increase in damage rate of up to ~10 times (as compared with distilled water) with 100 ppm aqueous polymer solutions of this type, in spite of the concomitant increase in ordinary viscosity of ~4 times. This is then taken to be strong evidence of the importance of the microjet damage mechanism in these tests because of the jet-stabilizing effect of the polymer. However, pertinent to the present discussion, it also shows that the major effect of the polymer addition is not its effect upon ordinary viscosity, since vibratory cavitation-damage tests with liquids of varying viscosity, such as glycerol-water solutions, show a strong decrease of damage rates with increased viscosity.

Only limited tests have so far been reported for the effects of polymer solution additions to water upon cavitation-inception sigma. The effect in one case (1000 ppm polymer addition) was a suppression of cavitation inception, i.e., a decrease in

---

† See ASTM Standard Method of Vibratory Cavitation Erosion Test, ANSI/ASTM, No. G 32-77, 1977.

inception sigma [180, 183]. Assuming that the main effect of the polymer addition is the suppression of cross-flow velocities, this suppression of bubble growth is to be expected. Of course, this trend would also be expected considering only the corresponding increase in ordinary viscosity.

In summary, information concerning the effects of polymer addition is thus too sketchy at present to draw any conclusions pertinent to the effects of viscosity change upon the cavitation behavior of ordinary liquids such as water.

Summary of viscosity-sigma effects   Various preliminary conclusions can be drawn from this consideration of the possible effects of liquid viscosity upon cavitation inception. These are listed below according to their probable order of importance:

1. It is likely that one of the more important effects relating liquid viscosity and cavitation-inception sigma will be found in the different sensitivities of sigma to viscosity changes to be found for separated flows (such as orifices, blunt obstacles, etc.) as compared with nominally nonseparated flows (venturis, streamlined objects, etc.). Such a difference should exist because of the much greater dependence of separated-flow cavitation-inception sigma upon the traveling vortices generated in the high-shear layer around the central jet. Increased viscosity may substantially increase inception sigma for separated flows by suppressing the strong and local underpressures generated by such vortices.

2. Reynolds number is probably the most important of the conventional non-dimensional flow parameters which should be considered for the correlation of inception-sigma changes which might be attributed to viscosity changes. However, past experience shows that it is not a completely sufficient correlating parameter for sigma. Probably the most important effect which prevents the complete correlation of sigma with Reynolds number is the effect of the time of exposure of bubbles to underpressure. Thus a pure velocity and size effect appears to exist above and beyond the conventional effect of Reynolds number changes, brought about only by changes in viscosity. Conventional Reynolds number effects include both effects upon turbulence level (and hence local turbulent underpressures) and those upon boundary-layer thickness, which influences the entrainment of microbubbles trapped in wall crevices. For both of these effects, increase in Reynolds number would increase cavitation-inception sigma.

3. A viscosity increase for constant Reynolds number would increase shear stresses both in the shear layer of separated flows (thus increasing the number and intensity of individual traveling vortices) as well as in the vicinity of micro-bubbles trapped in upstream wall crevices, which is probably more important for nonseparated flows. Both these effects would tend to increase cavitation probability and thus reduce inception sigma.

4. Bubble "slip velocities" (relative to the main flow) would be reduced in general by an increase of viscosity. This effect would reduce microbubble penetration into high-pressure regions, thus perhaps reducing damage, but it probably would not substantially affect inception. However, more complex effects due to

changed "slip velocities" (due to viscosity changes) for the entire loop may affect both the dissolving propensity and the entrainment capability for micro-bubbles by changing the "dwell time" in different pressure regimes around the circuit.

5. The probable effect of changes in viscosity, as seen through the analysis of individual bubble dynamics, is such that both bubble growth (inception) and collapse (damage) will be somewhat inhibited for more viscous liquids. However, existing numerical analyses indicate that these effects are probably small for liquids with viscosity similar to water.

Thermodynamic effects "Thermodynamic effects" upon cavitation have been discussed in several portions of this book, including the present chapter. In general, these "effects" exist with relation to cavitation damage as well as to performance and inception, and hence must be mentioned in various portions of this book. Since the present chapter concerns "Bubble Growth and Nucleation," thermodynamic effects must be considered from those viewpoints here. This has been done briefly in the foregoing in discussing the effects of Prandtl and Jacob numbers in either cavitation or boiling bubble dynamics. As indicated, the two phenomena become essentially identical for subcooled boiling applications, where from the cavitation viewpoint "thermodynamic effects" become important. From the same viewpoint, these are entirely controlling for saturated boiling applications, where inertial restraints to growth or collapse become negligible and only heat-transfer mechanisms control bubble behavior.

In general, since thermodynamic effects from the cavitation viewpoint add an additional restraint to growth or collapse, they must act in the direction of restraining either bubble growth or collapse. Thus they very substantially reduce damage in some cases, and little doubt on this score appears to exist. Similarly, they must also inhibit nucleation to some extent, so that an increase in inception sigma should be expected. However, the extent of this effect is difficult to predict in a given case. This aspect of the problem is discussed in detail in Chap. 4 on bubble dynamics, and will not be considered further here. No precise data relating the increase of sigma as referred to acoustic or first-visible initiation due to thermodynamic effects are known to the author at this point. Unless "nuclei" spectra were also measured precisely, any observed effect due to increased temperature might well be partially due also to changes in this very important nucleation parameter.

The thermodynamic effects relating to cavitating component performance can be substantial, and have already been discussed in this chapter concerning other machinery scale effects. Since thermodynamic effects can only be expected to restrict the growth of a cavitation region, their effect upon cavitating-machinery components must be assumed to be favorable, in the sense that the existence of any cavitation is a disadvantage. This is certainly a "truism" from the viewpoint of erosion, but may not always be true for "performance effects." It has sometimes been observed, particularly with centrifugal pumps, that a very small quantity of cavitation actually slightly improves the performance, i.e., there is sometimes a small rise in the head-NPSH curve before the overall fall-off of head when substantial cavitation occurs. The most plausible mechanism to explain this type of behavior, in the author's opinion, is that the streamline pattern allowed when a small cavity exists along the blading surface is more favorable than that formed by the blade itself.

3-4 GENERAL CONCLUSIONS

Bubble growth and nucleation in general have been covered in this chapter. While cavitation applications have been emphasized herein, some consideration has also been given to boiling situations, particularly in those regimes where boiling and cavitation overlap, such as highly subcooled boiling or cavitation cases where "thermodynamic effects" predominate. In these cases, cavitation and boiling are nearly identical phenomena.

Bubble growth and nucleation in general has been covered in Sec. 3-1, while Secs. 3-2 and 3-3 dealt with special features of major interest. Section 3-2 concerns particularly the effects of entrained gas "micronuclei" in generating the inception of conventional "macrobubbles," of which cavitation and boiling regimes are presumed to consist, as well as methods for their measurement and detection. Such micronuclei are presumably of primary importance for cavitation inception, and are also at least contributory to boiling inception in most cases. Inception mechanisms of primary importance only to boiling, such as stationary wall crevices, are not discussed in full detail, as they are beyond the scope of this book. The measurement and detection of traveling "nuclei" is believed by the writer to be the present necessary direction of research for further improvement of understanding and predictability of cavitation-inception sigma and related scale effects.

Section 3-3 covers actual cavitation-inception scale effects as they are observed in various machinery components. The level of understanding and predicting ability in this highly complex field is reviewed in detail, with relation to all pertinent fluid and flow parameters. A brief summary of the state of the art is provided there as well. In general, it can be stated that predicting ability with regard to cavitation-inception and performance scale effects (as well as damage) is extremely limited at this time, so that only highly empirical and highly uncertain approaches are as yet available.

REFERENCES

1. Knapp, R. T., J. W. Daily, and F. G. Hammitt: "Cavitation," McGraw-Hill, New York, 1970.
2. Holl, J. W.: Nuclei and Cavitation, *Proc. Symp. on Role of Nucleation in Boiling and Cavitation,* 1970, ASME; see also *Trans. ASME, J. Basic Engr.,* vol. 93, pp. 681–688, 1970.
3. Tong, L. S.: "Boiling Heat Transfer and Two-Phase Flow," John Wiley, New York, 1965.
4. Rohsenow, W. M.: Nucleation with Boiling Heat Transfer, paper no. 70-HT-18, *Proc. Symp. on Role of Nucleation in Boiling and Cavitation,* 1970, ASME.

5. Holl, J. W., and A. L. Treaster: Cavitation Hysteresis, *Trans. ASME, J. Basic Engr.*, vol. 88, ser. D, no. 1, pp. 199–212, March, 1966; see also J. W. Holl: Sources of Cavitation Nuclei, *Proc. Fifteenth Amer. Towing Tank Conf.*, June, 1968, Ottawa, Canada.

6. Harvey, E. N., D. K. Barnes, W. D. McElroy, A. H. Whiteley, D. C. Pease, and K. W. Cooper: Bubble Formation in Animals—I. Physical Factors, *J. Cellular and Comp. Physiol.*, vol. 24, no. 1, pp. 1–22, August, 1944.

7. Harvey, E. N., A. H. Whiteley, W. D. McElroy, D. C. Pease, and D. K. Barnes: Bubble Formation in Animals—II. Gas Nuclei and Their Distribution in Blood and Tissues, *J. Cellular and Comp. Physiol.*, vol. 24, no. 1, pp. 23–24, August, 1944.

8. Harvey, E. N., D. K. Barnes, W. D. McElroy, A. H. Whiteley, and D. C. Pease: Removal of Gas Nuclei from Liquids and Surfaces, *J. Amer. Chem. Soc.*, vol. 67, p. 156, 1945.

9. Harvey, E. N., W. D. McElroy, and A. H. Whiteley: On Cavity Formation in Water, *J. Appl. Phys.*, vol. 18, no. 2, pp. 162–172, 1947.

10. Sette, D., and F. Wanderlingh: Nucleation by Cosmic Rays in Ultrasonic Cavitation, *Physical Review*, vol. 125, pp. 409–417, 1962.

11. Coacci, R., P. Marietti, D. Sette, and F. Wanderlingh: On the Acoustic Study of Nucleation by Energetic Particles in Fluids, *J. Acoustical Soc. Amer.*, vol. 49, no. 1, pp. 246–252, 1971.

12. Knapp, R. T.: Cavitation and Nuclei, *Trans. ASME*, vol. 80, pp. 1315–1324, 1958.

13. Fox, F. E., and K. F. Herzfeld: Gas Bubbles with Organic Skin as Cavitation Nuclei, *J. Acoustical Soc. Amer.*, vol. 26, pp. 984–989, 1954.

14. Bernd, L. H.: Cavitation, Tensile Strength and the Surface Films of Gas Nuclei, paper 4, *Proc. Sixth Symp. on Naval Hydrodynamics*, 1966, Washington, D.C.

15. Plesset, M. S.: "Bubble Dynamics," CIT Rept. 5-23, Calif. Inst. of Tech., Pasadena, Calif., February, 1963.

16. Glaser, D. A.: Bubble Chamber Tracks of Penetrating Cosmic Ray Particles, *Physical Review*, vol. 91, pp. 762–763, 1953.

17. Bell, C. R.: "Radiation Induced Nucleation of Vapor Phase," Ph.D. thesis, Nuclear Engr. Dept., MIT, 1970.

18. Greenspan, M., and C. E. Tschiegg: Radiation-Induced Acoustic Cavitation; Apparatus and Some Results, *N.B.S., J. Research*, vol. 71, ser. C, no. 4, pp. 299–312, October–December, 1967.

19. Yilmaz, E., and F. G. Hammitt: Effects of Fast-Neutron Irradiation and High-Intensity Magnetic Fields upon Cavitation Thresholds and Nuclei Spectra in Water, *Nucl. Science and Engr.*, vol. 63, no. 3, pp. 319–329, July, 1977.

20. Pyun, J., F. G. Hammitt, and A. Keller: Microbubble Spectra and Superheat in Water and Sodium, Including Effect of Fast Neutron Irradiation, *Trans. ASME, J. Fluids Engr.*, vol. 99, pp. 87–97, March, 1976.

21. Holz, R. E.: "On the Incipient Boiling of Sodium and Its Application to Reactor Systems", Argonne National Lab. Rept. ANL-7884, Argonne, Ill., 1971.

22. Plesset, M. S., and D. Y. Hsieh: Theory of Gas Bubble Dynamics in Oscillating Pressure Fields, *Physics of Fluids*, vol. 3, pp. 882–892, 1960.

23. Daily, J. W., and V. E. Johnson Jr.: Turbulence and Boundary Layer Effects on Cavitation Inception from Gas Nuclei, *Trans. ASME*, vol. 78, pp. 1695–1706, 1956.

24. Yeh, H. C., and W. J. Yang: Dynamics of Bubbles Moving in Liquids with Pressure Gradient, *J. Appl. Phys.*, vol. 39, pp. 3156–3165, 1968.

25. Ivany, R. D., F. G. Hammitt, and T. M. Mitchell: Cavitation Bubble Collapse Observations in a Venturi, *Trans. ASME, J. Basic Engr.*, vol. 88, ser. D, no. 3, pp. 649–657, September, 1966.

26. Chincholle, L., and A. Sevastianov: Etude du Mouvement d'une Bulle de Gaz dans une Liquide à l'Interieur d'un Diffuseur, *La Houille Blanche*, no. 5, pp. 355–360, 1976.

27. Arndt, R., and A. T. Ippen: "Cavitation near Surfaces of Distributed Roughness," MIT Hydrodynamics Lab. Rept. 104, Lexington, Mass., 1967.

28. Arndt, R., and A. T. Ippen: Rough Surface Effects on Cavitation Inception, *Trans. ASME, J. Basic Engr.*, vol. 90, ser. D, pp. 249–261, 1968.

29. Holl, J. W.: "The Effect of Surface Irregularities on Incipient Cavitation," Ordnance Research Lab. Rept. TM 53410-03, Penn State University, 1958.

30. Holl, J. W.: The Inception of Cavitation on Isolated Surface Irregularities, *Trans. ASME, J. Basic Engr.*, vol. 82, ser. D, pp. 169–183, 1960.
31. Holl, J. W.: The Estimation of the Effects of Surface Irregularities on the Inception of Cavitation, in G. M. Wood et al. (eds.), *Symp. on Cavitation in Fluid Machinery*, 1965, pp. 3–15, ASME.
32. Birkhoff, G.: "Hydrodynamics," Princeton University Press, Princeton, N.J., 1950.
33. Oldenziel, D. M.: Measurements on the Cavitation Susceptibility of Water, *Proc. Fifth Budapest Conf. on Hydraulic Machinery*, 1975, vol. 2, pp. 737–748.
34. Karplus, H. B., R. B. Massow, and R. L. Williams: Transducer Design Considerations for Ultrasonic Inspection of Nuclear Reactors, *Trans. ANS Annual Meeting*, June, 1973, pp. 85–86.
35. Plesset, M. S., and T. P. Mitchell: On the Stability of a Spherical Shape of a Vapor Cavity in a Liquid, *Quart. Appl. Math.*, vol. 13, pp. 419–430, 1956.
36. Plesset, M. S.: Bubble Dynamics, in Robert Davis (ed.), "Cavitation in Real Fluids," pp. 1–18, Elsevier Publishing Company, Amsterdam, 1964.
37. Taylor, G. I.: *Phil. Trans. Roy. Soc.*, ser. A, vol. 223, pp. 289–343, 1923.
38. Goldstein, S.: "Modern Developments in Fluid Dynamics," vol. 1, pp. 196–197, Oxford Clarendon Press, Oxford, 1938.
39. Shalobasov, I. A., and K. K. Shal'nev: Effect of an External Magnetic Field on Cavitation and Erosion Damage, *Heat Transfer-Soviet Res.*, vol. 3, no. 6, 1971.
40. Shal'nev, K. K., I. A. Shalobasov, and Yu. S. Zvragincev: Influence of Direction of Magnetic Field Vector on Cavitation and Erosion, *Akad. Nauk., U.S.S.R.*, vol. 213, no. 3, pp. 574–576, 1973.
41. Shalobasov, I. A., K. K. Shal'nev, Yu. S. Zvragincev, S. P. Kozyrev, and E. V. Haldeev: Remarks Concerning Magnetic Fields in Liquids, *Electronic Machining of Materials, Akad. Nauk., Moldasky C.C.R.*, vol. 3 (57), pp. 56–59, 1974.
42. Kamiyama, S., and T. Yamasaki: Theory on Charged Bubble Growth, *1977 ASME Cavitation and Polyphase Flow Forum*, 1977, pp. 3–6.
43. Branover, G. G., A. S. Vasilyev, and J. M. Gelfgat: Investigation of Transverse Magnetic Field on the Flow in the Pipe with Sudden Expansion (in Russian), *Magnetohydrodynamics*, vol. 3, pp. 99–104, 1967.
44. Jones, T. B.: "Electrohydrodynamically Enhanced Heat Transfer in Liquids—A Review," Dept. Elec. Engr. Rept. NSF/ENG74-24113/RR2/77, Colorado State University, Ft. Collins, Colo., March, 1977.
45. Acosta, A. J., and B. R. Parkin: Cavitation Inception—A Selective Review, *J. Ship Res.*, vol. 19, no. 4, pp. 193–205, December, 1975.
46. Noordzij, L.: Some Experiments on Cavitation Inception with Propellors in the NSMB-Depressurized Towing Tank, *Int. Shipbuilding Prog.*, vol. 23, no. 265, pp. 300–306, September, 1976.
47. Arakeri, V. H., and A. J. Acosta: Cavitation Inception Observations on Axisymmetric Bodies at Supercritical Reynolds Numbers, *J. Ship Res.*, vol. 20, no. 1, pp. 40–50, March, 1976.
48. Kamiyama, S., and T. Yamasaki: On the Influence of Velocity on Gaseous Cavitation Occurrence in the Flow in Converging-Diverging Duct, *1977 Cavitation and Multiphase Flow Forum*, June, 1977, pp. 7–8, ASME; see also S. Kamiyama and T. Yamasaki: Prediction of Gaseous Cavitation Occurrence in Sodium and Water Based on Two-Phase Critical Flow Analogy, *1976 Cavitation and Polyphase Flow Forum*, 1976, pp. 46–48, ASME.
49. Wong, C. P. C., G. C. Vliet, and P. S. Schmidt: Magnetic Field Effects on Bubble Growth in Boiling Liquid Metals, *Proc. Second Topical Conf. on Tech. of Nucl. Fusion*, September, 1976, vol. 2, p. 407.
50. Hammitt, F. G.: "Cavitation Scale Effects between Model and Prototype," Working Group Rept. 1, 1970, available as University of Michigan UMICH Rept. 03371-3-I, July, 1970; see also Effects of Gas Content upon Cavitation Inception, Performance, and Damage, *J. Hyd. Research (IAHR)*, vol. 10, no. 3, pp. 259–290, 1972.
51. Numachi, F.: "Uber die Kavitationsentstehung mit besonderem Bezug auf den Luftgehalt des Wassers," Tech. Rept. of Tohoku Imp. University, vol. XII, no. 3, 1937.
52. Numachi, F., and T. Kurokawa: Tech. Rept. of Tohoku Imp. University, vol. XII, no. 4, 1938.
53. Numachi, F., and T. Kurokawa: "Uber den Einfluss des Luftgehalts auf die Kavitationsentstehung im Salzwasser," Tech. Rept. of Tohoku Imp. University, vol. XII, no. 4, 1938.

54. Numachi, F., and T. Kurokawa: "Uber den Einfluss des Luftgehalts auf die Kavitationsenstehung im Meerwasser," Tech. Rept. of Tohoku Imp. University, vol. XII, no. 4, 1938.
55. Numachi, F., and T. Kurokawa: Uber den Einfluss des Luftgehalts auf die Kavitationsentstehung, *Werft Reederei Hafen*, vol. XX, 1939.
56. Gutsche, F.: Hohlsog—Kavitations Bildung in Lufthaltigem Wasser, *Schiffbau Heft*, vol. II, 1939.
57. Edstrand, H.: "The Effect of the Air Content of Water on the Cavitation Point and upon the Characteristics of Ships' Propellors," no. 6, Publications of the Swedish State Shipbuilding Expt. Tank, Göteborg, Sweden, 1946.
58. Edstrand, H.: "Cavitation Tests with Model Propellors in Natural Sea Water with Regard to the Gas Content of the Water and its Effect upon Cavitation Point and Propellor Characteristics," no. 15, Publications of the Swedish State Shipbuilding Expt. Tank, 1950.
59. Lindgren, H., and C. A. Johnsson: "Cavitation Inception on Head Forms, ITTC Comparative Experiments," no. 58, Publications of the Swedish State Shipbuilding Expt. Tank, 1966; presented at the Eleventh Int. Towing Tank Conf., Tokyo, 1966.
60. Vuskovic, I.: Recherches Concernantes l'Influence de la Teneur en Air sur la Cavitation et la Corrosion, *Bull. Escher-Wyss*, vol. 19, pp. 83–90, 1940.
61. Crump, S. F.: "Determination of Critical Pressures for Inception of Cavitation in Fresh Water and Sea Water as Influenced by Air Content of the Water," DTMB (U.S. Navy) Rept. 575, 1949.
62. Crump, S. F.: "Critical Pressure for Inception of Cavitation in a Large Scale Numachi Nozzle as Influenced by Air Content of the Water," DTMB (U.S. Navy) Rept. 770, 1951.
63. Williams, E. E., and P. McNulty: Some Factors Affecting the Inception of Cavitation, paper 2, *Proc. 1955 NPL Symp. in Hydrodynamics*, HMSO, London, 1956.
64. Ziegler, G.: Tensile Stresses in Flowing Water, paper 3, *Proc. 1955 NPL Symp. in Hydrodynamics*, HMSO, London, 1956.
65. Kermeen, R. W., J. T. McGraw, and B. R. Parkin: Mechanism of Cavitation Inception and the Related Scale Effects Problem, *Trans. ASME*, vol. 77, pp. 533–541, 1955.
66. Parkin, B. R., and J. W. Holl: "Incipient Cavitation Scaling Experiments for Hemispherical and 1.5 Caliber Ogive-Nosed Bodies," Penn State University Rept. NORD 1958-264, 1953.
67. Holl, J. W.: An Effect of Air Content on the Occurrence of Cavitation, *Trans. ASME, J. Basic Engr.*, vol. 82, ser. D, pp. 941–946, 1960.
68. Ripken, J. F., and J. M. Killen: Gas Bubbles: Their Occurrence, Measurement and Influence in Cavitation Testing, *Proc. 1962 IAHR Symp. on Cavitation and Hydraulic Machinery*, pp. 37–57, Sendai, Japan, 1963.
69. Killen, J. M., and J. F. Ripken: "A Water Tunnel Air Content Meter," University of Minnesota, Hydr. Lab. Rept. 70, St. Anthony Falls, 1964.
70. Scheibe, F. R.: Cavitation Occurrence Counting—A New Technique in Inceptive Research, *ASME Cavitation Forum*, pp. 8–9, 1966.
71. Scheibe, F. R., and J. M. Killen: "New Instrument for the Investigation of Transient Cavitation in Water Tunnels," University of Minnesota, Hydr. Lab. Memo M-113, St. Anthony Falls, June, 1968.
72. Bindel, S., and R. Lombardo: Influence de la Vitesse et de la Teneur en Air de l'Eau sur l'Apparition de la Cavitation sur Modèle, *Proc. Assoc. Tech. Maritime Aero.*, 1964, Paris.
73. Bindel, S.: "Etude Experimentale de l'Influence de la Teneur en Air et de la Vitesse sur l'Apparition de la Cavitation en Tunnel," Colloque Euromech no. 7, Grenoble, 1965.
74. Bindel, S., and J. C. Riou: Influence de la Vitesse, de la Teneur en Air de l'Eau et de l'Echelle sur l'Apparition de la Cavitation sur Modèle. *Proc. Assoc. Tech. Maritime Aero.*, 1969, Paris.
75. Silverleaf, A., and L. W. Berry: "Propeller Cavitation as Influenced by the Air Content of the Water," Ship Division, National Physical Lab. Rept. 31, Teddington, England, August, 1962.
76. Tullis, J. Paul, and R. Govindarajan: Cavitation and Size Scale Effects for Orifices, *J. Hydraulics Div.*, ASCE, vol. 99, no. HY3, pp. 417–430, March, 1973.
77. Tullis, J. Paul: Cavitation Scale Effects for Valves, *J. Hydraulics Div.*, ASCE, vol. 99, no. HY7, pp. 1109–1128, July, 1973.
78. Bertrand, J. P.: "Cavitation de Mélange—Compte Rendu des Premiers Essais," SOGREAH Rept. R. 9093, DRME, June, 1966, Grenoble, France.

79. Bertrand, J. P.: "Cavitation de Mélange—Deuxième Compte Rendu d'Essais," SOGREAH Rept. R. 9285, DRME, June, 1966, Grenoble, France.

80. Bertrand, J. P.: "Cavitation de Mélange—Troisième Compte Rendu d'Essais," SOGREAH Rept. R. 9307, DRME, June, 1966, Grenoble, France.

81. Duport, J., and J. P. Bertrand: "Cavitation de Mélange Rapport Général de l'Etude," SOGREAH Rept. R. 9404, DRME, November, 1966, Grenoble, France.

82. Duport, J. P.: La Cavitation de Mélange, *Revue Francaise de Mécanique*, no. 24, pp. 79–87, 1967; also available as SOGREAH Rept. NT. 1370, January, 1966, Grenoble, France.

83. Nomarski, M., J. Bertrand, P. Danel, and J. Duport: "Méthode Optique de Mesure et de Dénombrement des Bulles de Gaz au Sein d'un Ecoulement," SOGREAH Rept. NT. 1399, Euromech, Grenoble, France, April, 1966.

84. Danel, F.: "Etude de la Cavitation: Mesures des Gaz Contenus dans les Liquides," SOGREAH DEM, 7 Apr., 1971, Grenoble, also in *La Houille Blanche*, no. 4, pp. 309–315, 1971.

85. Ahmed, O., and F. G. Hammitt: Determination of Particle Population Spectra from Water Tunnel using Coulter-Counter, *ASME 1969 Cavitation Forum*, pp. 26–28, 1969.

86. Hammitt, F. G.: Observations of Cavitation Scale and Thermodynamic Effects in Stationary and Rotating Components, *Trans. ASME, J. Basic Engr.*, vol. 85, ser. D, pp. 1–16, March, 1963.

87. Ericson Jr., D. M.: "Observations and Analysis of Cavitating Flow in Venturi Systems," Ph.D. thesis, Nuclear Engr. Dept. University of Michigan, Ann Arbor, Mich., June, 1969; also available as University of Michigan ORA Rept. 01357-23-T or U.S. Air Force Rept. AFLC-WPAF8-Jun. 69 35, Dayton, Ohio.

88. Hammitt, F. G., and D. M. Ericson Jr.: Scale Effects Including Gas Content upon Cavitation in a Flowing System, *Proc. Symp. on Pumps and Compressors*, 1970, Leipzig, E. Germany; also available as University of Michigan ORA Rept. 01357-11-T.

89. Ahmed, O.: "Bubble Nucleation in Flowing Stream," Ph.D. thesis, Nuclear Engr. Dept., University of Michigan, 1974.

90. Jekat, W.: A New Approach to Reduction of Pump Cavitation—Hubless Inducer, *Trans. ASME, J. Basic Engr.*, vol. 69, no. 1, 1967, and discussion by F. G. Hammitt of above, pp. 137–139.

91. Fallström, P. G.: Swedish State Power Admin., Stockholm, Sweden, personal letter to F. G. Hammitt, Oct. 20, 1969.

92. Gast, P.: "Experiment Untersuchungen uber den Beginn der Kavitation an unstromten Korpern," Fakultat fur Maschinenbau an der Technischen Hochschule Darmstadt zur Erlangung des Grades eines Doktor-Ingenieurs, December, 1971.

93. Richardson, E. G.: Detection of Gaseous Nuclei in Liquids Using an Ultrasonic Reverberation Chamber," Mech. Engr. Res. Lab., Fluid Mech. Div., Fluids Note no. 38, NEL, East Kilbride, Scotland, February, 1956.

94. Richardson, E. G., and M. A. K. Mahrous: "Ultrasonic Tests with Water Samples," Mech. Engr. Res. Lab., Fluid Mech. Div., Fluids Note no. 39, NEL, March, 1956.

95. Iyengar, K. S., and E. G. Richardson: "The Role of Cavitation Nuclei," Mech. Engr. Res. Lab., Fluid Mech. Div., Fluids Note no. 57, NEL, August, 1957.

96. Iyengar, K. S., and E. G. Richardson: "The Optical Detection of Cavitation Nuclei," Mech. Engr. Res. Lab., Fluid Mech. Div., Fluids Note no. 55, NEL, January, 1958.

97. Galloway, W. J.: Apparatus for the Study of Cavitation in Liquids, *J. Acoustic Soc. Amer.*, vol. 26, p. 149, 1954.

98. Hayward, A. T. J.: The Role of Stabilized Gas Nuclei in Hydrodynamic Cavitation Inception, *J. Phys., Appl. Phys.*, vol. 3, ser. D, pp. 574–579, 1970.

99. Ward, C. A., A. Balakrishnan, and F. C. Hooper: On the Thermodynamics of Nucleation in Weak Gas-Liquid Solutions, paper no. 70-FR-20, in "The Role of Nucleation in Boiling and Cavitation," ASME Symposium Booklet, 1970.

100. ASME Symposium Booklet, "The Role of Nucleation in Boiling and Cavitation," ASME, 1970; also ASME Discussion Booklet (same title), pp. 7–11, 1970.

101. Nystrom, R. E., and F. G. Hammitt: Behavior of Liquid Sodium in a Sinusoidal Pressure Field, paper no. 70-FE-20, in "The Role of Nucleation in Boiling and Cavitation," ASME, 1970; also available in *Trans. ASME, J. Basic Engr.*, vol. 92, ser. D, no. 4, pp. 671–680, December, 1970.

102. Hammitt, F. G.: Behavior of Liquid Sodium in a Sinusoidal Pressure Field Including Contained Gas Effects, *J. Acoust. Soc. Amer.*, vol. 51, no. 5, pt. 2, pp. 1725–1732, 1972.

103. Landa, I., and E. S. Tebay: The Measurement and Instantaneous Display of Bubble Size Distribution, Using Scattered Light, *1970 ASME Cavitation Forum*, pp. 36–37.

104. Landa, I., E. S. Tebay, V. Johnson, and J. Lawrence: "Measurement of Bubble Size Distribution Using Scattered Light," Tech. Rept. 707-4, June, 1970, Hydronautics, Laurel, Md.

105. Keller, A.: The Influence of the Cavitation Nucleus Spectrum Inception, Investigated with a Scattered Light Counting Method, *Trans. ASME, J. Basic Engr.*, vol. 94, no. 4, pp. 917–925, 1972.

106. Yilmaz, E., A. Keller, and F. G. Hammitt: "Comparative Investigations of Scattered Light Counting Methods for Registration of Cavitation Nuclei and the Coulter Counter," University of Michigan ORA Rept. UMICH 01357-36-T (Mod. 1), Ann Arbor, Mich.; see also *La Houille Blanche*, no. 1, p. 59, 1976.

107. Keller, A., F. G. Hammitt, and E. Yilmaz: Comparative Measurements by Scattered Light and Coulter Counter Method for Cavitation Nuclei Spectra, *1974 ASME Cavitation and Polyphase Flow Forum*, pp. 16–18; also *J. Acoust. Soc. Amer.*, pp. 324–333, February, 1976.

108. Arefiev, N. B., V. A. Bazin, and A. F. Pokhilko: Methods for Determining Size Distribution of Cavitation Nuclei in the Flow, Izvestia VNIIG, *Trans. Vedeneev, All-Union Res. Inst. of Hydraulic Engr.*, vol. 104, pp. 81–84, 1974.

109. Oldenziel, D. M.: Measurements on the Cavitation Susceptibility of Water, *Proc. Fifth Conf. on Fluid Machinery*, 1975, vol. 2, pp. 737–748, Budapest Technical University, Budapest.

110. Schiebe, F. R.: A Method for Determining the Relative Cavitation Susceptibility of Water, *Conf. on Cavitation, Fluid Mach. Group, Inst. Mech. Engrs.*, September, 1974, pp. 101–108, Heriot Watt University, Edinburgh.

111. Turner, W. R.: "Physics of Microbubbles," Vitro Lab. Repts. TN 01654.01-1, TN 01654.01-2, and TN 02242.01-1, Silver Spring, Md., July and August, 1963, and July, 1970.

112. Gavrilov, L. R.: Free Gas Content of a Liquid and Acoustical Technique for its Measurement, *Soviet Physics Acoustics*, vol. 15-3, January to March, 1970.

113. Lions, N.: Detection des Gaz Entraines dans le Sodium aux Surfaces Libres, in "Alkali Metal Coolants," International Atomic Energy Agency, Vienna, 1967.

114. Knight, J. A.: Determination of Gaseous Void Fractions by Measurement of the Velocity of Sound in Hot Flowing Sodium, *Proc. Fluid Dynamic Measurements in the Industrial and Medical Environment*, April, 1972, Leicester University, England.

115. Ahmed, O., and F. G. Hammitt: Determination of Particle Population Spectra from Water Tunnel Using Coulter Counter, *1969 ASME Cavitation Forum*, June, 1969, pp. 26–28.

116. Pyun, J.: "On the Use of Coulter Counter to Measure the Microbubble Spectrum in Water and its Effect on the Superheat of Water," Ph.D. thesis, Nuclear Engr. Dept., University of Michigan, April, 1973, Ann Arbor, Mich.

117. Pyun, J., F. G. Hammitt, and A. Keller: Role of Microbubble Spectra in Cavitation Threshold, *Trans. ASME, J. Fluids Engr.*, pp. 87–97, March, 1976.

118. Yilmaz, E.: "Comparison of Two Nucleus Spectrum Measuring Devices and the Influence of Several Variables on Cavitation Threshold in Water," Ph.D. thesis, Nuclear Engr. Dept., University of Michigan, 1974, Ann Arbor, Mich.

119. Peterson, F. B.: Monitoring Hydrodynamic Cavitation Light Emission as a Means to Study Cavitation Phenomena, *Symp. on Testing Techniques in Ship Cavitation Research*, May 31 to June 2, 1967, The Technical University of Norway, Trondheim, Norway.

120. Peterson, F. B.: Incipient and Desinent Cavitation on an ITTC Head Form in a Large Water Tunnel, *1971 ASME Cavitation Forum*, pp. 35–38, 1971.

121. Peterson, F. B., F. Danel, and A. Keller, Determination of Bubble and Particulate Spectra and Number Density in a Water Tunnel with Three Optical Techniques, App. 1, Cavitation Com. Rept., *Fourteenth Iht'l. Towing Tank Conf.*, September, 1975, Ottawa.

122. Hammitt, F. G., A. Keller, O. Ahmed, J. Pyun, and E. Yilmaz: Cavitation Threshold and Superheat in Various Fluids, *Conf. on Cavitation, Fluid Mach. Group, Inst. Mech. Engrs.*, September, 1974, pp. 341–354, Heriot Watt University, Edinburgh.

123. Nystrom, R. E.: "Ultrasonically Induced Sodium Superheat," Ph.D. thesis, Nuclear Engr. Dept., University of Michigan, 1969, Ann Arbor, Mich.
124. Hammitt, F. G.: Cavitation Damage and Performance Research Facilities, in J. W. Holl and G. M. Wood (eds.), *Symp. on Cavitation Research Facilities and Techniques*, May, 1964, pp. 175–184, ASME.
125. Plesset, M. S., and D. Y. Hsieh: Theory of Gas Bubble Dynamics in Oscillating Pressure Fields, *Physics of Fluids*, vol. 3, pp. 882–892, 1960.
126. Plesset, M. S.: Discussion of "Role of Microbubble Spectra in Cavitation Threshold," by J. Pyun, F. G. Hammitt, and A. Keller, *Trans. ASME, J. Fluids Engr.*, pp. 87–97, March, 1976.
127. Plesset, M. S.: Effect of Dissolved Gases on Cavitation in Liquids, *Zeitschrift fur Flugwissenschaften*, vol. 19 (Heft 3), p. 120, 1971.
128. Parkin, B. R., and R. W. Kermeen: The Roles of Convective Air Diffusion and Liquid Tensile Stresses during Cavitation Inception, *Proc. IAHR Symposium*, 1962, Sendai, Japan.
129. Gallant, H.: Research on Cavitation Bubbles (trans.), *Oesterreichische Ingenieur Zeitschrift*, no. 3, pp. 74–83, 1962; see also Electricité de France, Traduction no. 1190, Chatou.
130. Johnson, V. E., and T. Hsieh: The Influence of Entrained Gas Nuclei Trajectories on Cavitation Inception, *Proc. Sixth Naval Hydrodynamics Symp.*, 1966, Washington, D.C., Off. Nav. Res.
131. Ivany, R. D., and F. G. Hammitt: Cavitation Bubble Collapse in Viscous, Compressible Liquids—Numerical Analysis, *Trans. ASME, J. Basic Engr.*, vol. 87, ser. D, pp. 977–985, 1965.
132. Hickling, R., and M. S. Plesset: Collapse and Rebound of a Spherical Cavity in Water, *Physics of Fluids*, vol. 7, 1964, pp. 7–14, 1964.
133. Smith, R. H., and R. B. Mesler: A Photographic Study of the Effect of an Air Bubble on the Growth and Collapse of a Vapor Bubble near a Surface, *Trans. ASME, J. Basic Engr.*, vol. 94, ser. D, no. 4, pp. 933–942, December, 1972.
134. Mousson, J. M.: Pitting Resistance of Metals under Cavitation Conditions, *Trans. ASME*, vol. 59, pp. 399–408, 1937.
135. Rasmussen, R. E. H.: Some Experiments on Cavitation Erosion in Water Mixed with Air, paper 20, *Proc. 1955 NPL Symp. on Cavitation in Hydrodynamics*, HMSO, London, 1956.
136. Rasmussen, R. E. H.: Experiments on Flow with Cavitation in Water Mixed with Air, *Trans. Danish Acad. Tech. Sci.*, no. 1, 1949.
137. Hobbs, J. M., and A. Laird: Pressure, Temperature and Gas Content Effects in the Vibratory Cavitation Erosion Test, *1969 ASME Cavitation Forum*, pp. 3–4.
138. Hobbs, J. M., A. Laird, and W. C. Brunton, "Laboratory Evaluation of the Vibratory Cavitation Erosion Test," NEL Rept. 271, 1967, East Kilbride, Scotland.
139. Sirotyuk, M. G.: The Influence of Temperature and Gas Content in Liquids on the Cavitation Process, *Acoustics Journal*, vol. 12, no. 1, pp. 87–92, 1966, USSR.
140. Garcia, R., and F. G. Hammitt: Cavitation Damage and Correlations with Material and Fluid Properties, *Trans. ASME, J. Basic Engr.*, vol. 89, ser. D, pp. 753–763, 1967.
141. Devine, R., and M. S. Plesset: "Temperature Effects in Cavitation Damage," Calif. Inst. of Tech., Div. Engr. and Appl. Sci. Rept., pp. 85–87, 1964, Pasadena, Calif.
142. Petracchi, G.: Investigation of Cavitation Corrosion (in Italian), *Metallurgica Italiana*, vol. 41, pp. 1–6, 1944; English summary in *Engr. Digest*, vol. 10, pp. 314–316, 1949.
143. Plesset, M. S.: On Cathodic Protection in Cavitation Damage, *Trans. ASME, J. Basic Engr.*, vol. 82, ser. D, pp. 808–820, 1960.
144. Pearsall, I. S.: in J. Gordon Cook (ed.), "Cavitation," M & B Monographs, Mechanical Engineering, ME/10, Mills and Boon, London, or in the United States, Crane, Russak and Company, New York, 1972.
145. Pearsall, I. S.: The Supercavitating Pump, no. 54/73, and Design of Pump Impellers for Optimum Cavitation Performance, no. 55/73, *Proc. 1973, Institution of Mechanical Engineers*, vol. 187, pp. 649–665 and pp. 667–678, 1973.
146. Pearsall, I. S.: "A Review of Cavitation Scale Effects in Hydraulic Machines," IAHR Working Group no. 1, Cavitation Scale Effects, *Proc. IAHR Symp.*, January, 1974.
147. Hammitt, F. G.: Detailed Cavitation Flow Regimes for Centrifugal Pumps and Head vs NPSH

Curves, *1975 ASME Cavitation and Multiphase Flow Forum*, pp. 12–15; also a discussion of this paper by C. F. Wislicenus, ibid., 147.

148. Hammitt, F. G.: Observations of Cavitation Scale and Thermodynamic Effects in Stationary and Rotating Components, *Trans. ASME, J. Basic Engr.*, vol. 85, ser. D, pp. 1–16, March, 1963.

149. Hammitt, F. G., et al.: "Cavitation Performance of a Centrifugal Pump with Water and Mercury," University of Michigan ORA Rept. UMICH 03424-10-I, August, 1961, Ann Arbor, Mich.

150. Stahl, H. A., and A. J. Stepanoff: Thermodynamic Aspects of Cavitation in Centrifugal Pumps, *Trans. ASME*, vol. 78, pp. 1691–1693, 1956.

151. Holl, J. W., and A. L. Treaster: Cavitation Hysteresis, *Trans. ASME, J. Basic Engr.*, vol. 88, ser. D, pp. 385–398, 1961.

152. Jekat, W.: A New Approach to Reduction of Pump Cavitation—Hubless Inducer and Discussion by F. G. Hammitt, *Trans. ASME, J. Basic Engr.*, vol. 89, ser. D, no. 1, pp. 130–139, March, 1967.

153. Nechleba, M.: "Hydraulic Turbines," Constable, London, 1957.

154. Karelin, V. J.: "Kavitacionnye Javlenija V Centrobeznyel," Masgiz Moshva, 1963; see also "Cavitation Phenomena in Centrifugal and Axial Flow Pumps," Trans. R. J. Dobble, Nat'l. Lending Library, Boston Spa, Yorkshire, 1965.

155. Noskiewicz, J.: "Kavitace," Academia, Prague, 1969.

156. Hammitt, F. G., O. S. M. Ahmed, and J.-B. Hwang: Performance of Cavitating Venturi Depending on Geometry and Flow Parameters, *ASME Cavitation and Polyphase Flow Forum*, pp. 18–21, 1976.

157. Huebotter, P. R., et al.: Principle Results of U.S. Base Technology Program on Cavitation in LMFBR Plants, *Trans. ANS Winter Meeting*, Nov. 15–19, 1976; and T. J. Costello, R. L. Miller, and S. L. Schrock: "Cavitation Test," Westinghouse Advanced Reactor Div. Progress Rept. W-ARD XARA-52045, June, 1976.

158. Stepanoff, A. J.: Cavitation in Centrifugal Pumps with Liquids Other Than Water, *J. Engr. for Power, Trans. ASME*, vol. 83, ser. A, p. 79, January, 1961.

159. Ericson, D. M.: "Observation and Analysis of Cavitating Flow in Venturi Systems," University of Michigan ORA Rept. UMICH 01357-13-T, July, 1969, Ann Arbor, Mich.

160. Holl, J. W.: An Effect of Air Content on Occurrence of Cavitation, *Trans. ASME, J. Basic Engr.*, vol. 82, pp. 941–946, 1960.

161. Holl, J. W., and G. F. Wislicenus: Scale Effects of Cavitation, *Trans. ASME, J. Basic Engr.*, vol. 83, pp. 385–398, 1961.

162. Bonnin, J., R. Bonnafoux, and J. Gicquel: Comparaison des Seuils d'Apparition de la Cavitation dans un Tube de Venturi dans l'Eau et le Sodium Liquid, *EdF, Bull. de la Direction des Etudes et Recherches, Série A Nucléaire, Hydraulique*, Thermique no. 1, pp. 3–12, 1971.

163. Hammitt, F. G., et al.: Cavitation Threshold and Superheat in Various Fluids, *Proc. Conf. on Cavitation, Inst. Mech. Engrs.*, September 1974, pp. 341–354, Herriot-Watt University, Edinburgh.

164. Ardellier, A., and J. C. Duquesne: "Etude Experimentale de la Cavitation dans un Ecoulement de Sodium à Travers des Diaphragmes et une Tuyère," Commissariat à Energie Atomique Note Technique SDER/73/163, Service de Technologie des Reacteurs à Sodium, Dept. des Reacteurs à Neutrons Rapides, Cadarache, March 9, 1973.

165. Arndt, E. A., and A. P. Keller: Free Gas Content Effects on Cavitation Inception and Noise in a Free Shear Flow, *Proc. Grenoble IAHR Meeting*, April, 1976.

166. Tullis, J. P., and R. Govindarajan: Cavitation and Size Scale Effects for Orifices, no. HY3, *Proc. ASCE, J. Hydraulics Div.*, March, 1973, pp. 417–439.

167. Duport, J. P.: La Cavitation de Mélange, *Revue Française de Mécanique*, no. 24, pp. 79–88, 1967.

168. Bertrand, J. P.: "Cavitation de Mélange Compte Rendu Des Premiers Essais," Pts. 1, 2, and 3, Sogreah Papers, Grenoble, France.

169. Duquesne, J. C., S. Elie, and J. P. Constantin: Cavitation in Flow Distribution Devices of Fast Reactor Cores—Problems Related to Phénix, *Trans. ASME, Cavitation and Polyphase Flow Forum*, pp. 33–34, 1976.

170. Tullis, J. P., M. L. Albertson, and B. W. Marschner: Flow Characteristic of Valves, *Proc. IAHR Lausanne Symp.*, Oct. 8–11, 1968.

171. Tullis, J. P., and B. W. Marschner: Review of Cavitation Research on Valves, no. HY1, *Proc. ASCE, J. Hydraulics Div.*, January, 1968, pp. 1–16.
172. Tullis, J. P.: Choking and Supercavitating Valves, no. HY12, *Proc. ASCE, J. Hydraulics Div.*, December, 1971, pp. 1931–1945.
173. Tullis, J. P.: Cavitation Scale Effects from Valves, no. HY7, *Proc. ASCE, J. Hydraulics Div.*, July, 1973, pp. 1109–1128.
174. Hammitt, F. G., et al.: Cavitation Threshold and Superheat in Various Fluids, *Proc. Conf. on Cavitation, Fluid Mach. Group, Inst. Mech. Engrs.*, September, 1974, pp. 341–354, Heriot Watt University, Edinburgh.
175. Hammitt, F. G., and N. R. Bhatt: Sodium Cavitation Damage Tests in Vibratory Facility Temperature and Pressure Effects, *1975 ASME Polyphase Flow Forum*, pp. 22–25, 1975.
176. Canavelis, R.: "L'Erosion de Cavitation dans les Turbomachines Hydrauliques," Ph.D. thesis, Faculté des Sciences, L'Université de Paris, 1966.
177. Poritsky, H.: The Collapse or Growth of a Spherical Bubble or Cavity in a Viscous Fluid, *Proc. First Nat'l. Congress Appl. Mech.*, 1952, pp. 823–825, ASME.
178. Chincholle, L.: Visualisation des Ecoulements Relatifs dans les Machines Tournantes a Rotoscope, *La Houille Blanche*, no. 1, pp. 51–58, 1968.
179. Chincholle, L.: Etude du Microjet qui Suit une Bulle Animée d'un Double Mouvement de Translation et d'Implosion," *C. R. Acad. Sci. Paris*, vol. 265, sér. A, pp. 882–885, December, 1967.
180. Ellis, A. T., J. G. Waugh, and R. Y. Ting: Cavitation Suppression and Stress Effects in High-Speed Flows of Water with Dilute Macromolecule Additives, *Trans. ASME, J. Basic Engr.*, vol. 90, pp. 459–466, September, 1970.
181. Ellis, A. T., and J. W. Hoyt: Some Effects of Macromolecules on Cavitation Inception, *1968 Cavitation Forum, ASME*, pp. 2–3, June, 1968.
182. Ting, R. Y., and A. T. Ellis: Bubble Growth in Dilute Polymer Solutions, *Physics of Fluids*, vol. 17, no. 7, pp. 1461–1462, July, 1974.
183. Ting, R. Y.: Cavitation Suppression by Polymer Additives: Concentration Effect and Implication on Drag Reduction, *AiChE Journal*, vol. 20, no. 4, pp. 827–828, July, 1974.
184. Ting, R. Y.: Viscoelastic Effect on Polymers on Single Bubble Dynamics, *AiChE Journal*, vol. 21, no. 4, pp. 810–813, July, 1974.
185. Hoyt, J. W.: Jet Cavitation in Polymer Solutions, *1973 Polyphase Flow Forum, ASME*, pp. 44–47, June, 1973.
186. Hoyt, J. W.: "Influence of Natural Polymers on Fluid Friction and Cavitation" (preprint), Symposium on Testing Techniques in Ship Cavitation Research, 31 May–2 June, 1967, Trondheim, Norway.

FOUR

BUBBLE DYNAMICS IN MULTIPHASE FLOW

4-1 CAVITATION BUBBLE GROWTH AND COLLAPSE

4-1-1 General Background

Cavitation flow regimes are by definition multiphase flow regimes. Two phases are most importantly involved, i.e., a liquid and its own vapor. However, in almost all real cases, at least a trace quantity of noncondensable gas such as air is also involved significantly in both bubble collapse and inception, but particularly inception. Thus cavitation is generally a two-phase, three-component flow.

The nonliquid portion in general can be either in the form of quasifixed cavities or "traveling" bubbles.† It is generally agreed today that it is the collapse of these latter to which cavitation erosion can be attributed. Semifixed cavities are usually found in cases of relatively low cavitation "sigma."‡ Consider, for example, a cavitating hydrofoil, propellor blade, or similar component. For sufficiently high sigma, cavitation will not be present. However, for very much lower sigma for the same component, a large relatively fixed cavity will be formed. In the extreme case, the flow regime will become that of "supercavitation," i.e., the cavity termination will be downstream of the cavitating body (hydrofoil, etc.). If sigma is then raised from the supercavitating condition, the cavity length will be reduced so that its termination point will move upstream. Cavity attachment

---

† Nomenclature introduced by Knapp, Daily, and Hammitt [1].
‡ "Cavitation sigma," $\sigma = (p_\infty - p_v)/\rho V^2/2$.

will then occur along the body. A further increase in sigma will cause the fixed cavity to disappear completely. However, after complete disappearance of the fixed cavity, there will still remain small cavitation "bubbles" in the region of minimum pressure. These generally will be entrained in the flow, and hence are "traveling" cavities. It is the collapse of these which presumably produces cavitation damage.

Even for those cavitation regimes dominated by a fixed or semifixed cavity, it is generally true that traveling bubbles exist in the interface region between the cavity and the main liquid flow. One important cavitation-damage flow regime, studied originally by Knapp [1, 2], is that of the traveling bubbles penetrating the region of stagnation pressure, sometimes found at the closing end of a semi-static cavity along a submerged body such as an ogive. This flow regime is shown schematically in Fig. 4-1 (reprinted for convenience from Ref. 1). The collapse in the region of cavity termination of the traveling bubbles is presumed responsible for the cavitation damage observed in this region. Thus, in most cases, it appears that traveling bubbles, the dynamics of which are the main subject of this chapter, are responsible for the cavitation damage which often occurs with such flows.

Other possibilities for damaging mechanisms exist. For example, the partial collapse of large semifixed cavities, i.e., not traveling bubbles, can create large forces, even if insufficient local instantaneous pressures for "pitting" are created. These forces can cause damage to adjacent structures, possibly through damaging vibrations. However, this is not the typical cavitation pitting mechanism, which in general must involve bubble collapse. A flow regime such as that shown in Fig. 4-1 can be responsible for such large and possibly damaging vibrational forces. High-speed motion pictures by Knapp [1, 2] indicated that the cavity in the particular case studied (Fig. 4-1) was by no means steady-state. Rather, it exhibited a cycle of growth and partial collapse due to the action of a reentry liquid jet, penetrating into the cavity periodically from the stagnation region at the downstream termination point. The time period of this cavity collapse is much greater than that

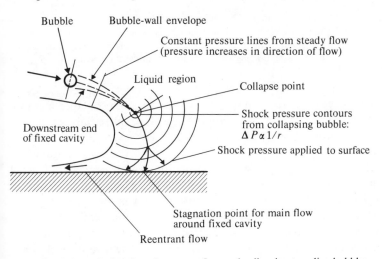

Figure 4-1 Schematic drawing of reentrant flow and collapsing traveling bubbles.

associated with the collapse of small individual traveling bubbles, so that the vibration frequencies thereby induced have a much greater period than that associated with the collapse of individual bubbles.

It is the growth and collapse of individual bubbles which is the primary subject of this chapter. This subject can be considered from many viewpoints, depending upon the degree of complexity desired in the mathematical models. The following sections will consider these various possible analytical approaches, commencing with the most simple.

4-1-2 Ideal Treatment

General: Rayleigh and Besant Historically, the earliest analyses of bubble growth and collapse considered mathematical models idealized to the maximum extent possible. The earliest of these was that by Besant [3] and was published in 1859. However, this work considerably predated the appearance of bubble dynamics as an important engineering problem and was merely a mathematical textbook treatment of an interesting and tractible mathematical problem. Hence, it was not followed up until Lord Rayleigh's analysis [4] in 1917, which laid the foundation for much of the analytical work which continues to the present. Lord Rayleigh, associated with the British Admiralty at that time, was concerned with the first important engineering manifestations of liquid "cavitation" (so named in that era). The engineering problem was associated with new and prematurely high-speed marine propellors, used essentially because of the earlier shift from reciprocating steam-engine to steam-turbine marine drives. Rayleigh quoted Besant's 1859 formulation of the problem of the dynamics and kinematics of an empty spherical cavity in a constant density liquid with constant pressure at infinity:

> An infinite mass of homogeous incompressible fluid acted upon by no forces is at rest, and a spherical portion of the fluid is suddenly annihilated; it is required to find the instantaneous alteration of pressure at any point of the mass, and the time in which the cavity will be filled up, the pressure at an infinite distance being supposed to remain constant (Knapp, Daily, and Hammitt, Ref. 1, p. 98).

Rayleigh then proceeded to solve the problem using an energy balance method, whereas Besant had used a straightforward application of the spherically symmetric equations of conservation of mass and momentum. However, Besant did not elaborate upon his solution or apply it to the cavitation case, as did Rayleigh. Full details of the Rayleigh analysis are given in Ref. 1 and are generally well known to cavitation researchers. Hence they will not be repeated here, except in brief summary. However, it is important to note that in many ways Rayleigh "solved" a most significant portion of the overall cavitation bubble-dynamics problems. However, various aspects, which at first glance might appear relatively minor and which were definitely beyond the scope of his "ideal" approach, have assumed considerable later-day importance and have been the subject of numerous much more complex analyses continuing to the present time. Since it is not so well known, and since its more conventional approach utilizing the conventional fluid momentum and mass conservation relations forms the basis for many of the later more complete analyses, the Besant solution will be included in this chapter.

The foregoing discussion illustrates the meaning in this context of "ideal treatment." Both the Rayleigh and Besant solutions assumed "ideal-fluid" conditions, i.e., incompressible, inviscid liquid; neither assumption is entirely valid. In addition, it was assumed at first that the cavity was empty or contained a fluid (presumably vapor) which remained at constant pressure (less than liquid pressure) throughout the bubble collapse. Such a "constant vapor pressure" assumption is fully equivalent to that of an empty cavity. Also, surface tension was neglected. Rayleigh [4] did extend his solution to consider a perfect gas within the cavity, which could be assumed to be compressed during the collapse either isothermally or adiabatically. However, the coupled heat-transfer problem, generated by the adiabatic compression of such a gas or by the necessary condensation of the vapor within, if the vapor pressure were to remain constant, was not considered.

In addition to the "ideal-fluid" assumptions discussed above, there are also important questions relating to the geometry of the problem, which were not considered in these "ideal" treatments. The most important of these is the lack of spherical symmetry if bubble collapse is to occur near a solid object and thus cause damage. Spherical symmetry is also destroyed in many real cavitation flow regimes by the existence of velocity and pressure gradients, and also by the presence of other nearby bubbles. The consequences of these assumptions will be discussed in this chapter, along with the research investigations leading to their evaluation. First, however, the Besant solution will be presented and the Rayleigh results summarized.

Besant analysis Historically, the earliest analysis of the growth or collapse of a vapor or gas bubble, or void, in a continuous liquid medium known to the author is that of Besant [3]. This is included here in detail because of its simplicity and general applicability. The same result, but using a different approach, was also obtained by Lord Rayleigh [4], as already mentioned.

Consider the case of an expanding or contracting spherical bubble in an inviscid fluid. The basic equations of motion and continuity can be written in polar coordinates for spherical symmetry as follows:

$$\frac{\partial u}{\partial t} + u\frac{\partial u}{\partial r} = -\frac{1}{\rho}\frac{\partial p}{\partial r} \tag{4-1}$$

and

$$\frac{\partial}{\partial r}r^2u = 0 \tag{4-2}†$$

---

† This form of the conservation of mass equation for spherical symmetry with an incompressible fluid is obvious if one considers that $4\pi\rho r^2 u$ is the mass flow rate for a source or sink.

where u = radial component of velocity
r = radius measured from center of bubble
ρ = liquid density
p = pressure

From Eq. (4-2):

$$r^2 u = \text{constant} = R^2 U = R^2 \dot{R}$$

(4-3)

or

$$u = \frac{\dot{R}R^2}{r^2}$$

where the capital letters refer to the same quantities as measured at the liquid-void interface, i.e., the bubble wall, and the dot indicates the time derivative. The last version of Eq. (4-3) is, of course, directly obvious from physical considerations for incompressible, spherically symmetric flow. By substitution of Eq. (4-3) into Eq. (4-1):

$$\frac{\partial}{\partial t}\frac{\dot{R}R^2}{r^2} + \frac{\dot{R}R^2}{r^2}\frac{\partial}{\partial r}\frac{\dot{R}R^2}{r^2} = -\frac{1}{\rho}\frac{\partial p}{\partial r}$$

or

$$\frac{R^2\ddot{R}}{r^2} + \frac{2R\dot{R}^2}{r^2} - \frac{2\dot{R}^2 R^4}{r^2} = -\frac{1}{\rho}\frac{\partial p}{\partial r}$$

Integrating between $r = r$ and $r = \infty$:

$$R^2\ddot{R}\int_r^\infty \frac{dr}{r^2} + 2R\dot{R}^2\int_r^\infty \frac{dr}{r^2} - 2\dot{R}^2 R^4 \int_r^\infty \frac{dr}{r} = -\frac{1}{\rho}\int_p^{p_\infty} dp$$

(4-4)

or

$$\frac{R^2\ddot{R}}{r} + \frac{2R\dot{R}^2}{r} - \frac{\dot{R}^2 R^4}{2r^4} = -\frac{1}{\rho}(p_\infty - p)$$

At the bubble wall, $r = R$ and $p = P$, so that Eq. (4-4) becomes

$$R\ddot{R} + 2\dot{R}^2 - \frac{\dot{R}^2}{2} = -\frac{1}{\rho}(p_\infty - P)$$

(4-5)

or

$$R\ddot{R} + \tfrac{3}{2}\dot{R}^2 = \frac{1}{\rho}(P - p_\infty)$$

which is sometimes called the extended Rayleigh equation for the bubble-wall motion.

Significant results: Rayleigh analysis The Rayleigh analysis [4], essentially solving the equations earlier formulated by Besant and discussed in the previous section, leads to various highly significant results from the viewpoint of cavitation damage. Essentially, the Rayleigh analysis makes plausible the idea that cavitation damage is primarily a result of fluid-mechanical rather than corrosive effects. In most cases this is apparently true, though sometimes corrosion also makes a significant contribution. Due to its "ideal" nature, this analysis does not show in detail the mechanisms by which mechanical cavitation pitting may occur. However, it does show the potential for the generation of very high fluid pressures and velocities by cavitation bubble collapse. In fact, according to the Rayleigh model, these would become infinite at the completion of bubble collapse, when the orginally finite bubble has become a mathematical point. Infinite energy is then ascribed to zero mass, so that no violation of the energy-conservation law is involved. The existence of much higher pressures within the liquid during collapse than the "stagnation pressure" for the process is possible, since the essence of the collapse problem is its nonsteady nature. In a sense, such bubble collapse provides a very good mechanical amplifier, due primarily to the spherical geometry. The low-intensity energy, originally spread throughout the liquid and related to p_∞, is concentrated by the collapse into a much smaller mass and then exists at much higher levels of intensity. While the Rayleigh model demonstrates the potential for very high, local, and transient pressures and velocities in the liquid as a result of bubble collapse, it does not provide a plausible mechanism for the transmission of the pressures and velocities to an eroded surface. It is the search for such mechanisms which has been the subject of much of the cavitation-damage research since Rayleigh.

Figure 4-2 (reprinted from Ref. 1 for convenience) shows the essentials concerning the pressure distribution in the liquid during bubble collapse according to the Rayleigh analysis. While the pressure at large distances remains p_∞ during the collapse and that at the bubble wall zero (if vapor pressure, internal gas, surface tension, and viscosity are neglected), there is an increasing rise in liquid pressure near the bubble wall as the collapse proceeds. The radial position of this pressure peak approaches 1.56 times the bubble-wall radius as the collapse proceeds toward completion. The amplitude of the pressure peak approaches infinity at this time, as does the wall velocity of the bubble. However, as will be shown in a later section, if the existence of either liquid compressibility or viscosity is assumed, these "infinities" vanish as would be expected.

The Rayleigh analysis also allows calculation of the time of collapse [1, 4, 5] in the form of a characteristic time

$$\tau = 0.91468R_0\sqrt{\frac{\rho}{p}} \qquad (4\text{-}6)$$

Table 4-1 (reprinted from Ref. 1 for convenience; originally from calculations in Ref. 5) shows precise values of the nondimensional time as the collapse proceeds in terms of $\beta = R/R_0$. Figure 4-3 (reprinted from Ref. 1 for convenience) shows the collapse curve of Rayleigh compared with photographic results of Knapp and Hollander [6]. The slight disagreement as the collapse proceeds toward completion may be due to the lack of complete symmetry of the experimental bubble and/or to a small quantity of air entrapped within the bubble. The fact that the experimental bubble commences a "rebound" at the completion of collapse tends to confirm the probable existence of an important amount of air, from the viewpoint of bubble collapse, within the bubble.

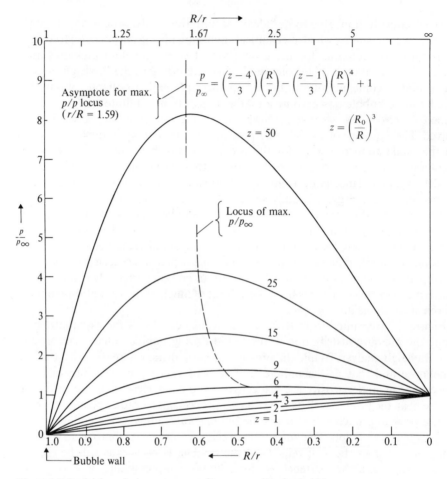

Figure 4-2 Rayleigh analysis: pressure profile near a collapsing bubble.

However, Fig. 4-3 was not introduced here primarily to illustrate the approximate experimental confirmation of the Rayleigh collapse model, but rather to illustrate the form of the radius versus time curve there generated. For an appreciable part of the total collapse time the radius remains approximately constant, but then the collapse accelerates rapidly toward completion. Since the high velocities and pressures also occur only during this final portion of the collapse, its detailed study becomes extremely important in attaining an understanding of cavitation-damage mechanisms. As noted from Fig. 4-3, the duration of complete collapse for a bubble of initial radius of 3.5 mm, in ~1 atm pressure field, is ~0.7 ms. However, the final portion of the collapse, important from the viewpoint of detailed damage mechanisms, requires only a few microseconds, in a typical case such as that shown in Fig. 4-3. Since the time duration is extremely short and the bubble very small during this final collapse period, it is extremely difficult to obtain precise experimental information for this portion of the bubble-collapse process. Primarily for this reason, there is still room for doubt concerning detailed collapse mechanisms in various cases, so that substantial research efforts continue in the cavitation-damage field.

Table 4-1 Values of the dimensionless time $t' = t/R_0\sqrt{\rho/p_\infty}$ from Eq. (4-5). (Error less than 10^{-6} for $0 \le \beta \le 0.96$)

β	$t\dfrac{\sqrt{p_\infty/\rho}}{R_0}$	β	$t\dfrac{\sqrt{p_\infty/\rho}}{R_0}$	β	$t\dfrac{\sqrt{p_\infty/\rho}}{R_0}$
0.99	0.016145	0.64	0.733436	0.29	0.892245
0.98	0.079522	0.63	0.741436	0.28	0.894153
0.97	0.130400	0.62	0.749154	0.27	0.895956
0.96	0.174063	0.61	0.756599	0.26	0.897658
0.95	0.212764	0.60	0.763782	0.25	0.899262
0.94	0.247733	0.59	0.770712	0.24	0.900769
0.93	0.279736	0.58	0.777398	0.23	0.902182
0.92	0.309297	0.57	0.783847	0.22	0.903505
0.91	0.336793	0.56	0.790068	0.21	0.904738
0.90	0.362507	0.55	0.796068	0.20	0.905885
0.89	0.386662	0.54	0.801854	0.19	0.906947
0.88	0.409433	0.53	0.807433	0.18	0.907928
0.87	0.430965	0.52	0.812810	0.17	0.908829
0.86	0.451377	0.51	0.817993	0.16	0.909654
0.85	0.470770	0.50	0.822988	0.15	0.910404
0.84	0.489229	0.49	0.827798	0.14	0.911083
0.83	0.506830	0.48	0.832431	0.13	0.911692
0.82	0.523635	0.47	0.836890	0.12	0.912234
0.81	0.539701	0.46	0.841181	0.11	0.912713
0.80	0.555078	0.45	0.845308	0.10	0.913130
0.79	0.569810	0.44	0.849277	0.09	0.913489
0.78	0.583937	0.43	0.853090	0.08	0.913793
0.77	0.597495	0.42	0.856752	0.07	0.914045
0.76	0.610515	0.41	0.860268	0.06	0.914248
0.75	0.623027	0.40	0.863640	0.05	0.914406
0.74	0.635059	0.39	0.866872	0.04	0.914523
0.73	0.646633	0.38	0.869969	0.03	0.914604
0.72	0.657773	0.37	0.872933	0.02	0.914652
0.71	0.668498	0.36	0.875768	0.01	0.914675
0.70	0.678830	0.35	0.878477	0.00	0.91468
0.69	0.688784	0.34	0.881062		
0.68	0.698377	0.33	0.883528		
0.67	0.707625	0.32	0.885876		
0.66	0.716542	0.31	0.888110		
0.65	0.725142	0.30	0.890232		

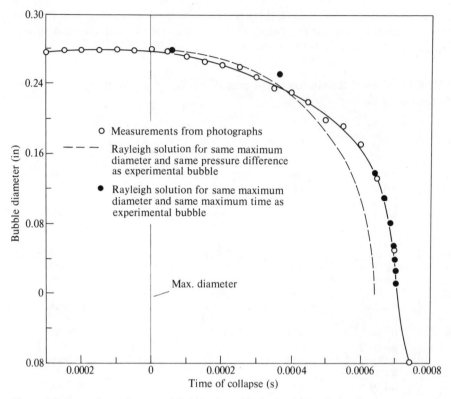

Figure 4-3 Comparison of measured bubble size with the Rayleigh solution for an empty cavity in an incompressible liquid with a constant pressure field.

Inclusion of internal gas and surface tension in Rayleigh model The basic Besant-Rayleigh differential equation for bubble collapse [Eq. (4-5)] can be easily modified to include both surface tension and the effect of an internal gas in addition to vapor pressure:

$$\ddot{R} - \tfrac{3}{2}\dot{R}^2 = \frac{P_1 - p_\infty}{\rho} \tag{4-7}$$

where
$$P_1 = p_v - \frac{2\sigma}{R} + \frac{NT}{R^3} \tag{4-8}$$

N = constant including the gas properties
T = absolute temperature

In the original Besant and Rayleigh formulations, it was assumed that a "void" in the liquid was suddenly annihilated. From this classical problem, the differential equation (4-7) results, but $P_1 = 0$ for this case, as previously shown. In all cases the pressure term making up the right-hand side of the equation is actually the pressure differential between the liquid pressure at a large distance, p_∞, and that at the bubble, i.e., the liquid pressure at the bubble wall, P_1. In the classical Besant-Rayleigh problem $P_1 = 0$.

It is obvious that this model can be easily generalized to include the effects of internal bubble pressure and surface tension. The simplest model, beyond the original formulation, is the assumption of a constant pressure within the bubble such as, for example, the vapor pressure p_v. Then $P_1 = p_v$ in Eqs. (4-7) and (4-8) and the solution of the differential equation is in no way complicated from the version in which $P_1 = 0$.

The next-simplest version, which was in fact also considered by Rayleigh [4], is the assumption of an ideal gas within the bubble, which is compressed according to the ideal-gas laws as the bubble collapses. Rayleigh solved this problem for the isothermal case, introducing the last term in Eq. (4-8), that is, NT/R^3, to represent internal gas pressure. This formulation obviously follows Boyle's law for isothermal compression. For most cases of bubble collapse or growth, it is a reasonably good approximation for gas-filled bubbles. However, it does often "lose" its validity for the cavitation bubble-collapse process near termination, where the bubble-wall velocities are too great and the time too short for adequate heat transfer to occur. Thus, the final portion of the collapse is often modeled more closely by the adiabatic assumption. In this case the final term in Eq. (4-8) would become NT^γ/R^3, where γ is the ratio of specific heats of the gas.

The second term on the right-hand side of Eq. (4-8) represents the effect of surface tension. For the static case, we have

$$P_1 - p_i = \frac{2\sigma}{R} \tag{4-9}$$

where $2\sigma/R$ is the pressure differential between the inside and outside of a static bubble and p_i is the pressure within the bubble. It will be noted that this equation also applies for the stress in a thin-walled spherical pressure vessel, i.e., the stress and the surface tension play identical roles in these two different physical situations. Equation (4-9) is derived by a static-force balance between pressure forces acting on the projected area of the spherical surface and surface tension, or stress, acting on the circumference.

It will be noted from Eqs. (4-7) and (4-8) that the term representing internal gas pressure and vapor pressure are simply additive. Both then act to restrain collapse and to accelerate growth. The term representing surface tension [Eq. (4-8)] is of the opposite sign to the terms representing gas and vapor pressure. Since these latter act to restrain collapse, the surface-tension effect will be to accelerate collapse. However, it will restrain (or prevent, in many cases) growth. Thus surface tension always acts opposite to gas and/or vapor-pressure effects. Note that all the terms in Eq. (4-8) must always have the signs indicated externally, since all the parameters included in these terms are inherently positive.

It should be noted at this point that the original differential equations (4-7) and (4-8) apply equally well to either collapse or growth,† though bubble collapse was the problem for which they were originally generated. These equations have been used, in fact, to study growth and collapse for the case of a cavitation

† Also true of Besant analysis.

bubble subjected to an oscillating pressure field. Of course, the use of an oscillating pressure in an incompressible liquid presents no special problem, at least in the formulation rather than the solution stage, since simply $p_\infty = p_\infty(t)$. Of course, if internal gas pressure or surface tension are included, P is already $P(t)$. The problem of an oscillating pressure at "infinity" presents little problem for an *incompressible* liquid, since the speed of sound therein is infinite. The problems involved with the introduction of compressibility and viscosity will be considered later.

While Eqs. (4-7) and (4-8) are equally applicable to either bubble collapse or growth, as explained in the foregoing, they are essentially applicable to "cavitation" rather than "heat-transfer" bubbles. They apply to cases of either growth or collapse wherein the principle restraints against bubble-wall motion are inertial, i.e., the conventional case of cavitation. If, instead, conventional boiling is considered, the restraints against bubble growth are of thermal rather than inertial nature, i.e., bubble growth can accurately be considered as limited to the rate of evaporation of the liquid to fill the growing bubble and maintain an internal pressure equal to vapor pressure. The fact that this internal pressure is larger than the pressure at large distances is, of course, the mechanism by which the bubble will grow. However, the controlling differential equation is that of heat transfer rather than momentum conservation, as in the Besant-Rayleigh formulation. The inertial resistance to the growth of the heat-transfer bubble is essentially negligible, so that it will grow as rapidly as the inward supply of heat for evaporation of the necessary vapor will permit. Thus the Besant-Rayleigh differential equation is pertinent to neither growth nor collapse for a heat-transfer bubble. Formulations suitable for such heat-transfer bubbles are treated in a later section of this chapter.

Of course, there are many cases of interest where both inertial and heat-transfer effects are important. In the cavitation literature, such a combined case is one in which "thermodynamic effects" are important. These cases will be treated in Sec. 4-1-5. Such a case is that of cavitation in "hot" water as opposed to conventional cold water cavitation. Another case is that of cavitation in cryogenic liquids such as liquid hydrogen or oxygen, important from the viewpoint of liquid-propellant rocket pumps.

A case of boiling in the intermediate range, where both inertial and thermal effects must be considered, is "subcooled" boiling. In fact, highly subcooled boiling and thermodynamic-effect cavitation are essentially identical from the viewpoint of bubble dynamics, if not from their method of generation.

4-1-3 Real-Fluid Parameters

General The "real-fluid" effects in probable order of importance from the viewpoint of bubble collapse are: thermal effects, liquid compressibility, viscosity, and surface tension. From the viewpoint of bubble growth, the order might be: thermal effects, surface tension, and viscosity. For growth, compressibility is not an important parameter, since the bubble-wall velocities, or more appropriately the liquid Mach numbers, are not sufficient to make liquid compressibility an important effect. On the other hand, in the final stages of collapse, if spherical symmetry were maintained and bubble gas content not large, velocities and Mach numbers would become very large. In this case, the collapse velocities would be limited primarily by compressibility effects, or by viscosity, if very viscous liquids such as, for example, lubricating oils were involved.

In collapse, surface tension is not of major importance if it is in the range of that for ordinary water. On the other hand, surface tension can be of controlling importance in bubble growth and nucleation, as indicated by the static balance relation between surface tension and pressure-differential effects [Eq. (4-9)], which controls commencement of growth, i.e., nucleation. In collapse, in most typical cases of interest in the study of cavitation damage, when the radius becomes sufficiently small for the surface-tension term to become important, wall velocities are very large and the inertial term dominates. Of course, for the collapse of sufficiently small bubbles, surface tension would dominate, but such bubbles are too small to be of interest in the study of cavitation damage.

Thermal effects, i.e., "thermodynamic effects" according to the cavitation literature, can be of major importance in either growth or collapse in certain instances, and in other cases they are entirely negligible. They are essentially negligible for most cases of cavitation in cold water, which was the physical situation considered by Besant and Rayleigh in their original formulations. On the other hand, for "hot" water the thermal effects can become predominant in both collapse and growth, i.e., they substantially affect both the cavitation-damage rate and the other effects of cavitation upon component performance. These questions will be discussed in much greater detail in a later section. At this point it is difficult to delineate more precisely the relative importance of "thermodynamic" effects compared to the other "real-fluid" effects already discussed, since as yet there is no generally accepted parameter grouping by which the thermodynamic effects can be evaluated and there is no generalized comprehensive mathematical study which can be utilized for this purpose. The thermal effects differ in kind from other pertinent real-fluid parameters such as compressibility, viscosity, and surface tension, for example, in that these involve only single, easily describable, fluid properties, the variations of which have been studied in the cavitation literature. This has not been comprehensively done for thermodynamic effects, since the parameter groups, the variations of which must be studied, are not as yet fully known.

Viscosity In general, the effect of viscosity is clear in that it must at least to some extent reduce rates of growth or collapse compared to those attained in an inviscid liquid for otherwise identical problem parameters. The effect of viscosity can be considered as a damping effect, and provides a degradation of mechanical energy to thermal energy. The rate of such energy degradation in a given case can be computed by integrating the "dissipation function" over the liquid region. Since viscosity varies over a very wide range, as compared to other liquid properties such as surface tension or density, between liquids of interest, or for the same liquid within the temperature range of interest, its effect in cavitation bubble dynamics can vary from negligible to very important in some instances. Table 4-2 (reprinted for convenience from Ref. 1) lists typical values of viscosity and other pertinent fluid properties for some liquids of interest over a range of pertinent temperatures.

The introduction of the viscosity effect into the analysis is surprisingly easy, in that it appears only as a "boundary condition" in the description of the liquid pressure at the bubble wall [Eq. (4-8) with an additional term], so that the original Besant-Rayleigh differential equation [Eq. (4-7)] still applies, as apparently first pointed out in 1952 by Poritsky [7]. His work resolved what might well have been termed the "viscosity paradox." The basic Navier-Stokes momentum balance equation for an incompressible fluid under conditions of spherical symmetry can be so arranged that no term involving viscosity is present. It would thus appear at first glance that viscous effects do not influence the bubble-dynamics problem, but the simple energy consideration of degradation of mechanical energy through viscous effects indicates that this is not a tenable hypothesis. Poritsky [7] pointed out in his analysis of the problem that the viscous effects do alter the pressure at the bubble wall and thus act to reduce the effective pressure differential in such a way as to reduce rates of either bubble growth or collapse. Thus, though not realized earlier, the Besant-Rayleigh formulation is not limited to the inviscid case. These points are developed in more detail in the following. [See also Eq. (3-4).]

The "viscosity paradox" mentioned above arises from the fact that viscosity can be eliminated from the equation of motion for an incompressible source or sink flow such as that describing spherically symmetric bubble collapse or growth. In its simplest and most conventional form this is Eq. (4-1). It is shown as Eq. (23) in Goldstein's classical book [8] and is presented below for convenience in the form here pertinent:

$$\frac{\partial \mathbf{V}}{\partial t} - \mathbf{V} \times \omega = -\text{grad}\left(\frac{p}{\rho} + \Omega + \tfrac{1}{2}\mathbf{V}^2\right) - v\,\text{curl}\,\omega \qquad (4\text{-}10)$$

where ω is vorticity, i.e., curl $\mathbf{V}$, Ω is the body force potential, and v is the kinematic viscosity. For source or sink flow of an incompressible fluid it can easily be shown that curl $\mathbf{V} = 0$. Therefore, the equation of motion for this case reduces to

$$\frac{\partial \mathbf{V}}{\partial t} = -\text{grad}\left(\frac{p}{\rho} + \mathbf{V}^2/2\right) \qquad (4\text{-}11)$$

even for a viscous fluid, if body force is neglected, which is certainly appropriate for most cases of bubble collapse or growth. Equation (4-11) can, of course, easily be reduced to the form of Eq. (4-1). Hence we see the bubble dynamics "viscosity paradox," since it seems intuitively obvious that bubble-growth or -collapse rates must depend to some extent upon the viscosity of the liquid.

The "paradox" was resolved initially by Poritsky [7], by noticing that the normal stresses in both liquid and gas (or vapor) at the bubble wall must be balanced rather than the pressures. The essentials of the analysis are given below.

Table 4-2 Properties of various liquids

Property	Water 70°F	Water 300°F	Mercury 70°F	Mercury 500°F	Lead-bismuth† 500°F	Lead-bismuth† 1500°F	Sodium 500°F	Sodium 1500°F	Potassium 500°F	Potassium 1500°F	Lithium 500°F	Lithium 1500°F	Ethanal 68°F	Glycerin 68°F
Density, slugs/ft^3	1.939	1.779	26.27	25.17	20.13	18.69	1.724	1.474	1.518	1.249	0.97	0.891	1.53	2.45
Specific weight, γ(lb/ft^3)	62.43	57.3	845.9	810.3	648.0	601.8	55.50	47.45	48.88	40.20	31.21	28.35	49.3	78.8
Surface tension, σ(lb/ft) × 10^2	0.5015	0.322	3.187	2.875	2.735	2.528	1.202	0.83	0.545	0.425	2.625	2.122	0.153	
Bulk modulus, (lb/in^2) × 10^{-6}	0.31	0.248	4.11	3.94	3.16	2.92	0.765	—	—	—	0.160	0.140	—	0.630
Specific heat, Btu/(lbm-°F)	1.00	1.03	0.033	0.032	0.035	0.035	0.3155	0.3030	0.1864	0.1891	1.03	0.991	0.581 (77°F)	0.57
Thermal diffusivity, (ft^2/s) × 10^5	0.147	0.186	4.694	7.03	12.69	14.5	71.4	63.6	73.1	62.8	19.44	21.42		
Heat of vaporization, Btu/lbm	1054	910	127.7	126.5	113.3	113.3	1970	—	932.0	812.4	2600.0	2600.0		
Vapor pressure, lb/in^2	0.36	67.62	2.5 × 10^{-5}	1.93	<0.001	<0.001	~10^{-5}	7.737	<0.001	24.66	<0.001	0.1	0.85	2 × 10^{-6}
Prandtl number (dimensionless)	6.8	1.18	0.026	-0.0091	0.014	0.009	0.0065	0.0038	0.0047	0.0032	0.05	0.022	—	12.5
Kinematic viscosity, ν(ft^2/s) × 10^6	10.8	2.17	1.174	0.75	1.78	1.306	4.6	2.34	3.46	2.15	10.8	6.17	16.4	12,700

† Eutectic mixture (44.5 weight percent Pb).

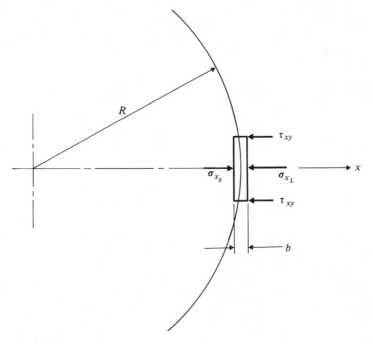

Figure 4-4 Pill-box balance for bubble wall.

Refer to Fig. 4-4 for the appropriate force balance for a "pill-box" control volume cut from the bubble wall. The thickness of the control volume b is arbitrarily thin so that shear forces acting upon its circumference can be neglected. However, its radius is finite, so that the normal stresses in the X direction acting thereon must be balanced. For this simple one-dimensional incompressible case, the pertinent relations between normal stresses and pressure are

$$\sigma_x = -p + \mu \frac{2\partial u}{\partial x} \tag{4-12}$$

where

$$-p = \bar{\sigma} = \tfrac{1}{3}(\sigma_x + \sigma_y + \sigma_z) \tag{4-13}$$

If the gas or vapor viscosity are neglected for simplicity, assuming that they are typically small compared to the liquid viscosity, then

$$p_g = -\sigma_x = P - \mu_L \cdot \frac{2\partial u}{\partial x} \tag{4-14}$$

where P is liquid pressure at the bubble wall, as also given by Eq. (3-4). Continuity [Eq. (4-2)] gives

$$\frac{\partial}{\partial r} r^2 u = 2ru + r^2 \frac{\partial u}{\partial r} = 0 \quad \text{so that} \quad \frac{\partial u}{\partial r} = \frac{\partial u}{\partial x} = \frac{-2u}{r} \tag{4-15}$$

Combining these relations:

$$P = p_g + 2\mu_L\left(\frac{-2u}{r}\right)_R$$

or

$$P = p_g + 4\mu_L\frac{\dot{R}}{R}$$

Since $\dot{R}$ is negative for a collapsing bubble, if we use the absolute value of $\dot{R}$ and also include the surface-tension term,† we obtain

$$P = p_g - \frac{2\sigma}{R} + 4\mu_L\frac{[\dot{R}]}{R} \tag{4-16}$$

where p_g is total pressure within the bubble, whether gas, vapor, or both. It is thus seen that for a collapsing bubble the effect of viscosity is opposite to that of surface tension, as would be expected. Since surface tension increases collapse velocity, viscosity decreases it, as intuitively expected. For bubble growth, the sign of $\dot{R}$ is reversed so that both surface tension and viscosity restrain growth, as also would be intuitively expected.

If the effect of the vapor or gas viscosity within the bubble is also considered, in an analogous manner, the resultant equation is

$$P = p_g - \frac{2\sigma}{R} + 4(\mu_L + \mu_g)\frac{\dot{R}}{R} \tag{4-17}$$

The energy dissipation due to viscous-liquid effects in bubble growth or collapse can also be easily evaluated [7] through application of the "dissipation function" for the fluid. In this case:

$$P_{\text{diss}} = \int_R^\infty \Phi\, 4\pi r^2\, dr = 48\pi\mu_L \int_R^\infty u^2\, dr \tag{4-18}$$

where Φ is the dissipation function and P_{diss} the power dissipated by viscosity. Substituting the continuity relation Eq. (4-2) into Eq. (4-18) and integrating, one obtains

$$P_{\text{diss}} = 16\mu_L\pi\dot{R}^2R \tag{4-19}$$

From this it is obvious that the Rayleigh energy balance analysis [4] could not be applied correctly to a viscous fluid unless P_{diss} were included in the energy balance.

Numerical calculations by Ivany [9, 10] show that, as predicted by Poritsky's precomputer analysis [7], the effect of viscosity in bubble collapse is essentially negligible for fluids such as cold water, but reduces collapse velocity to virtually zero for fluids such as lubricating oils (~ 1470 times cold water viscosity) for a "standard bubble" collapsing under a pressure differential to infinity of 1 atm (Fig. 4-5).

† To be discussed in detail in a later section.

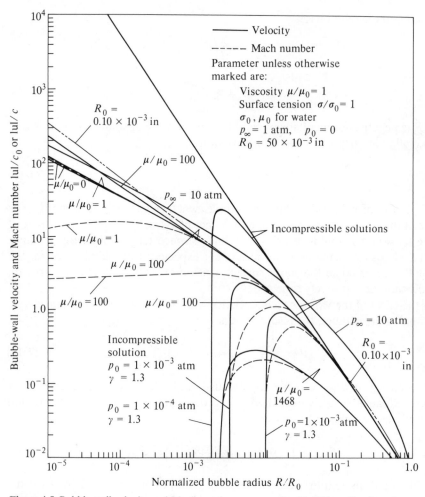

Figure 4-5 Bubble-wall velocity and Mach number vs. normalized bubble radius for reference bubble parameters except where noted otherwise on individual curves.

Compressibility, viscosity, and surface-tension effects and results

General The effects of liquid compressibility are important, as for all fluid-flow problems, only when the Mach number is appreciable. Since the sonic velocity in water is ~1500 m/s, this is not likely to be the case except in relatively unusual circumstances. For example, it is certainly not in general a consideration in bubble-growth problems, except perhaps in cases involving actual underwater explosions. The World War II investigations of underwater-explosion problems [11, 12] are perhaps the first instance where it was necessary to consider liquid-compressibility effects. The Kirkwood-Bethe approximation [12] did in fact lead to the earliest relatively comprehensive and accurate consideration of liquid-compressibility effects in cavitation bubble collapse, i.e., the study of Gilmore [13]. This model has since been utilized in various computerized numerical studies of bubble collapse.

As is well known, the original ideal-fluid incompressible Rayleigh analysis leads to an infinite bubble-wall collapse velocity as the radius approaches zero. However, the liquid Mach number remains zero, since sonic velocity is infinite in an incompressible fluid. If spherical symmetry were maintained, as assumed by Rayleigh [4], for the entire collapse, it is clear that liquid Mach number in a liquid such as water would also achieve very large (supersonic) values as the collapse proceeded toward zero radius. Thus it appears intuitively probable that liquid-compressibility effects would become important for such a collapse as it proceeded toward completion. More recent research shows that in most cases spherical symmetry is not maintained through a sufficient radius ratio for such large velocities to be attained, and this feature is discussed in detail in a later section. However, to obtain a realistic evaluation of the collapse behavior for hypothesized spherical symmetry, it is necessary to investigate the effects of both liquid viscosity and compressibility. The viscosity effect was discussed in the previous section; that of compressibility will be summarized here.

Analysis The general Navier-Stokes equations of motion and the conservation of mass equation are of course the basic tools for the bubble-collapse (or growth) analysis considering liquid compressibility. These were most suitably arranged for that purpose by Gilmore [13], and his approach, also used by Ivany [9, 10], will be followed here. Related material is also presented in Ref. 1. The pertinent version of the Navier-Stokes equation is

$$\frac{D\mathbf{V}}{Dt} = -\frac{1}{\rho}\,\text{grad }p + \frac{4}{3}\frac{\mu}{\rho}\,[\text{grad (div }\mathbf{V})] \tag{4-20}$$

If this is combined with the conservation of mass relation, the result is

$$\frac{D\mathbf{V}}{Dt} = -\frac{1}{\rho}\,\text{grad }p + \frac{4}{3}\frac{\mu}{\rho}\left[\text{grad}\left(-\frac{1}{\rho}\frac{D\rho}{Dt}\right)\right] \tag{4-21}$$

At this point several assumptions were made by Gilmore, and also by the others who followed the same procedure. First, the viscous effect is assumed to be small, as is the compressibility effect, so that their product, which is represented by the last term of Eq. (4-21), is assumed to be negligible. However, the compressibility effect is still included through the continuity relation [Eq. (4-22)] which is its most important aspect:

$$\frac{1}{\rho}\frac{D\rho}{Dt} + \text{div }\mathbf{V} = 0 \tag{4-22}$$

The major effect of viscosity can still be included through the boundary condition at the bubble wall, as explained in the previous section, following the pioneering Poritsky analysis [7]. This effect was not computed by Gilmore or any of the subsequent investigators, to my knowledge, with the exception of Ivany [9, 10].

Another assumption made by Gilmore and followed by various later investigators was the assumption that the liquid was "barotropic," i.e., density is a function only of pressure. Perhaps the most suitable liquid equation of state is that of Tait [14]. This form, shown below, has been used by most investigators in this field:

$$\left(\frac{p + B}{p_r + B}\right) = \left(\frac{\rho}{\rho_r}\right)^n \tag{4-23}$$

where p_r and ρ_r are any reference pressure and density. For cold water, values of $B = 3000$ bar and $n = 7$ were used.

Gilmore now defines a new quantity h, which may be thought of simply as a form of enthalpy pertinent to the barotropic fluid assumed, i.e.,

$$h(p) = \int_{p_\infty}^{p} \frac{dp}{\rho} \tag{4-24}$$

Then

$$\text{grad } h = \text{grad} \int_{p_\infty}^{p} \frac{dp}{\rho} \tag{4-25}$$

Equation (4-21) is now written neglecting the last term involving the crossproduct of viscous and compressibility effects, as previously explained, and Eq. (4-25) substituted for the pressure term. Converting to spherical coordinates:

$$\frac{D\mathbf{V}}{Dt} \frac{\partial h}{\partial r} = -\frac{\partial h}{\partial r} \tag{4-26}$$

The compressibility effect is now introduced through substituting the sonic velocity, i.e.,

$$c^2 = \frac{dp}{d\rho} \quad \text{and} \quad \frac{dh}{dp} = \frac{1}{\rho}$$

Then the equation of motion becomes

$$\frac{D\rho}{Dt} = \frac{d\rho}{dh} \frac{Dh}{Dt} = \frac{Dh}{Dt} \frac{d\rho}{dp} \frac{dp}{dh} = \frac{Dh}{Dt} \frac{\rho}{c^2} \tag{4-27}$$

and the contintinuity equation becomes

$$-\frac{1}{c^2} \frac{Dh}{Dt} = \text{div } \mathbf{V} \tag{4-28}$$

There are now two partial differential equations [Eqs. (4-27) and (4-28)] with three dependent variables, V, h, and C, and two independent variables, r and t.

By using the Tait equation of state [Eq. (4-23)], c and h can be reduced to one variable, leaving two equations with two dependent and two independent variables, which can be solved simultaneously.

Various solutions to these equations exist in the literature prior to those of

Ivany and Hammitt [9, 10] already discussed. Flynn [15] presented an approximate analytical solution using a computed table of values of pressure and density rather than an explicit equation of state [such as Eq. (4-23)]. Mellen [16, 17] computed the bubble-wall velocity using Gilmore's method, and used this to compute the pressure in the liquid at a fixed distance far from the point of bubble collapse. He computed the propagation of the shock resulting from the complete collapse of an empty bubble to zero radius. He included the shock pressure as a function of distance from the center of collapse, with some experimental verification from spark-induced cavitation bubbles of 1 to 2 cm initial radius. Schneider [18] used the method of characteristics in a hand calculation to obtain a graphical solution of the compressible flow equations. Brand [19, 20] made a similar calculation to that of Schneider, except that a computerized finite difference computation of the characteristics was made rather than a graphical solution. As did Schneider, he found that shock waves resulted from collapse upon a rigid sphere. Hickling and Plesset [21] present the most thorough solutions to the compressible equations for bubbles collapsing and rebounding upon contained gas, which is assumed to be compressed adiabatically. The computer solution was continued to the point where a shock wave formed from the rebound pressure wave. They did not use the Kirkwood-Bethe approximation [12], as did Gilmore [13] and other subsequent investigators. Since the results agree closely with the other investigations, the validity of this approximation for the bubble-collapse problem is thus verified. The key results of Hickling and Plesset [21] are presented in Figs. 4-6 to 4-8 (reprinted from Ref. 1 for convenience).

Except for Ivany [9, 10], none of the above investigators included the effect of viscosity. Ivany discussed it in relation to the boundary condition at the bubble wall. The resulting relation at the bubble wall is

$$P(R) = P_i(R) - \frac{2\sigma}{R} - \frac{4\mu U}{R} - \frac{4\mu U}{3c^2}\frac{dH}{dR} \qquad (4\text{-}29)$$

where the capital letters signify values at the bubble wall, small letters having been used for the same parameters within the liquid. P_i is the total vapor and gas pressure within the bubble. The effects of gas viscosity are assumed to be negligible compared to those in the liquid. According to Gilmore [13], the last term is of the same order of significance as the term involving the crossproduct between compressibility and viscous effects neglected earlier. Hence this term was also neglected in the Ivany analysis [9,·10].

Results   As already mentioned, Figs. 4-6 to 4-8 present the key results of the Hickling-Plesset study [21], which is much more comprehensive in scope than the preceding, primarily precomputer, studies mentioned in the foregoing. Hickling and Plesset show the effects of different quantities of entrapped permanent gas. Such gas inevitably causes a "rebound" since time for its solution in the liquid is not available. It thus prevents the collapse from going to completion. In almost all real cases with almost any fluid, the presence of some entrapped permanent gas seems inevitable, since the cavitation bubbles are generally supposed to

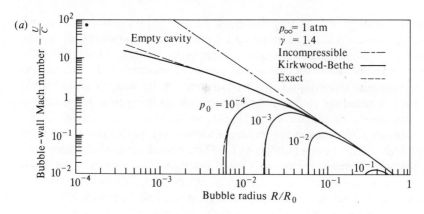

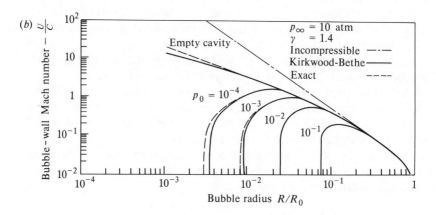

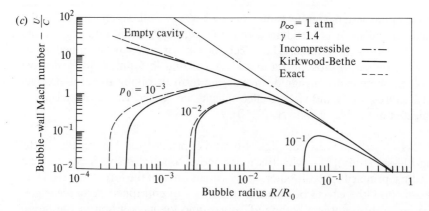

Figure 4-6 Bubble-wall velocity vs. bubble radius for decreasing gas content. Compressible liquid without viscosity or surface tension; gas content determined by initial pressure p_0. (a) Gas constant $\gamma = 1.4$; ambient pressure $p_\infty = 1$ atm. (b) Gas constant $\gamma = 1.4$; ambient pressure $p_\infty = 10$ atm. (c) Gas constant $\gamma = 1.0$; ambient pressure $p_\infty = 1$ atm.

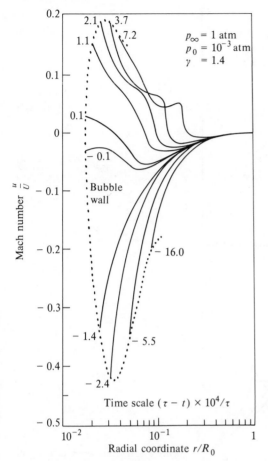

Figure **4-7** Instantaneous velocity in the liquid vs. distance from the bubble wall during collapse and rebound. Compressible liquid without viscosity or surface tension. Gas constant $\gamma = 1.4$; ambient pressure $p_\infty = 1$ atm; initial pressure $p_0 = 10^{-3}$ atm.

nucleate originally from gas microbubbles. In addition, it is quite probable that vapor would act as a permanent gas during the final phases of collapse, since time for its condensation and removal of the resultant latent heat would not be possible in the final phases of collapse. In typical cases, time durations for this phase are ~1 μs. Later numerical studies by Hickling [22] and Mitchell and Hammitt [23], as well as various others, tend to confirm this conclusion, as does the experimentally observed "sonoluminescence" which is generally ascribed to very high temperatures [22].

The Hickling-Plesset results (Figs. 4-6 to 4-8) and those of Ivany-Hammitt [9, 10] indicate bubble-wall liquid Mach numbers between 10 and 100, if spherical symmetry could be maintained through radius ratios of the order 10^3 to 10^4. This certainly verifies the necessity for the consideration of compressibility effects. Significant resultant reductions in wall velocity from the Rayleigh case are shown in these figures. However, as previously indicated, this maintenance of symmetry appears to be most unlikely in real cases. This point is discussed in detail in the section concerning asymmetrical bubble collapse.

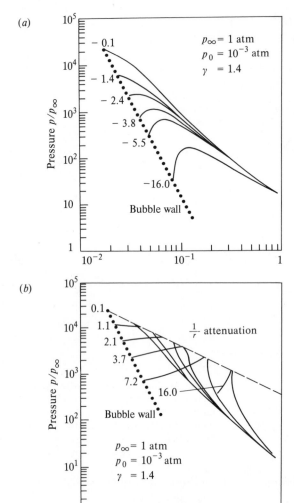

(a)

(b)

Figure 4-8 Instantaneous pressure in the liquid vs. distance from the bubble wall. Compressible liquid without viscosity or surface tension. Gas constant $\gamma = 1.4$; ambient pressure $p_\infty = 1$ atm; initial pressure $p_0 = 10^{-3}$ atm. (a) Bubble collapse. (b) Bubble rebound.

Figure 4-8b shows the envelope of maximum pressure within the liquid during rebound. This is generally considerably greater than the liquid pressure during collapse at distances from the center of collapse greater than the initial radius R_0. Assuming as a first approximation that the bubble center remains stationary during collapse, R_0 would be the closest possible distance to an adjacent material surface which might be damaged by the collapse. Of course, a symmetrical collapse would be out of the question in such close proximity to a wall. However, it can easily be shown for ideal-fluid considerations [11] that the bubble-collapse center would approach a solid wall during collapse, and conversely be repelled from a free surface. Since an analysis of this situation assuming maintenance of spherical symmetry would obviously not be physically meaningful, these questions will only be discussed in detail in the section considering asymmetrical-collapse phenomena.

As shown in Fig. 4-8, the maximum pressure at a distance of R_0 from the collapse center is ~ 500 bar for a typical case. The maximum pressure decreases approximately as r^{-1}. Obviously this pressure would not be sufficient to damage most structural materials of the sort, which are in fact known to be readily damageable by cavitation. Figure 4-9 from Ivany's study [9, 10] shows the maximum-pressure envelope around the bubble at various distances during collapse. In this case, the maximum pressure at a distance R_0 is about 2000 bar, which of course is also not damaging to most materials. Hence, it would appear from these results that while the pressure wave upon rebound may be an important contributory factor in the damage mechanism the wave upon collapse is not, and other contributing factors must also be involved, as, for example, the high-velocity microjet resulting from asymmetric bubble collapse (discussed in a later section), as well as the migration of the bubble center toward the wall.

Figure 4-5, already mentioned concerning viscosity, summarizes Ivany's results [9, 10] concerning the effects of viscosity and compressibility. As in the Hickling-Plesset study, the existence of highly supersonic liquid bubble-wall Mach numbers is verified upon collapse, if spherical symmetry is maintained through large radius ratios. The effects of surface tension and viscosity are included in these calculations, omitted from the Hickling-Plesset study. The effect of variation of surface tension was not included; rather, a surface-tension value pertinent to cold water was used. It was found, however, that the surface-tension effect was not important in any portion of the collapse (even though it approaches infinity as collapse approaches

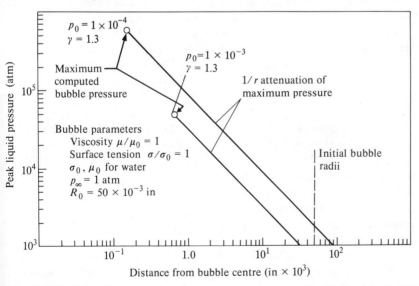

Figure 4.9 Peak liquid pressure on rebounding pressure wave vs. distance from bubble center, assuming $1/r$ attenuation.

completion), since the inertial term grows faster. Surface-tension effects are, of course, important in nucleation, as discussed in Chap. 3.

The effects of liquid compressibility in strongly reducing collapse velocities as compared to the ideal-fluid Rayleigh analysis [1] for radius ratios below $\sim 10^{-2}$, also shown by the Hickling-Plesset analysis [21], are confirmed (Fig. 4-5). The effects of viscosity are also shown in Fig. 4-5. The effects for viscosities similar to that of cold water become significant only for radius ratios of 10^{-3} to 10^{-4}, and are more important for liquid Mach number than velocity. For a viscosity of ~ 100 times that of cold water, the effect becomes significant for radius ratios in the range 10^{-2} to 10^{-3}. For the critical viscosity predicted by Poritsky [7], that is, 1468 times cold water viscosity, the collapse is brought to a virtual standstill. This is a viscosity about equivalent to that of a residual crude oil.

The effects of various quantities of entrapped gas, studied in the Hickling-Plesset analysis [21], were investigated also by Ivany [9, 10] and are shown in Fig. 4-5. The results are essentially in agreement with those of Hickling and Plesset.

In brief summary of the results from Hickling-Plesset [21] and Ivany-Hammitt [9, 10], it can be stated that for cavitation bubble collapse:

1. The effects of surface tension are probably never of substantial significance in bubble collapse for any liquid of engineering importance.
2. The effects of viscosity can only be substantial for liquids with viscosities in the range of lubricating oils.
3. The effects of liquid compressibility for liquids with bulk moduli of the range of water are important only if spherical symmetry is maintained through very large radius ratios, which present knowledge indicates is not probable for engineering applications. The effects of liquid compressibility are thus probably negligible for such fluids as liquid metals (sodium, mercury, etc.). However, to my knowledge, no investigation has been made for highly compressible liquids of engineering importance such as, for example, petroleum products and other organic liquids. It is conceivable that for such liquids the effects of both compressibility and viscosity would be substantial in some engineering applications.

Surface tension, gas content, and miscellaneous   The effects of very small permanent gas content within the bubble upon bubble collapse are shown in Figs. 4-5 to 4-8, already discussed, and also in Figs. 4-9 and 4-10 from Ivany [9, 10]. As indicated in all these cases, the collapse is stopped at a finite radius ratio by the compression of the internal gas, assumed to be adiabatically compressed in all cases. The minimum radius ratio attained decreases, as would be expected for greater initial gas contents. Initial internal gas pressures of 10^{-1} to 10^{-4} bar were investigated in these studies.

Internal gas contents of these general magnitudes were investigated, since it is believed that in most cases cavitation bubbles "nucleate" from either entrapped or entrained gas "microbubbles," the sizes of which depend primarily upon surface-tension effects, as discussed in detail in Chap. 3. While, as stated above, surface

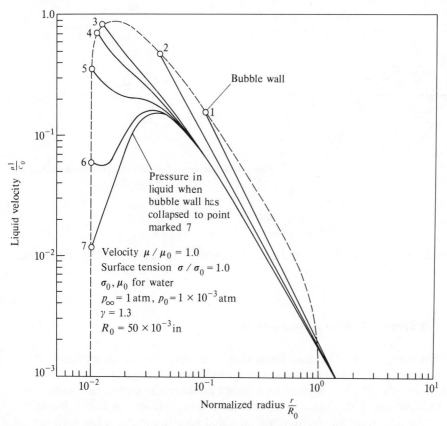

Figure 4-10 Liquid velocity vs. normalized radius for bubble containing gas.

tension is not of substantial importance directly in bubble collapse, it is of importance through its influence on nucleation in affecting the entire aspect of the cavitation field, and hence the initial size from which the bubble collapses.

The quantitative relation, sometimes called the "Rayleigh equation," particularly in the heat-transfer literature, through which surface tension enters the bubble equation of motion has been already included as Eq. (4-9). Since its derivation is very simple and also pertinent in other fields, such as the calculation of wall stresses in a thin-walled spherical pressure vessel, it is included next, even though it has been assumed and used previously in this book.

Consider a static-force balance (Fig. 4-11) applied to the hypothetical "free body" of the upper half of a typical bubble between the pressure differential between internal gas pressure P_i and external liquid pressure P_L, attempting to pull the bubble apart, and the surface tension σ applied to the circumference, attempting to hold the bubble together. Then

$$\pi R^2(P_i - P_L) = 2\pi R\sigma \tag{4-30}$$

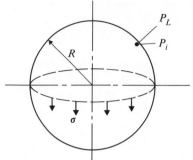

Figure 4-11 Schematic for surface-tension term derivation.

so that
$$P_i - P_L = \frac{2\sigma}{R}$$
(4-31)

Capital letters are used since the quantities apply at the bubble wall rather than within the fluid.

4-1-4 Effects of Bubble Asymmetries

General background   The most important "real-fluid" deviation, especially from the viewpoint of cavitation damage, from the classical Rayleigh analysis is undoubtedly that of bubble asymmetries, which predominate particularly during the collapse process. This situation is discussed relatively briefly in Ref. 1, but more recent and important information will be added here. In general, it has been shown theoretically [24] that the collapse process of an initially spherical bubble is basically unstable, so that small perturbations may grow into major asymmetries, while the growth process is essentially stable. The physical mechanisms involved are somewhat similar to those of the classical Taylor instability, involving the acceleration of a lighter fluid away from the interface between lighter and heavier fluids. However, in the bubble case, the instability occurs in the opposite sense because of the controlling effects of the spherical geometry. Photographic information in general confirms these theoretical expectations [23] in that, in most engineering situations, growing bubbles in most cases remain essentially spherical while collapsing bubbles do not.

The asymmetry of major importance from the viewpoint of cavitation damage is obviously the necessary proximity of the wall to be damaged. However, pressure gradients, relative velocity ("slip"), or the presence of a body force such as gravity produce relatively similar results in the actual shape of bubble-collapse profiles. These results are in general confirmed both by high-speed photographs and numerical analyses, as will now be reviewed.

Analytical approaches   The numerical analysis computer problem is complicated by the fact that the moving boundary, of indeterminate shape at a given instant, between liquid and vapor must be followed mathematically during the progress of the bubble collapse. Procedures for doing this in various other similarly complex problems have been developed under various pseudonyms such as MAC ("marker and cell" technique), PIC ("particle in cell"), etc. [25–29]. The former (MAC) is written for incompressible fluids and was hence adopted at the author's laboratory [30–32] for the bubble-collapse problem. The latter (PIC) is for highly compressible fluids only, and hence not applicable for the liquid regime in bubble collapse. Since MAC allows the inclusion of viscosity, its effect was also investigated by Mitchell [30–32], who included the effects of pressure gradient and relative velocity as well as wall proximity. These were investigated in separate problems. No combined-effect analysis has yet been reported. Spherical coordinates were used, since they were assumed to be least likely to lead to numerical instabilities.

During approximately the same time period as the Mitchell analysis at Michigan, the problem was also investigated at the California Institute of Technology (CIT) by Chapman and Plesset [33, 34]. They used an entirely different numerical technique based upon potential flow assumptions, and hence were unable to include the effects of viscosity. These, at least for water, are probably not of major importance. Since they only considered the wall-proximity problem, of primary importance from the viewpoint of cavitation damage, they used cylindrical coordinates. This allowed them to follow the progress of the microjet further than was possible in Mitchell's analysis with spherical coordinates, since in that case the numerical procedure must be terminated as the microjet approaches the initial bubble center. The results of these numerical studies, the only ones so far reported to the author's knowledge, are discussed later.

The general mechanism of collapse and microjet formation in a direction determined by the initial asymmetries is easily explicable from relatively simple arguments, involving the conservation of linear momentum [35]. Assuming some initial linear momentum of the bubble virtual mass, i.e., including a portion of the surrounding liquid, this momentum must be conserved during collapse, since no external forces other than viscous effects are applied to the bubble during this period. Hence, as the collapse proceeds and the bubble virtual mass decreases, its linear velocity must increase proportionately. As the bubble size approaches zero, this velocity would approach infinity. The initial linear momentum is a direct result of the initial asymmetry such as, for example, relative velocity, pressure gradient, gravity or other asymmetrical body force, wall proximity, etc. A simplified analysis considering this effect was also reported by Chincholle [36, 37]. Termed the "rocket effect," his investigation did not consider the actual bubble-shape changes during collapse, but showed the trajectory of bubble centroid.

The wall effect, i.e., the "attraction" of a bubble toward a rigid wall, and also its repulsion from a free surface was predicted by studies of underwater explosions during World War II [11], which were based on potential flow analyses and used the concept of "mirror images." Later work shows that a sufficiently flexible boundary [38] is similar to a free surface in this respect, thus partially explaining the experimental observation of the remarkable cavitation-damage resistance of soft rubbers and other elastomeric materials, including even Plexiglas,† in some applications.

In very simple and naive terms, the wall attraction and free-surface repulsion can be explained as follows. Consider a spherical bubble near a rigid wall starting to collapse. If the collapse were to be spherical, the radial liquid velocity in the inward direction would need to be uniform at all points of the bubble periphery. However, liquid "access" to the region between bubble and wall would be relatively restricted compared to that on the side away from the wall. Thus the radial-wall collapsing velocity on the wall side would be less than that elsewhere, so that a motion of the bubble centroid toward the wall would be generated. Thus a linear momentum of the bubble centroid in this direction would be created. Its necessary conservation (neglecting viscous effects) would lead to an acceleration of the bubble virtual mass toward the wall as the collapse continued, resulting in the formation of the eventual high-velocity microjet to which an important portion of cavitation damage is usually attributed.

In the case of a highly resilient wall or free surface, the above argument would apply in an inverse sense. Acceleration of the liquid on the side of the bubble away from such an interface is essentially restrained by the inertia of the liquid column extending theoretically from the bubble wall to infinity in that direction. Similar inward acceleration of the bubble wall near a free surface would involve only the inertia of the restricted liquid column between the bubble wall and free surface, plus the inertia of the adjoining air column, which is also, as a first approximation, considered to infinity, i.e., to a "long distance." Hence the inertia restraint for bubble collapse on the side toward the free surface would be much less than that on the opposite side, assuming that the bubble was in reasonable proximity to the free surface. Thus acceleration on the free-surface side of the bubble would be greater, so that the bubble centroid would move away from the free surface. As the collapse proceeded, this acceleration of the bubble centroid would increase. Microjet formation would then occur away from the surface, rather than toward it as in the case of a rigid boundary. Precisely the same arguments can be made for a highly resilient interface such as, for example, a thin rubber diaphragm [39, 40]. Figures 4-12 and 4-13 show this effect for a study here involving spark-generated bubbles. Similar results were obtained in a more comprehensive study of this effect at Cambridge University [38]. A sufficiently resilient surface would then be expected to repel collapsing bubbles, which is consistent with the already observed good cavitation-damage resistance of various elastomeric materials [40, 41].

During bubble growth the opposite effects occur for entirely analogous reasons, i.e., a growing bubble is repelled from a solid surface and attracted toward a highly resilient or free surface, as discussed in more detail later.

† Venturi tests at author's laboratory.

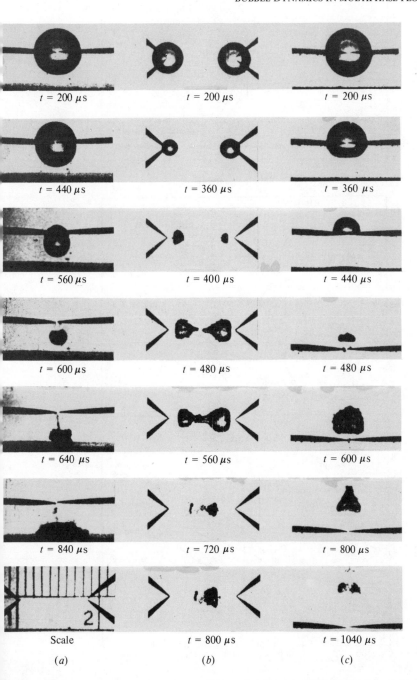

$t = 200\ \mu s$ $t = 200\ \mu s$ $t = 200\ \mu s$

$t = 440\ \mu s$ $t = 360\ \mu s$ $t = 360\ \mu s$

$t = 560\ \mu s$ $t = 400\ \mu s$ $t = 440\ \mu s$

$t = 600\ \mu s$ $t = 480\ \mu s$ $t = 480\ \mu s$

$t = 640\ \mu s$ $t = 560\ \mu s$ $t = 600\ \mu s$

$t = 840\ \mu s$ $t = 720\ \mu s$ $t = 800\ \mu s$

Scale $t = 800\ \mu s$ $t = 1040\ \mu s$

(*a*)   (*b*)   (*c*)

Figure 4-12 Various static spark-generated bubble collapse cases. (*a*) Bubble collapse adjacent to a brass plate. (*b*) Bubble collapse adjacent to another bubble. (*c*) Bubble collapse adjacent to a rubber membrane.

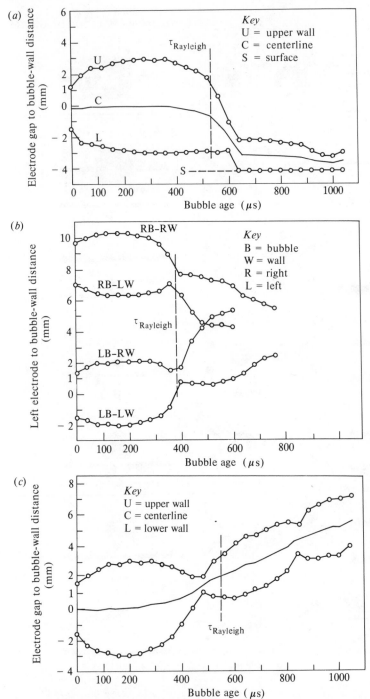

Figure 4-13 Apparent wall position for bubble collapse. (*a*) Apparent wall position along the axis of symmetry as a function of time for a bubble collapse next to a brass plate. (*b*) Apparent wall position along the axis of symmetry as a function of time for a bubble collapse adjacent to another bubble. (*c*) Apparent wall position along the axis of symmetry as a function of time for a bubble collapse adjacent to a rubber membrane.

Specific cases

Wall proximity—simplified analyses of double-bubble and other cases As pre-viously indicated, wall proximity is the most important asymmetry from the view-point of cavitation damage. Also, as previously discussed, the problem of bubble-centroid motion was initially approximately solved [11] without considering the details of asymmetric bubble-collapse profiles. This type of solution also has been applied much more recently [36, 37]. Since the original solution [11] relied upon a method of mirror images, considering a hypothetical "mirror-image" bubble behind the wall, the wall was then considered a plane of symmetry for a double-bubble problem. This model is essentially exact, if the effects of viscosity are neglected, and hence it seems reasonable to investigate the double-bubble problem as being of interest in itself, as well as shedding light on the single-bubble wall-proximity problem. This has been done [39, 40] photographically (Figs. 4-12 and 4-13), and the results are in fact very similar to the wall problem with microjet formation along the axis connecting the bubbles and in the direction toward the system centroid.

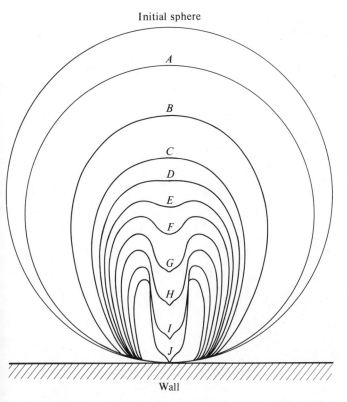

Figure 4-14 Computed bubble-collapse surfaces from case 1, Plesset-Chapman [34].

Comprehensive wall-proximity numerical analyses

1. *Chapman and Plesset* [33, 34]. The most complete numerical investigation of the wall-proximity bubble-collapse problem, in terms of following the collapse to completion, is that of Chapman and Plesset [33, 34]. Their results showing the initial substantial collapse of the sides normal to the wall, followed by the accelerating collapse of the side away from the wall, and the development of the expected microjet are shown in Figs. 4-14 and 4-15. Figure 4-15*b* provides a very good verification of these results [42].

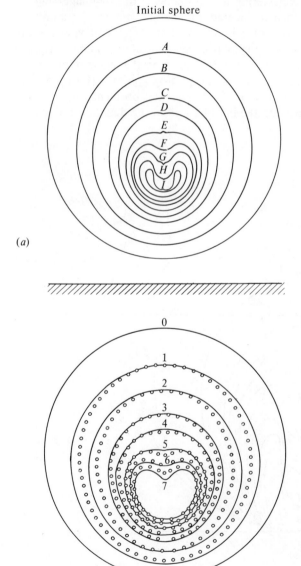

(*a*)

(*b*)

Figure 4-15 Computed bubble collapse from case 2, Plesset-Chapman [34, 42].

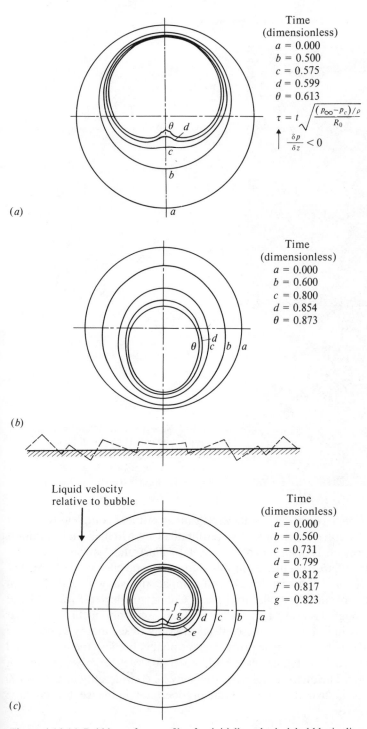

Time
(dimensionless)
$a = 0.000$
$b = 0.500$
$c = 0.575$
$d = 0.599$
$\theta = 0.613$

$$\tau = t \sqrt{\frac{(p_\infty - p_c)/\rho}{R_0}}$$

$\dfrac{\delta p}{\delta z} < 0$

(a)

Time
(dimensionless)
$a = 0.000$
$b = 0.600$
$c = 0.800$
$d = 0.854$
$\theta = 0.873$

(b)

Liquid velocity
relative to bubble

Time
(dimensionless)
$a = 0.000$
$b = 0.560$
$c = 0.731$
$d = 0.799$
$e = 0.812$
$f = 0.817$
$g = 0.823$

(c)

Figure 4-16 (*a*) Bubble surface profiles for initially spherical bubble in linear pressure gradient, $\sigma = 0.57$, and defines pressure gradient magnitude (Mitchell). (*b*) Bubble surface profiles for initially spherical bubble with center 1.5 R_0 for rigid wall (Mitchell). (*c*) Bubble surface profiles for initially spherical bubble moving relative to surrounding liquid, $V_\infty = 0.1515$, Mitchell Hammitt [30–32].

169

Figure 4-17 Nonsymmetric collapse of a bubble in a pressure gradient. High-speed photographs of collapse in a two-dimensional venturi with a $\frac{1}{4}$ inch throat in water at 74.6 ft/s. Air content 2.35 % by volume, 132 μs between frames, 1 μs [46].

2. *Mitchell and Hammitt* [30–32]. Somewhat similar but less complete results in terms of the microjet development were obtained by Mitchell and Hammitt [30–32], as previously mentioned. Figure 4-16 illustrates typical computer-output bubble-collapse profiles from this study, showing effects (*a*) of pressure gradient and (*b*) of wall proximity. Careful comparisons show that those from wall proximity are entirely similar to those of Chapman-Plesset (Figs. 4-14 and 4-15), even though viscosity and wall roughness were not included in that study. It is thus indicated that these effects are only of secondary importance in this particular bubble-collapse problem, presumably because the bubble-wall velocities attained in a typical case with water for such a collapse are not sufficiently large. The microjet velocity itself may be considerably larger (~ 100 to 1000 m/s is indicated),[†] but this factor does not influence the collapse behavior otherwise.

[†] Plesset and Chapman [33, 34] estimate 130 and 170 m/s for two typical cases with water, but up to 1000 m/s has been estimated by Brunton [43]. Photographs (Fig. 4-19) show ~ 100 m/s [44, 45].

Wall roughness was necessarily included in the Mitchell-Hammitt study because of the form of the spherical-coordinate computing cells. This was not involved in the Chapman-Plesset case because of their use of cylindrical coordinates. The wall roughness used was approximately that of a cast-steel wall for a ~2-mm bubble initial diameter, so the roughness result is of practical significance.

Kling, Timm, and Hammitt [40, 44, 45]. The Chapman-Plesset and Mitchell-Hammitt computer results were confirmed at about the same time by the high-speed cinematographic results of Kling and Hammitt [44, 45], obtained in a venturi with spark-generated bubbles in water using framing rates ~10^6 Hz. Camera speeds of this order are necessary, since the final important portion of the bubble collapse in pertinent cases occurs in only a few microseconds. Figure 4-17 (reprinted for convenience from Ref. 1) and Ref. 46 show bubble collapse in a venturi pressure gradient, indicating bubble flattening normal to the pressure gradient. Figure 4-18 (reprinted from Ref. 1 for convenience) shows schematically the various forms of bubble collapse expected, including external

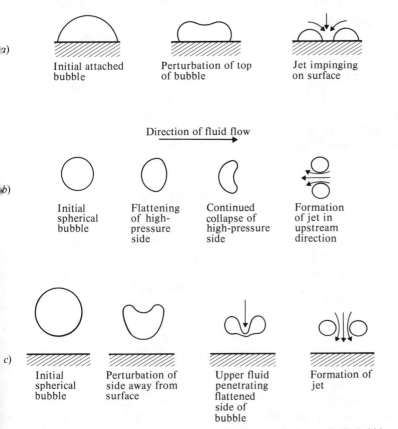

Figure 4-18 Jet-collapse models. (*a*) Hemispherical bubble attached to wall. (*b*) Bubble moving into pressure gradient (such as venturi diffuser flow). (*c*) Bubble collapsing near wall.

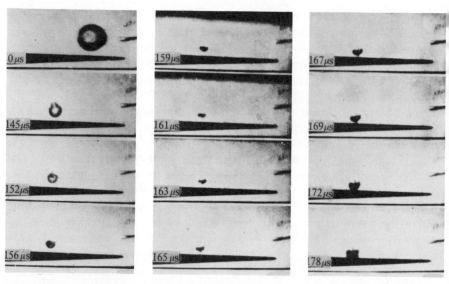

Figure 4-19 Selected frames from a sequence taken at 550,000 frames/s of a spark-generated cavita tion bubble near a splitter in a venturi, exposure 1.8 μs per frame, flow right to left, magnification × [44, 45].

velocity and pressure gradient (Fig. 4-17) and wall proximity. Figure 4-19 is a sequence of high-speed motion pictures [44, 45] showing the collapse of a spark generated bubble in a two-dimensional venturi near a thin flat plate which i aligned parallel to the flow along the venturi axis. The complete expected sequence of events is observed: first, flattening of sides normal to the wall, then of sides parallel to and distant from the wall, followed by microjet generation and its impingement upon the wall.

The final event is that of bubble "rebound," i.e., regrowth of the vapor and gaseous mass. Such rebounds are typical of bubble collapses in flowing water systems, perhaps being due primarily to entrapped gas and noncondensing vapor in the short time available. These features are discussed in detail in Sec. 4-1-5. Rebound is important in the damage process, since the strength of pressure pulses in the liquid due to a growing bubble are considerably greater than those due to collapse, as shown by Hickling and Plesset [21] for spherical collapse and growth.

Figures 4-20 and 4-21 show detailed experimental bubble-collapse profiles from the Kling-Hammitt study taken from photographic enlargements of the individual frames of high-speed motion pictures (Fig. 4-19). These show all the details of intermediate bubble shapes predicted numerically [30–34] and discussed previously. From these photos, for an initial ∼4-mm diameter bubble, the eventual microjet diameter is ∼80 μm and the jet was estimated to have an impact velocity of ∼100 m/s. However, in later experimental and photographic work by Brunton [43] and his colleagues at Cambridge University it was estimated that velocities up to 1000 m/s were possible. An

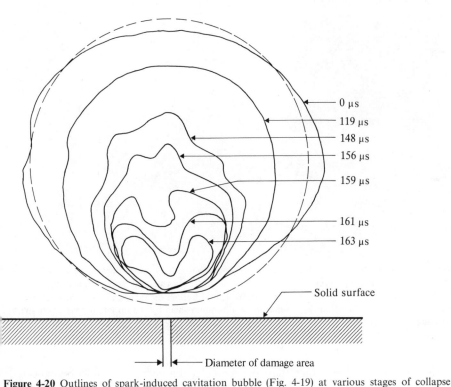

0 µs
119 µs
148 µs
156 µs
159 µs
161 µs
163 µs

Solid surface

Diameter of damage area

Figure 4-20 Outlines of spark-induced cavitation bubble (Fig. 4-19) at various stages of collapse showing the mode of deformation [44, 45].

intermediate jet velocity was predicted by the Chapman-Plesset study [33, 34], as previously stated.

The Kling-Hammitt study also allowed a one-to-one correspondence to be obtained between bubble collapses and craters in the soft aluminum wall. In normal cavitation [47, 48] this ratio appears to range up to 10^6, so that the one-to-one correspondence in these tests results only from nearly complete control of the bubble parameters achieved using spark-generated bubbles. Figure 4-22 shows details of a crater in soft aluminum so created, indicating

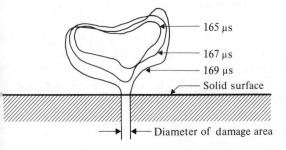

165 µs
167 µs
169 µs
Solid surface
Diameter of damage area

Figure 4-21 Outlines of spark-induced cavitation bubble (Fig. 4-19) at various stages of rebound showing the bubble impinging on the nearby solid surface [44, 45].

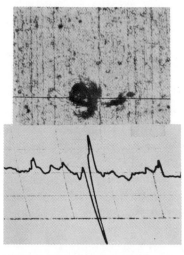
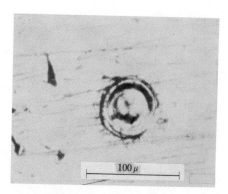

Figure 4-22 (*a*) Photomicrograph and accompanying proficorder trace of damage produced on the 0.5 mm thick aluminum sheet by collapse with an initial normalized wall distance, b_0, of 1.15. Scale divisions 66.0 μ horizontal, 0.635 μ vertical. Initial bubble diameter $\cong 2$ mm. (*b*) Photomicrograph of damage produced on the 0.5 mm thick aluminum sheet by the initiating spark of a spark-induced cavitation bubble with an initial normalized wall distance, b_0, of 1.15. Initial bubble diameter $\cong 2$ mm.

that it is somewhat larger than the microjet diameter, as would be expected from consideration of liquid jet impact characteristics discussed in Chap. 6.

The high-speed bubble-collapse photos obtained by Kling [44, 45], for example, provide information from one viewing direction only, so that the three-dimensional form of the collapse profiles can only be assumed from considerations of symmetry. To investigate further the three-dimensional form of bubble collapse, Timm [39, 40] obtained high-speed motion pictures of spark-generated venturi bubble collapse, using the same flow geometry as that of Kling [44, 45], for two perpendicular viewing directions simultaneously, i.e., side pictures were combined with top pictures of venturi bubble collapse adjacent to a wall (Fig. 4-23). These simultaneous pictures (Fig. 4-23) were obtained using the same high-speed camera employed by Kling, but using a mirror system to obtain pictures from perpendicular directions together on each frame. In general, the side-view bubble pictures (lower pictures in each frame) are quite similar to those of Kling (Fig. 4-19), showing again the attraction of the bubble mass toward the wall, then the eventual impinging microjet, followed by "rebound" of the vapor-gas mass. The top-view pictures are interesting in showing a rapid breakdown of the circular profile but no pronounced systematic asymmetry, even though there is a significant axial pressure profile (toward the right in the pictures). The top-view pictures indicate only slight rebound compared to side view. In final picture, bubble has virtually disappeared from both views. The total sequence, including collapse, rebound, and final disappearance, requires only about 0.5 ms.

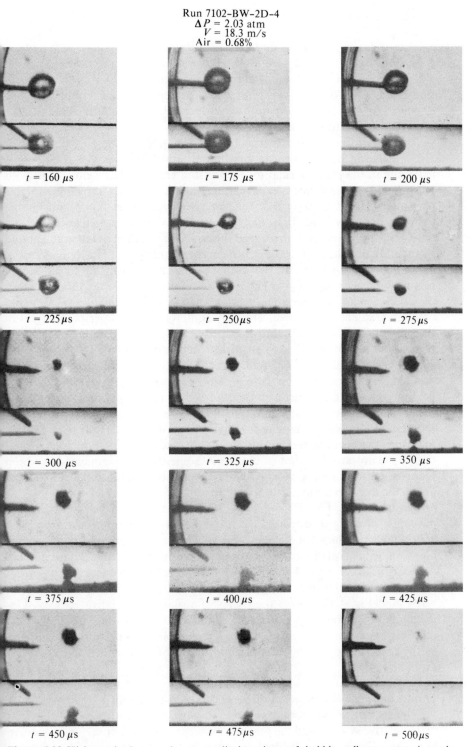

Run 7102–BW–2D–4
$\Delta P = 2.03$ atm
$V = 18.3$ m/s
Air = 0.68%

$t = 160\ \mu s$

$t = 175\ \mu s$

$t = 200\ \mu s$

$t = 225\ \mu s$

$t = 250\ \mu s$

$t = 275\ \mu s$

$t = 300\ \mu s$

$t = 325\ \mu s$

$t = 350\ \mu s$

$t = 375\ \mu s$

$t = 400\ \mu s$

$t = 425\ \mu s$

$t = 450\ \mu s$

$t = 475\ \mu s$

$t = 500\ \mu s$

Figure 4-23 High-speed photographs, perpendicular views of bubble collapse, venturi spark-generation (Timm [39, 40]).

Pressure or velocity gradient, body force, and relative velocity

1. *General.* The effects of pressure or velocity gradients, asymmetrical body force, and relative velocity between bubble centroid and external liquid are very similar to those of wall proximity, or a second identical bubble, already discussed. In all cases, an eventual microjet results, with intermediate profiles relatively similar to the wall-proximity case.

There is little precise experimental information on these effects, though Fig. 4-17 shows the effect of a venturi diffuser pressure gradient, causing bubble flattening in the plane normal to the pressure gradient, i.e., venturi axis, with presumably eventual microjet formation in the upstream direction. In this particular case, it is likely that relative velocity between liquid and bubble ("slip") is also involved. Presumably this slip would be negative in a rising pressure region such as the venturi diffuser, so that both pressure gradient and slip would cause the microjet to be in the upstream direction. This venturi case in particular was analyzed at the author's laboratory using a linear small-perturbation technique [49]. The possibility of microjet generation in either direction, depending upon the direction and strength of relative velocity and pressure gradient, was shown. The only comprehensive numerical analyses of the pressure gradient and relative velocity cases, even separately considered, to the author's knowledge, is that of Mitchell and Hammitt [30–32, 41]. No comprehensive analysis apparently exists of the asymmetrical body-force problem, such as gravity, but also possibly imposed magnetic or electrostatic fields, which may be of interest for various fusion reactor concepts using liquid metal coolants. As a first approximation, however, it would seem that the problem of a bubble rising due to gravity in an otherwise static liquid would involve both a relative velocity and a pressure gradient, so that microjet formation in the upward direction would result. Some crude experiments have been reported to the author, involving the final "jumping" of the collapsing bubble above the free surface in a simple beaker test.

2. *Mitchell analyses* [30–32, 41]. The Mitchell-Hammitt study of the wall-proximity case has already been discussed (Fig. 4-16). In his dissertation [32], Mitchell also included the effects of different pressure gradients and relative liquid velocity (slip) in separate problems. The qualitative results have already been discussed. Figure 4-16a shows typical computer results for a typical pressure gradient, which is approximately that produced by gravity with water, and Fig. 4-16c shows the effects of an imposed relative velocity. Unfortunately, the computer outputs do not proceed far enough to show final microjet development because of the necessary termination of the calculation as the microjet approaches the original origin of coordinates, as previously explained.

Figure 4-24 compares the effects of these various asymmetries upon bubble profiles during collapse, including the effects of two different pressure gradients, the stronger of which is approximately that of gravity with water. The effects of the relative velocity and the pressure gradients are qualitatively quite similar, but that of wall proximity causes initial bubble flattening in the opposite

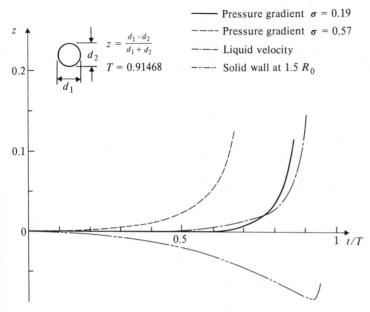

Figure 4-24 Comparison of distortions of initially spherical bubbles collapsing under the various asymmetric conditions (Mitchell [30–32]).

sense, i.e., the bubble dimensions in planes parallel to the wall diminish initially (Fig. 4-16*b*). The effects upon bubble-centroid migration of these same asymmetries were found also [30–32]. Again they are similar, and in this case in the same direction. For relative velocity, the eventual microjet is in the direction of the initial bubble velocity relative to the liquid (thus preserving initial linear momentum), and for the pressure gradients it is in the direction of decreasing pressure, essentially for the same reason.

Figure 4-25 compares the calculated jet velocities, i.e., maximum wall velocity, for two pressure gradients with the bubble-wall velocity from the classical Rayleigh spherical analysis [4] for the same radius ratios, R/R_0. The asymmetrical maximum wall velocities exceed those for the spherical Rayleigh collapse model for relatively high radius ratios ($R/R_0 \leq 0.35$), i.e., near the start of collapse for these asymmetrical cases. However, for smaller radius ratios, i.e., near the end of collapse, the Rayleigh velocity exceeds the asymmetric jet velocities, which are the maximum wall velocities. Although these results are available only for the pressure-gradient asymmetric collapse case, it seems intuitively obvious that qualitatively similar results would be found for the other cases. Even though the jet velocities are less than the Rayleigh wall velocities near the end of collapse, they are still sufficient to provide possibly a major component of the damaging mechanism, as previously discussed.

Compliant boundary effects: Lauterborn, Gibson, and Chahine studies [42, 50–53]
The effects of compliant boundaries in the vicinity of a collapsing bubble can be

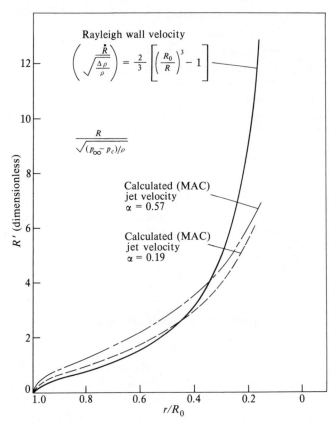

Figure 4-25 Jet velocity as function of radial position for bubbles collapsing in linear pressure gradients (Mitchell).

provided either by other neighboring bubbles, which are certainly very often present in many real cases, or compliant boundaries such as free or resilient surfaces. These effects in general are probably the most important asymmetrical effects to which little comprehensive analysis has yet been applied. Presumably this is due to the extreme complexities involved, making straightforward computer analyses highly expensive. Nevertheless, some pertinent information is available. The simplest cases have already been discussed (see Figs. 4-12 and 4-13, and Ref. 11, for example), but other pertinent information will now be reviewed.

The most comprehensive study to date of the compliant boundary problem known to the author is that of Gibson [50, 51], working in Brunton's laboratory at Cambridge University. Both theoretical and photographic studies were made with spark-generated bubbles, collapsing near a soft rubber wall. The theoretical profiles were derived from a perturbation model, and hence may not be as accurate as the numerical studies of either Mitchell or Chapman, discussed earlier. Bubble repulsion from the wall during its collapse was observed as expected and also as confirmed by the Timm photos (Figs. 4-12 and 4-13). Very recent work by Chahine [52] also confirms this bubble behavior.

Gibson also confirmed in great detail the bubble-collapse profiles in a pressure gradient, again using spark-generated bubbles in a static beaker. Figure 4-26 shows these results.

Some very recent and excellent results concerning single-bubble behavior using laser-generated bubbles and high-speed cinematography have also been reported by Lauterborn and Bolle [42, 53] at the University of Göttingen. The use of a concentrated laser beam for single-bubble generation for research purposes

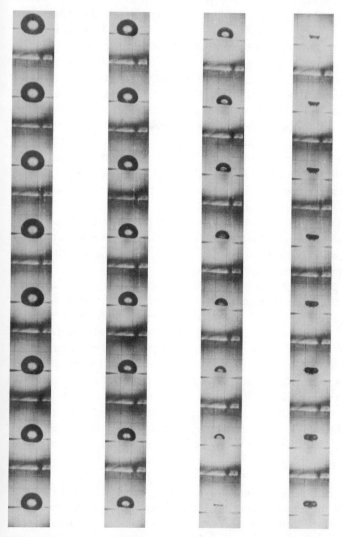

Figure 4-26 High-speed (2500 frames/s) photographs of initially spherical bubble collapsing and rebounding in linear pressure gradient, $\sigma = 0.186$ (Gibson [50, 51]).

represents some improvement over the spark-bubble technique, in that the liquid field is not perturbed by the presence of the electrodes and also there is no introduction of noncondensable electrolytic gases. Of course, the high-power pulse laser equipment is also much more expensive than that required for spark-bubble generation.

As already indicated, multiple bubbles may provide the most general source of compliant boundary effects in most cavitation (or boiling) situations. However, due to the complexity of the problem, very little comprehensive research on this problem has yet been attempted. Also, as previously indicated, the problem of the collapse of two identical bubbles in close proximity (Figs. 4-12 and 4-13) is essentially that of collapse of a single bubble near a rigid wall, and hence has already been solved to a major extent. Results similar to those of Timm (Figs. 4-12 and 4-13) have also been reported more recently by Lauterborn and Bolle [42, 53].

Comprehensive numerical analyses to consider multiple-bubble effects, other than that of two identical bubbles in close proximity, have not yet been reported, to the author's knowledge. This multiple-bubble problem appears to be nearly beyond the state of the art, within reasonable cost limits, for present-day large computers. Of course, all asymmetrical collapse analyses so far have neglected most real-fluid effects, except for viscosity [30–32, 41], which is relatively easy to include. These were included in spherically symmetric collapse studies, such as those of Hickling and Plesset [21, 22] and Ivany and Hammitt [9, 10]. In the asymmetrical analyses so far reported, the neglect of compressibility, which proved relatively to be most important in the spherical collapse analyses for low-viscosity liquids such as water [9, 10, 21, 22], is perhaps reasonably permissible, because liquid Mach numbers do not become significant (as opposed to spherical analyses) by the time the collapse is terminated by the microjet impact upon either the opposite bubble wall or adjacent material surfaces.

The effects of the growth or collapse of nearby bubbles during the collapse of a given bubble could be either primarily somewhat similar to the double-bubble case already discussed (Figs. 4-12 and 4-13), involving a mutual attraction between the collapsing "voids" (which are essentially "sink" flows), or alternatively the triggering of asymmetrical collapse in a given bubble by shock waves generated by the collapse or growth of adjacent bubbles. Bubble microjet formation could sometimes be triggered in this way, much like the formation of a "Monroe jet" in a shaped-charge explosive used in mining or the "Bazooka" antitank weapon used in World War II. This possibility was suggested and studied quantitatively by Kozyrev in the U.S.S.R. [54]. The concept represents admittedly an over-simplified model, but nevertheless it may be physically quite instructive in explaining the otherwise rather mysterious microjet collapse behavior.

Rebound phenomena   Bubble "rebound," or regrowth, is, as indicated by high-speed cinematography, quite common in cavitating flow regimes. It is the present general conception, in the author's opinion, that the liquid "shock waves" emanating from bubble rebound are one of the major mechanisms involved in mechanical cavitation damage, along with microjet impact, as already discussed.

This conception is confirmed in numerical analyses [9, 10, 21, 22] showing that the shock-wave strength on rebound is much greater than that occurring during Rayleigh collapse [4], and appears in fact to be of sufficient magnitude to contribute importantly to the observed damage. Nevertheless, no comprehensive numerical analyses of asymmetrical bubble collapse have yet been carried beyond the point of microjet impact [30–34]. This appears then to be a "next logical step" which is probably not beyond the capability of present computer techniques. Droplet-impact studies (Chap. 6) are certainly pertinent to some extent.

Figure 4-27 presents excellent Schlieren photographs of spherical shock waves in the liquid emanating from the collapse of a single cavitation bubble [114], obtained from the laboratory at Göttingen. These and other photos existing in the literature strengthen belief in the importance of the shock-wave mechanism of cavitation erosion. Bubble rebound is presumably primarily due to the compression of noncondensable gases within the bubble. Since cavitation bubbles presumably nucleate in most cases from entrained gas microbubbles, the presence of some gas within the collapsing bubble is inevitable. However, in some cases the internal pressure, which eventually causes rebound, may be augmented by vapor compression near the end of collapse. This effect is further discussed in Sec. 4-1-5. In the case of asymmetric collapse, a toroidal vortex is typically formed, along the axis by which the high-velocity microjet passes. The fluid centrifugal forces associated with the toroidal vortex, i.e., essentially a "smoke ring," also provide a mechanism for rebound in addition to that from internal gas or vapor. This is not the case for a spherical collapse.

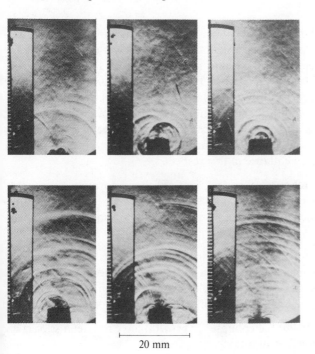

|——————| 20 mm

Figure 4-27 Schlieren photographs of spherical shock waves in the liquid emanating from the collapse of a single cavitation bubble [114].

General summarization of asymmetrical bubble-collapse studies Several general observations concerning the effects of various asymmetrical conditions, consistent with the previous discussions, seem useful.

1. Bubble collapse in most engineering situations does not even approximately follow the spherical collapse model first proposed by Rayleigh [4]. Rather, the "microjet mode" of collapse seems almost universally observed whether the dominant asymmetry is that of wall or free-surface proximity, relative velocity, pressure gradient, body force, or the presence of neighboring bubbles.

2. On both theoretical and experimental grounds, bubble growth is much more symmetrical than is collapse, so that the assumption of spherical symmetry provides a reasonably good model in that case.

3. The direction of the eventual microjet is always in the direction of the original linear momentum caused by the asymmetry itself. In the case of wall or surface proximity, or neighboring bubbles, this initial force unbalance must be due to differences in forces upon the neighboring bodies as compared to that with the liquid in the directions where no intermediate bodies exist.

4. Since linear momentum for the bubble virtual mass tends to be preserved during collapse, the associated linear velocity increases as the bubble virtual mass decreases with the collapse of bubble volume. This can be assumed to lead to the eventual high microjet velocity existing within a very small fluid mass.

5. Rigid surfaces "attract" the bubble centroid during bubble collapse, i.e., cause migration of the bubble "virtual mass" toward the surface, resulting sometimes in the eventual impingement of the liquid microjet thereon. Free surfaces, or sufficiently resilient solid surfaces, result in bubble virtual-mass repulsion from the surface during collapse and eventual microjet formation in the direction away from the wall. This bubble-centroid motion away from a resilient surface may be partially responsible for the good cavitation resistance of some rubber and other elastomeric materials. Effects during bubble growth are essentially the inverse of those during collapse in this respect.

6. One or several bubble "rebounds," i.e., regrowth of the vaporous volume, are common in flow systems. The resultant shock waves in the liquid constitute an important damage mechanism probably ranking in importance with microjet impact.

7. The collapse of two symmetrical bubbles in close proximity to each other is essentially the same problem as that of a single bubble near a rigid wall. Similarly, the growth of two identical bubbles in close proximity is essentially the same problem as that of growth of a single bubble near a free surface.

8. The problems of multiple-bubble collapse (or growth) have not been solved in detail, even for "ideal-flow" assumptions.

4-1-5 Thermodynamic Effects for Bubble Collapse

General The so-called cavitation "thermodynamic effect" [1] can be extremely important in both cavitation performance and damage. It is commonly understood to encompass all those features by which bubble growth or collapse are affected by factors other than those encountered in ordinary flow problems governed by the Navier-Stokes equations of motion and the mass conservation relations, but neglecting conservation of energy considerations such as heat transport or transfer and condensation or evaporation. Thus the "thermodynamic effect" could better be called a "thermodynamic heat-transfer effect." The original Rayleigh model [4] and subsequent improvements thereto, neglecting also these thermodynamic effects, did not consider the restraints upon bubble collapse or growth imposed by changes in internal pressure due to heat-transfer restraints. This neglect is relatively appropriate for liquids such as cold water, where the vapor density, and hence vapor mass within the bubble, is very small. For "hot water" or other similar liquids such as petroleum products, for example, this rapidly ceases to be the case as the temperature is raised, since vapor pressure, or density, is an exponential function of temperature. If vapor density within the bubble is sufficient, the flow regime becomes similar to that of subcooled boiling, where, to a first approximation, the restraints upon bubble growth or collapse are those of heat transfer and transport of the latent heat component released or absorbed by condensation or evaporation. Thus there are many important cases lying between classical cavitation and boiling in this respect. This situation was first recognized for cavitation by Stepanoff [55–57], and many recent publications upon the problem have been provided by Bonnin [58–60]. These and later specific analyses will now be discussed.

Stepanoff [55–57] It has been generally observed industrially [55–57] that the classical scaling factors for pump cavitation, such as suction specific speed S or the Thoma parameter σ_T, based primarily upon NPSH modeling, did not successfully model cavitation performance of turbomachines between liquids differing in the "thermodynamic" parameters; i.e., designs for hot water pumps could be considerably more aggressive from the viewpoint of satisfactory cavitation performance than for cold water pumps, considering only the conventional modeling parameters such as S, σ_T, etc. The same was found to be the case for pumps handling petroleum products, cryogenic liquids, etc. These trends appeared to apply for cavitation damage as well as performance. This general industrial observation was apparently first formalized in the open literature by Stahl and Stepanoff [55] in 1956. They formulated a "thermodynamic parameter" B defined as

$$B = \frac{V_v}{V_L} = \frac{v_v\,\Delta h_f}{v_L h_{fg}} \tag{4-32}$$

where B = thermodynamic factor = "B-factor"
        V = fluid total volume

v = fluid specific volume

Δh_f = increment of liquid enthalpy corresponding to a reduction of NPSH below saturation conditions

h_{fg} = heat of vaporization of liquid

This parameter essentially represents the ratio of the vapor volume formed under steady-state equilibrium conditions in a closed-system analysis to the liquid volume from which it was formed for a given small decrease of suppression pressure. It is thus a fluid "property" which can be computed and tabulated for various liquids at various saturation temperatures, and represents in a sense the cavitability of the liquid. This tabulation was in fact made by Stahl and Stepanoff [55]; correlated with empirical data, they hoped that a quantitative prediction could be made of the attainable improvement in machinery cavitation design parameters for different liquids (as compared to cold water) based on differences in the B-factor. The details of these analyses are given elsewhere in the literature [1, 55–57], so it is not necessary to repeat them in any greater detail here. However, much later experience and tests have shown that good quantitative predictions are not possible on the basis of this apparently oversimplified model. This may be primarily due to its assumption of equilibrium thermodynamics for determining the relative cavity size and its effect upon machine performance.

Many relatively comprehensive and sophisticated complex analyses have since been made to provide the framework for the desired quantitative predicting of "thermodynamic effects," both for damage and performance. In many cases, these investigators have developed modified B-factors designed to take into consideration various factors not considered in the original Stepanoff approach. One such approach is that of Bonnin [58–60] discussed next.

Bonnin [58–60] Bonnin in France has attempted [58–60] to improve the predicting capability for machinery cavitation thermodynamic effects which, as already indicated, is not good in terms of either the Stepanoff B-factor alone or other proposed modifications to it. Bonnin introduced a second B-factor to be used together with the Stepanoff factor to correlate existing experimental data better than would be otherwise possible. He presumes, on the basis of dimensional analysis, that a single B-factor cannot be adequate, and thus he proposes two parameters, B' and B''. On the basis of these two independent B-factors, he maps the total regime of cavitating flows and divides this into regions where inertial effects only are predominant, as in the Rayleigh model [4]: regions where only heat-transfer restraints are important, as in conventional boiling analyses, and regions where both inertial and heat-transfer effects must be considered, i.e., highly subcooled boiling or cavitation where thermodynamic effects must be considered. Some improvement in correlating ability is certainly achieved by the addition of a second correlating parameter, but it is not clear at this time whether the B-factors utilized in the Bonnin analyses are in fact those most appropriate.

Individual bubble-collapse analyses Several basic analyses based upon individual bubble-collapse dynamics, but investigating primarily those cases where thermodynamic as well as inertial restraints are involved, have contributed considerably to the understanding of and ability to utilize effectively "thermodynamic parameters" in the design of machinery subject to cavitation restraints. Some of the more significant of these will be reviewed in the following.

Florscheutz, Chao, and Wittke [61, 62]   Florscheutz, Chao, and Wittke [61, 62] were perhaps the first to analyze theoretically the collapse of cavitation bubbles† in cases where both inertial and thermodynamic restraints were important. They derived a new *B*-factor, B_{eff}, which included rate considerations. This allowed delineation on a more realistic theoretical basis of those flow regimes where thermodynamic effects could be neglected and those where they could not. This parameter contained time-dependent effects to reflect heat-transfer rates of latent heat of condensation away from the bubble, as well as the steady-state thermodynamic equilibrium considerations which alone were reflected by the original Stepanoff parameter [55–57]. Correlation of bubble-collapse regimes in terms of this revised parameter with high-speed photographic observations of individual bubble collapse were made, so that a more precise basis for the correlation of thermodynamic effects in bubble collapse was provided.

Hickling, Plesset, Zwick, and Hsieh [21, 22, 63–66]   The thermodynamic effect from a primarily macroscopic viewpoint was discussed in the foregoing. However, various microscopic studies of individual-bubble dynamics have contributed significantly to the understanding of the related phenomena. The work of Florscheutz, Chao, and Wittke [61, 62] was in that category. However, that of Hickling, Plesset, Zwick, and Hsieh [21, 22, 63–66] provides considerably more detail on individual collapse behavior in cases where gas or vapor effects within the bubble are important, i.e., cases where the thermodynamic effect is significant.

Hickling and Plesset [21, 22, 63, 64] investigated the actual heat-transfer rates occurring between bubble contents and surrounding liquid. In addition, they calculated the temperature profiles within a collapsing bubble, confirming the existence of very high internal temperatures upon bubble collapse which were sufficient to explain the existence of "luminescence," observed experimentally and discussed also in Sec. 6-3. Since this phenomenon has generally been observed in acoustically induced cavitation,‡ it has been called "sonoluminescence" in the cavitation literature [63, 64, 67, 68]. Various explanations for this phenomenon have been advanced in the past, including ionization effects, chemical reactions, etc. However, it now seems most probable that the luminescence is primarily the result of the very high temperatures induced by bubble collapse, particularly upon various impurities within the bubbles. Of course, this mechanism would require collapse through a relatively large volume ratio, which may not occur in most cases of flowing cavitation. This is consistent with the fact that the luminescence

---

† Applied originally to boiling bubbles.

‡ Also observed in laboratory and field flowing systems (such as dam tail-races), but less intense than for acoustic systems.

is much more intense in "ultrasonic cavitation" as opposed to cavitation in flowing systems, although it has been reported [68] in both cases.

The existence of high gas and vapor temperatures within collapsing bubbles might be thought to contribute in some way to cavitation damage, and some observations [69] of damaged surfaces appear to indicate the previous existence there of high temperatures. For example, cavitation "bluing" of steels, similar in appearance to oxide bluing at high temperatures, has been observed in the author's laboratory and elsewhere. However, it is probably due to oxide deposition from effects other than bubble collapse. Consideration of the presumed bubble-collapse mechanisms, already discussed, indicates the lack of a plausible mechanism for sufficiently rapid heat transfer between the bubble contents and the wall, the lack of sufficient heat capacity of the gas and vapor within the bubble, and the difficulty of obtaining a sufficiently high material-surface temperature to be damaging, considering the high thermal conductivity of the metals involved. Hence, in the opinion of the writer the existence of such localized high-temperature damage effects is unlikely. Of course, the high temperature and corresponding high pressure of the bubble contents is in fact instrumental in the reduction of bubble-collapse wall velocities, generally associated with thermodynamic effects, and thus causes an important reduction in damage in some cases.

Mitchell-Hammitt investigations [32, 70]   Numerical results relatively similar to those of Hickling and Plesset for the collapse of a spherical bubble of initial radius R_0 containing both noncondensable (and nondiffusing) gas and saturated vapor were obtained at the author's laboratory [32, 70]. Gas-diffusion effects were neglected in this study, as in that of Hickling discussed above, since there is insufficient time during such a collapse for diffusion effects to become important. Nonetheless, this is an important consideration in ultrasonic cavitation through the mechanism of rectified diffusion [66]. The problem to be considered, along with presumed temperature profiles within the bubble and in the liquid, is shown schematically in Fig. 4-28.

Figure 4-29 shows the resultant temperature profiles for a typical case compared with those of Hickling [63, 64]. Approximate agreement between the two studies is noted. The maximum ratio of absolute temperatures computed is ~9, indicating a maximum temperature within the bubble at the conclusion of collapse of ~2400°C for bubble collapse in room-temperature water. Thus the probability of luminescence due to high temperatures in such an event is confirmed.

The Mitchell-Hammitt study included the effect of an "evaporation coefficient" α, ranging from 0 to 1.0, to evaluate the effect of the nonequilibrium temperature differential at the bubble wall due to possible impurities on the wall, etc. This factor was not included in the Hickling-Plesset work, but was studied in considerable detail by Theofanous et al. [71, 72] and Bornhorst and Hatsopoulos [73, 74]. Three values of the evaporation coefficient were investigated in the Mitchell analysis, ranging from 0.01 to the maximum possible theoretical value corresponding to zero thermal resistance at the bubble wall, that is, $\alpha = 1.0$. The theoretical analysis of Theofanous [71, 72] indicates that the most probable value

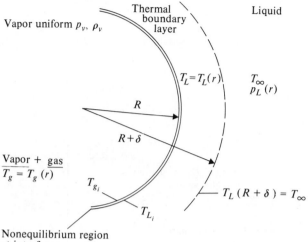

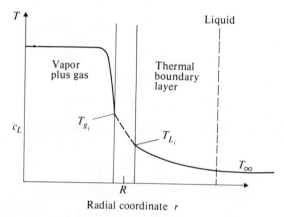

Figure 4-28 (a) Location of variables for thermodynamic effects study. (b) Sample temperature profile during collapse [32, 70].

for α lies between 0.05 and 0.1. The latter value was used in the calculations for Fig. 4-29. Figure 4-30 shows the effect of the condensation coefficient α upon the portion of original vapor mass remaining within the bubble at completion of collapse for α ranging from 0.01 to 1.0, indicating that for very low α the portion of vapor remaining is ~ 60 percent, whereas it is virtually zero for $\alpha = 1.0$. Thus for low α the vapor contributes significantly to the noncondensable gas-compression effect in restraining bubble collapse and producing an early rebound, but for an ideal condensation coefficient ($\alpha = 1.0$) the vapor effect in this respect is negligible. There is very little pertinent experimental information to guide the necessary numerical selection of α.

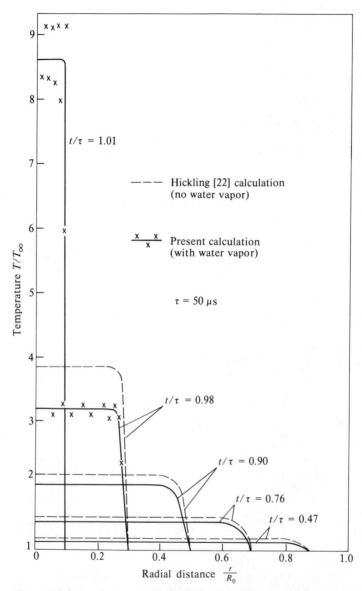

Figure 4-29 Internal temperature distributions at different radii for bubble containing nitrogen gas with and without water vapor; $\alpha = 0.1$, $pi = 0.075$ atm, $Pv = 0.0277$ atm, $\Delta p = 2.925$ atm, $R_0 = 0.1$ cm, $T_0 = 528°R$ [32, 70].

Figure 4-31 shows the effect of the full range of α upon the bubble-radius ratio (referred to R_0) as a function of time (referred to the classical Rayleigh collapse time [4], i.e., collapse velocity is the slope of this curve). In these terms, the assumed values of the condensation coefficient have little effect until the radius ratio is less than 0.9; i.e., this parameter would probably have only negligible

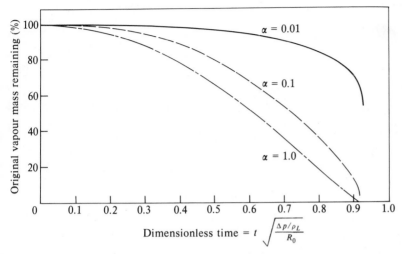

Figure 4-30 Effect for evaporation coefficient $\propto$ on quantity of vapor condensed during bubble collapse; "standard" bubble; $R_0 = 50 \times 10^{-3}$ inch, $T_0 = 537\,R$, $Pi_0 = 10^{-3}$ atm, $Pv_0 = 0.0277$ atm, $P_\infty = 1$ atm [32, 70].

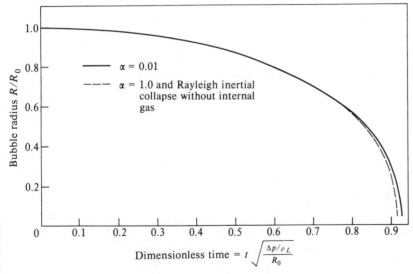

Figure 4-31 Effect of evaporation coefficient $\propto$ on bubble radius change with time for "standard" bubble (*see* Fig. 4.30) [32, 70].

effect for asymmetric bubble collapse, discussed before, since both photographic evidence and numerical analyses show that in most cases in flowing systems only relatively moderate collapse ratios will be attained before the formation of the microjet.

Figure 4-32 shows the effect of the condensation coefficient upon internal

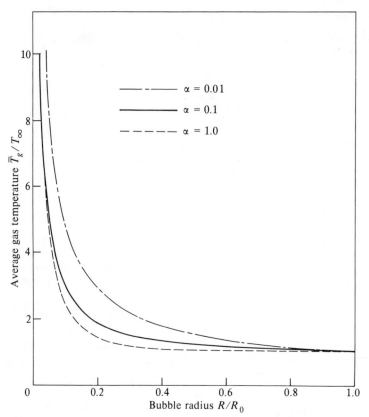

Figure 4-32 Effect of evaporation coefficient α on average internal temperature change with radius for "standard" bubble (*see* Fig. 4.30) [32, 70].

gas-vapor temperature, and Fig. 4-33, on internal pressure. The effect of α upon internal pressure (Fig. 4-33) is substantial throughout collapse, but that upon temperature (Fig. 4-32), only near its completion. Both effects are very substantial for radius ratios below ~0.2.

Figure 4-34 shows the effect upon internal pressure of the assumption of constant liquid temperature outside the bubble as opposed to variable liquid temperature. This effect is relatively small, though certainly not negligible.

Figure 4-35 shows the effect of the initial bubble radius upon internal gas-vapor temperature, indicating that maximum internal temperature is increased by a factor of ~ ×1.5 for the larger initial bubble radius compared to the smaller, i.e., 50 versus 5 mils (1.27 versus 0.127 mm). For smaller initial bubble radii, the collapse process is expected to be more nearly isothermal, rather than adiabatic, since the ratio of surface area to volume becomes larger for smaller size, so that outward heat transfer to maintain constant internal temperature is enhanced. Thus the maximum internal temperature attained is less for the bubble with the smaller initial radius. However, the internal gas pressure is very little influenced by

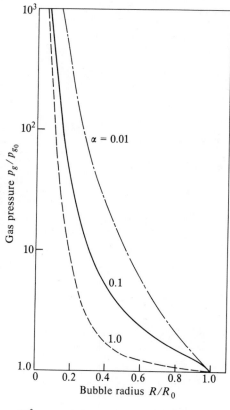

Figure 4-33 Effect of evaporation coefficient α on internal (gas + vapor) pressure change with radius for "standard" bubble (*see* Fig. 4.30) [32, 70].

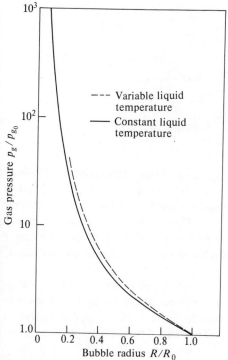

Figure 4-34 Effect of assumption of constant temperature in liquid on internal (gas + vapor) pressure change with radius for "standard" bubble (*see* Fig. 4.30), $\alpha = 0.1$ [32, 70].

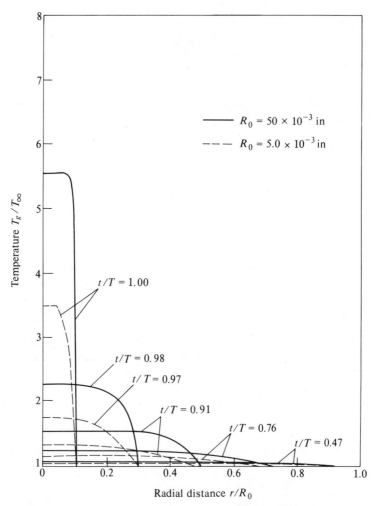

Figure 4-35 Internal temperature distribution at different stages of collapse for "standard" bubbles (*see* Fig. 4.30) with different initial radii, $\propto = 0.1$ [32, 70].

initial radius, as shown in Fig. 4-36. The corresponding calculations [32] show that the polytropic exponent γ for the mixture of water vapor and air takes values between ~ 1.10 and 1.20 during the earlier stages of collapse for the smaller bubble, but when R/R_0 reaches 0.05, γ is ~ 1.30. Thus these collapses are essentially isothermal at first, but approach adiabatic near conclusion. However, the larger bubble approaches the adiabatic condition more quickly, so that the gas-vapor temperatures therein are higher.

The effects of internal gas conductivity and liquid overpressure, i.e., "subcooling" from the viewpoint of boiling analyses, were also investigated. The effect of gas conductivity was essentially as expected, i.e., higher gas conductivities decrease internal gas-vapor temperature. This effect was also investigated by

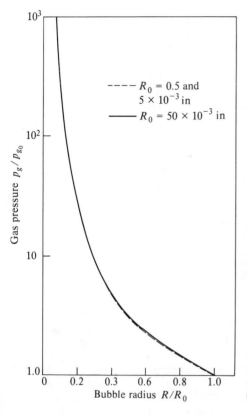

Figure 4-36 Effect of initial bubble radius on the internal (gas + vapor) pressure history during collapse of otherwise "standard" bubbles (*see* Fig. 4.30), $\propto = 0.1$ [32, 70].

Hickling [63, 64] with approximately similar results. This effect is taken to explain the reduced sonoluminescence [63, 64] for gases with higher thermal conductivity.

The effect of overpressure upon internal gas-vapor pressure is shown in Fig. 4-37. It is shown that the proportional gas-vapor pressure within the bubble is less for increased overpressure. Thus, for these conditions, spherical collapse would proceed through a greater radius ratio before rebound commences, so presumably damage would be increased. This is, of course, consistent with experimental observations in this respect.

Garcia, Hammitt, and Bhatt [75–79]   "Thermodynamic" effects with respect to cavitation damage are most easily demonstrated with a vibratory damage facility (discussed further in Chap. 5). The specimen to be damaged is vibrated at high frequency but low amplitude, while submerged in the test liquid, so that the test is essentially "static" in that liquid velocities are small and unimportant while accelerations are controlling. For this type of test it has been known for some time, and perhaps first reported by Devine and Plesset [80], that there was presumably for any liquid a maximum damage-rate temperature, lying (very roughly) midway between the melting point and boiling point for the test static pressure. For any given type of system, there is also presumably a maximum damage

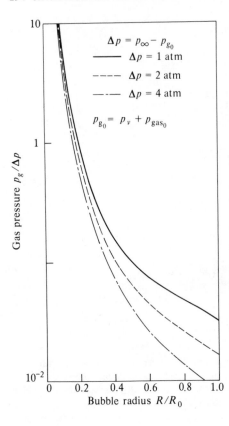

Figure 4-37 Effect of overpressure (Δp) on the dimensionless internal (gas + vapor) pressure history of collapsing bubbles, $R_0 = 50 \times 10^{-3}$ inch, $T_0 = 537°R$, $Pi_0 = 10^{-3}$ atm, $pv_0 = 0.0277$ atm, $\propto = 0.1$ [32, 70].

suppression pressure, i.e., there is no damage either for boiling or for too high a suppression pressure, which would suppress cavitation completely. This situation is discussed in more detail in Sec. 4-3.

Figure 4-38 illustrates the effects upon damage rate in our vibratory facility with water for changes in suppression pressure (NPSH · ρ) and temperature [77–79]. While the reason for the decrease in damage rate for decreasing temperature below the maximum damage temperature is not clear, that for temperature increase above the maximum damage temperature is presumably the thermodynamic effect. In essence, it is due to the increase in gas-vapor pressure within the bubble at collapse termination due to increase in vapor pressure and density, as well as to increased temperature of the bubble contents due to the inability of heat-transfer mechanisms to maintain isothermal conditions, as discussed in the previous section. This internal-pressure increase terminates collapse sooner than would otherwise be the case, resulting in reduced collapse velocities and radiated shock waves, thus reducing the cavitation-damage rate.

A series of tests in our laboratory using a vibratory facility have been conducted originally by Garcia and Hammitt [75, 76], using water and also various liquid metals as test liquids. The Florscheutz-Chao parameter B_{eff} [61, 62], previously discussed, was used to correlate the damage results. Our method of use of this

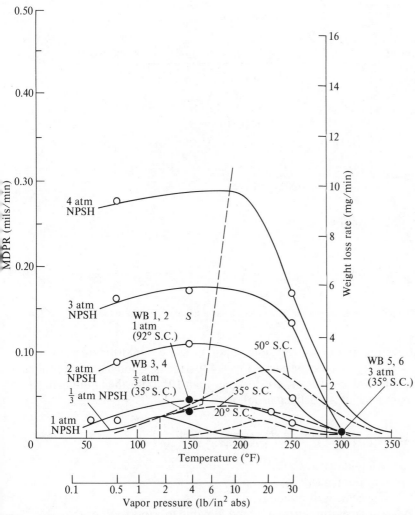

Figure 4-38 Maximum MDPR vs. temperature and vapor pressure for bearing brass (SAE-660), [77].

parameter differed somewhat from the original intention of Florscheutz and Chao [61] in that it was necessary to use the static suppression pressure for simplicity rather than the oscillating pressure of the vibratory facility, since this was not accurately known. Also, it was necessary to assume an initial bubble size (also actually unknown) and assume that it was the same for all tests. Nonetheless, the initial liquid-metal results (for bismuth, lithium, sodium, and mercury) all correlated well against B_{eff} as so computed (Fig. 4-39), showing a large region where thermodynamic effects were not important ($B_{eff} \geq 10^3$) and a region for lower values of B_{eff} where damage depended strongly on the thermodynamic effect. The relatively lower temperature liquid-metal results were in the region of nondependence upon thermodynamic effects, as was also molten bismuth at all

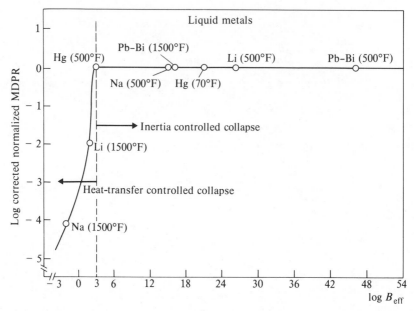

Figure 4-39 Effect of thermodynamic parameter on cavitation damage—liquid metals [75, 76].

temperatures because of its very low vapor pressure. Unfortunately, the water data did not fall upon the same curve (Fig. 4-40), since thermodynamic dependence commenced for values of $B_{\text{eff}} \le 10^{-1}$ (rather than 10^3 as for the liquid metals).

The parameter B_{eff} is defined as follows:

$$B_{\text{eff}} = \frac{J_a}{(\rho_L C_L \,\Delta T / \rho_v L)^2} \frac{\kappa_L}{R_0} \left(\frac{\rho_L}{\Delta p}\right)^{1/2}$$

where ρ_L = liquid density
ρ_v = vapor density
C_L = specific heat of liquid
ΔT = temperature drop in liquid film due to vaporization
J_a = Jacob number
L = latent heat of vaporization
κ_L = thermal diffusivity of liquid = $k_L / \rho_L C_L$
k_L = thermal conductivity of liquid
R_0 = equilibrium bubble radius
Δp = decrease in liquid pressure corresponding to decrease ΔT in liquid temperature

The parameter B_{eff} can be expressed in terms of the conventional Stepanoff B-factor [55–57] as

$$B_{\text{eff}} = B^2 \frac{\kappa_L}{R_0} (\text{NPSH})^{3/2}$$

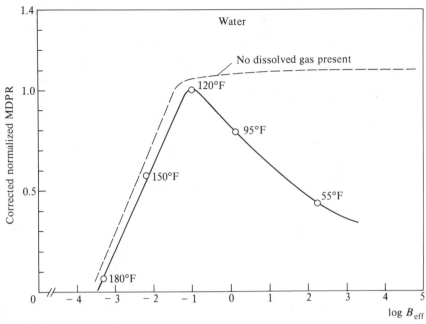

Figure 4-40 Effect of thermodynamic parameter on cavitation damage—water [75, 76].

It was shown that for low B_{eff} bubble growth and collapse are heat-transfer controlled, i.e., thermodynamic effects are important, and for high B_{eff} (fluids such as cold water) these processes are inertia controlled.

Later more comprehensive vibratory facility tests in sodium were conducted at the author's laboratory [79] and elsewhere, and are also included as Fig. 4-41. Unfortunately, these also do not fall upon the original Garcia curve (Fig. 4-39). From this disagreement of the later sodium data, as well as that of the water data, it must be concluded that B_{eff} as here used is not in itself a sufficient correlating parameter. This is no doubt partly due to the fact that test-facility parameters (specimen size, frequency, and amplitude) differ to some extent between the data sets, making our previously mentioned assumption of a constant NPSH·ρ and fixed initial bubble size for all cases particularly incorrect. These results tend to confirm Bonnin's previously discussed hypothesis [58–60] that it is in fact impossible to correlate thermodynamic effects of this kind in terms of a single correlating parameter, such as B_{eff}, and that two independent parameters are actually required.

Fixed-cavity theories

1. *General.* The original published conception of the thermodynamic effect by Stahl and Stepanoff [55–57] assumed the formation, for example, of a relatively fixed cavity in the inlet region of a cavitating pump which influenced the flow pattern in such a manner as to reduce the head, or other measurable parameter, by a prescribed amount, then defined as the "cavitation-inception"

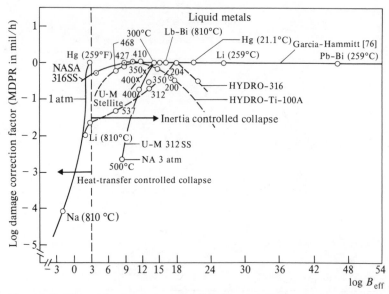

Figure 4-41 Effect of the thermodynamic parameter on cavitation damage—liquid metals [79].

condition. It was assumed that the same geometrical cavity would have been formed with other liquids, when the same head (or other parameter) decrease occurred. However, quantitatively different changes in inlet NPSH would be required to produce this result according to the B-factor for the liquid. Thus the concept of a relatively steady cavity in the sensitive region, responsible for the general cavitation effect upon the machine performance, is accepted as being valid, at least in many cases [1]. Damage in such flow regimes is presumed due to the collapse of "traveling bubbles" which skirt the cavity region and are impelled into the wall region by the reentry jet at the trailing edge of the cavity, the position of which is often found to oscillate at relatively high frequency [1, 81].

Since the above-described cavity model is reasonably valid and general, and also sufficiently simple for more detailed and meaningful analysis, it has in fact been utilized by several groups for this purpose. A specially designed venturi geometry with very sharp inlet geometry, and hence well-localized cavitation (Fig. 4-42), has been used by groups at NASA-Lewis Laboratory [82–84] with water and Freon-114. At the Bureau of Standards (BuStd), Boulder [85–89], under contract to NASA, the same cavitating venturi geometry was used, but with the cryogenic liquids, nitrogen and hydrogen. Hydrogen is one of the liquids for which thermodynamic effects appear to be most important.

Somewhat-related investigations during the same general time period were carried out by Holl and his colleagues [90–97] at Pennsylvania State University, using test bodies (ogives, etc.) submerged in a small tunnel with both

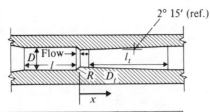

	Venturi scale	
	1.0	0.7
Free-stream diameter D	1.743	1.232
Throat diameter D_t	1.377	0.976
Radius of curvature R	0.183	0.128
Approach section length l	4.198	2.964
Throat length l_t	0.75	0.52
Diffuser length l_d	4.66	3.26

(*a*)

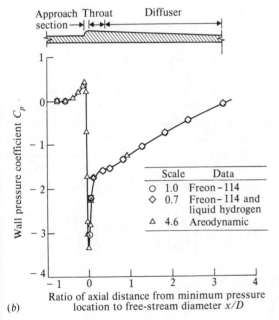

(*b*)

Scale	Data
○ 1.0	Freon-114
◇ 0.7	Freon-114 and liquid hydrogen
△ 4.6	Areodynamic

Wall pressure coefficient C_p

Ratio of axial distance from minimum pressure location to free-stream diameter x/D

Figure 4-42 (*a*) Venturi test sections. All dimensions in inches. (*b*) Wall pressure distribution for venturis [82–89].

water and freon as test liquids. Both groups proposed theoretical models to allow computation of the cavity pressure under various conditions, which is necessary for the a priori prediction of the thermodynamic effects in given machines. These studies will be reviewed in the following sections.

2. *NASA studies* [82–84]. The NASA studies involved water and Freon-114, while those at BuStd used liquid nitrogen and hydrogen, all in the same venturi geometry (Fig. 4-42). Thus a full range of cavitation thermodynamic-effect

parameters was investigated. The tests did not include damage effects, but rather performance effects including inception. The results of work under this program, performed primarily during the 1960s, are summarized by Ruggeri [84]. A method for predicting thermodynamic effects, i.e., changes in cavity-pressure depressions relative to stream vapor pressure, is presented. The prediction method accounts for changes in liquid, liquid temperature, flow velocity, and body scale, based upon theoretical and experimental studies. The method appeared [84] to provide good agreement between predicted and experimental results for geometrically similar venturis over the range of test liquids. Use of the method requires geometrical similarity for a body and cavitated region, and a known reference cavity pressure depression at some operating condition. The method can be applied to rotating machines such as pumps and inducers [83] as well as to stationary devices such as venturis [84].

Figure 4-43 shows typical vapor-head depression (feet of liquid) as a function of the vapor-to-liquid volume ratio in the cavitating region for different liquid temperatures for (a) water, liquid nitrogen, and Freon-114, and (b) liquid hydrogen. These results cannot be included on the same curve because of the much larger Δh_v values obtained with hydrogen for the same volume ratios.

Figure 4-44 shows cavity-head depressions for Freon-114 as a function of the distance from the position of minimum pressure for various liquid temperatures (Fig. 4-44a) and velocities (Fig. 4-44b), whereas Fig. 4-45 shows the effects of cavity size (Fig. 4-45a) for hydrogen, and venturi size (Fig. 4-45b) for Freon-114. These curves at least indicate the strong complexity of the situation, caused partly by the probable nonsteady nature of the cavity, the important effects of turbulence, and the significant mass transfer between the cavity and main liquid stream. This complexity is further indicated by the proposed modeling equation (4-33) below, showing the parameters which must be scaled to allow use of this predicting method. This relation shows the ratio between vapor and liquid volume, i.e., the thermodynamic parameter B introduced by Stahl and Stepanoff [55]:

$$\left(\frac{v_v}{v_l}\right)_{\text{pred}} = \left(\frac{v_v}{v_l}\right)_{\text{ref}} \left(\frac{\alpha_{\text{ref}}}{\alpha}\right)^m \left(\frac{V_0}{V_{0,\text{ref}}}\right)^n \left(\frac{D}{D_{\text{ref}}}\right)^{1-n} \left[\frac{(\Delta x/D)_{\text{ref}}}{\Delta x/D}\right]^p \qquad (4\text{-}33)$$

where v_l = volume of saturated liquid
v_v = volume of saturated vapor
α = thermal diffusivity
V_0 = free-stream velocity upstream of throat
D = upstream pipe diameter
Δx = axial length of cavitated region

The exponents m, n, and p depend, at least in part, on the heat-transfer process involved and must be determined by experiment. The derivation and use of Eq. (4-33) requires that (a) geometric similarity of cavitating flow is maintained for various liquids and flow conditions and (b) thermodynamic equilibrium

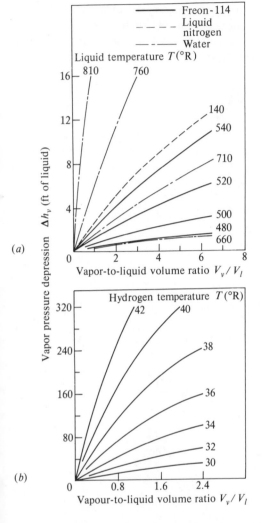

(a)

(b)

Figure 4-43 Vapor pressure depressions for various liquids and liquid temperature [82–84].

conditions exist within the cavitated region. At present, these two requirements can only be assumed for untested liquids.

3. *BuStd studies.* The BuStd work by Hord and his colleagues [85–89] were at first intended to provide checkpoints in cryogenic liquids for the tests at NASA [82–84] in water and Freon-114, all in geometrically similar venturi flow-paths. The flow-paths investigated, in addition to the NASA venturi, included a hydrofoil and three geometrically similar ogives. Thus crosscorrelation of the developed cavity data for five different hydrodynamic bodies was possible to achieve eventual best-fit correlations. Test fluids were liquid hydrogen and nitrogen. A new theoretical model was developed and applied to data from these five stationary models, as well as to the venturi and cavitating pump and axial inducer data from NASA. The techniques for pre-

dicting cavitation performance of pumping machinery are extended to include variations in flow coefficient, cavitation parameter (such as suction specific speed or inlet cavitation number K_c, for example), and equipment geometry. Hopefully these new predictive formulations can be used as a pump-design

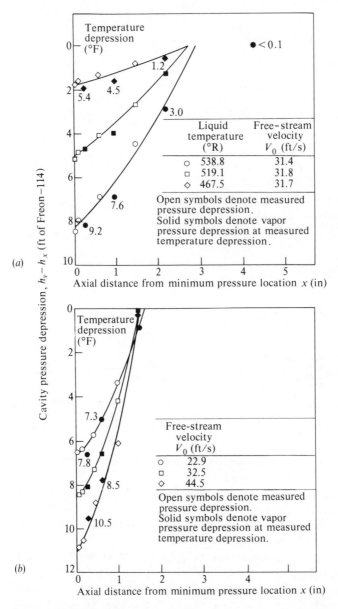

(a)

(b)

Figure 4-44 (a) Effect of free-stream liquid temperature on cavity pressure and temperature depressions. Freon-114; 1.0-scale venturi; nominal cavity length, $2\frac{3}{4}$ inches. (b) Effect of free-stream velocity on pressure and temperature depressions within cavitated regions of Freon-114. Venturi scale 0.7; liquid temperature 540°R; nominal cavity length, 1.6 inches [82–84].

tool and for the eventual development of a universal method for correlating pumping-machinery performance. Application of these predictive formulas requires prescribed cavitation test data, or an independent as-yet unavailable method for estimating the cavitation performance of each pump. If such a

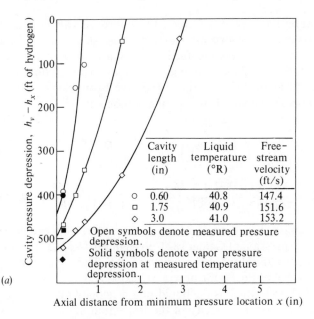

(a)

Axial distance from minimum pressure location x (in)

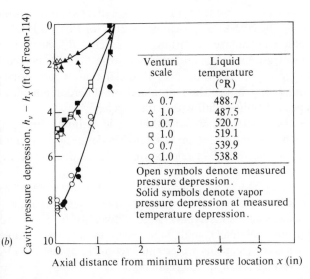

(b)

Axial distance from minimum pressure location x (in)

Figure 4-45 (a) Effect of cavity pressure depression. Liquid hydrogen; 0.7-scale venturi. (b) Effect of venturi scale on pressure and temperature depressions within cavitated regions of Freon-114. Cavity length, 1.4 free-stream diameters.

method were available, such as a proven detailed numerical flow model for the relatively complex rotating geometries involved, it would then be possible to predict detailed cavitation performance of new machinery designs without testing. Unfortunately, the present state of the art does not as yet include this capability.

The proposed correlations are based upon the improved calculation of B-factors, assuming isentropic vaporization of the liquid. Convection heat transfer and two-phase mass flux were included in the analysis. The relevance to pumps and other rotating machinery of analysis and experimentation with stationary bodies was strengthened, i.e., the correlative expressions devised for stationary bodies appear extendable to rotating equipment.

The proposed correlating parameters are

$$\frac{B}{B_{ref}} = \left(\frac{\alpha_{ref}}{\alpha}\right)^{E1}\left(\frac{V_0}{V_{0,ref}}\right)^{E2}\left(\frac{l}{l_{ref}}\right)^{E3}\left(\frac{v_{ref}}{v}\right)^{E4}\left(\frac{\sigma_{ref}}{\sigma}\right)^{E5}\left(\frac{D}{D_{ref}}\right)^{E6} \tag{4-34}$$

$$\frac{B}{B_{ref}} = \left(\frac{\alpha_{ref}}{\alpha}\right)^{E1}\left(\frac{MTWO}{MTWO_{ref}}\right)^{E2}\left(\frac{l}{l_{ref}}\right)^{E3}\left(\frac{v_{ref}}{v}\right)^{E4}\left(\frac{\sigma_{ref}}{\sigma}\right)^{E5}\left(\frac{D}{D_{ref}}\right)^{E6} \tag{4-35}$$

where l is the cavity length. Equation (4-34) is an improved and extended version of the simplified expression of Gelder [82], given in Eq. (4-33). Equation (4-35) was derived as a part of the Hord et al. studies [85–89]. The nomenclature is essentially the same as that of Eq. (4-33), previously explained, and except for the new term MTWO [82, 88], which is the liquid-phase velocity ratio; that is, V_0/V_1 where V_0 is the bulk-stream velocity of the test liquid at inlet and V_1 is the characteristic liquid velocity component which is normal to the cavity liquid-vapor interface. Details for the calculation of MTWO are given in Ref. 82.

Table 4-3 [85] summarizes the results for the various hydrodynamic shapes and liquids tested, showing the best-fit empirical exponents for Eqs. (4-34) and (4-35), as well as the predicted standard deviations in predicted B-factors. The percentage of standard deviations so computed range from 5 to 16 percent, allowing a much more precise prediction of this parameter than would otherwise be possible, assuming that future data obtained on different machines and hydrodynamic shapes continue to fall within this range.

Pertinent values of the cavitation B-factor for liquid helium, nitrogen, flourine and oxygen, Freon-114, and water are tabulated in detail by Hord and Voth [89], thus allowing the relatively easy calculation of the necessary parameters for these liquids.

4. *Penn State studies: Holl et al.* [90–97]. During roughly the same time period as that for NASA-Lewis and BuStd-Boulder, a comprehensive series of relatively similar studies was underway at Pennsylvania State University by Holl and his colleagues [90–97] using a relatively small, high-speed tunnel over a range of temperatures, velocities, and size, using water and Freon-113 as test liquids. Test bodies were submerged ogives of various calibers.

The Penn State group recommend a predicting model based upon gas-vapor entrainment considerations rather than the B-factor methods discussed

Table 4-3 Summary of correlative results for developed cavity data—venturi, hydrofoil, ogives, and combined venturi-hydrofoil-ogives (VHO) [85]

Line no.	Hydrodynamic body	Test fluids	Correlative equation	Exponents					Ref. run no.	Standard deviation† in B-factor	Mean percentage difference‡ in B-factor	$\overline{K}_{c,\min}$
				E1	E2	E3	E4	E6				
1	Venturi	H_2	(4-35)	(0.10)	0.59	0.18	—	—	071C	0.2234	5.5	2.459
2	Hydrofoil	H_2 and N_2	(4-35)	(−0.13)	0.59	0.27	—	—	255B	0.2565	9.9	1.833
3	Ogives	H_2 and N_2	(4-35)	(−0.05)	0.43	0.25	—	0.59	338B	0.2126	11.1	0.531
4	VHO	H_2 and N_2	(4-35)	—	0.51	0.28	—	0.43	338B	0.2618	11.7	—
5	Venturi	H_2	(4-34)	−1.92	0.74	0.31	—	—	071C	0.3466	9.0	2.459
6	Hydrofoil	H_2 and N_2	(4-34)	0.80	0.64	0.45	−1.00	—	255B	0.3717	12.7	1.833
7	Ogives	H_2 and N_2	(4-34)	0.32	0.21	0.34	−0.84	0.60	338B	0.2620	14.3	0.531
8	VHO	H_2 and N_2	(4-34)	—	0.11	0.36	—	0.58	338B	0.4311	16.3	—
9	Venturi	H_2 and F-114	(4-34)	1.0	0.8	0.3	—	−0.10	—	—	—	2.47

† Standard deviation $\equiv \sqrt{\Sigma(B - B_t)^2/(\text{NPTS} - 1)}$, where NPTS = number of data points (including "ref" data point), B_t = BFLASH and is computed from isentropic flashing theory, and B is computed from Eqs. (4-34) or (4-35).

‡ Mean percentage difference $\equiv \Sigma[|B - B_t|(100/B_t)]/(\text{NPTS} - 1)$.

in the last section. They believe that a more realistic analytical approach to the prediction of the thermodynamic effect can be formulated in this way. The probability of success, they believe, is greater than that with the B-factor approach, since the model is more closely based upon the actual physical phenomena observed.

The proposed "entrainment method" is a semiempirical approach for correlating temperature depression data [92]. Experimental values of temperature are compared with values predicted by the correlation equation. It is assumed that the vaporous cavity is continuously supplied with vapor from the cavity walls. The vaporization process requires energy in the form of heat which is transferred at the rate $\dot{q} = \lambda \dot{m}_v$, where λ is the latent heat of vaporization and $\dot{m}_v = \rho_v V_v A_v$, where V_v is vapor velocity and A_v is cavity cross-sectional area. Then $\dot{m}_v = \rho_v D^2 V_\infty C_Q$. In this equation D is the model diameter and C_Q is a flow coefficient defined as $C_Q = Q_v/D^2 V_\infty$, where Q_v is the volume flow rate of vapor into the cavity. The rate $\dot{q}$ is then evaluated using a film heat-transfer coefficient h between the cavity and main liquid stream. Combining all this, one obtains an expression for the cavity temperature depression,

$$\Delta T = \frac{C_Q}{h}\frac{D^2}{A_W} V_\infty \lambda \rho_v \tag{4-36}$$

This relation can also be expressed in terms of dimensionless coefficients:

$$\frac{C_p \Delta T}{\lambda} = \frac{C_Q}{C_A}\frac{\text{Pe}}{\text{Nu}}\frac{\rho_v}{\rho_L} \tag{4-37}$$

The equation can also be arranged so that the left-hand side corresponds to the Jacob number J. Pe and Nu are Peclet and Nusselt numbers, respectively, and C_p is specific heat of the liquid. The subscripts v and L refer to vapor and liquid. Equation (4-37) is similar to the relationship derived much earlier by Holl and Wislicenus [93], but corresponds more closely to that proposed by Acosta and Parkin [94]. It was then assumed that the volume flow rate of vapor required to sustain a vaporous cavity is equal to that required to maintain a ventilated cavity. The determination of C_Q was based upon experimental investigations at Penn State by Billet and Weir [95, 96]. The flow coefficient is then expressed in the form:

$$C_Q = C_1 \text{Re}^a \text{Fr}^b \left(\frac{L}{D}\right)^c \tag{4-38}$$

where C_1, a, b, and c are constants to be determined empirically, and Re and Fr are Reynolds and Froude numbers, respectively. The area coefficient C_A (Eq. 4-37) was determined from photographs by Billet, Holl, and Weir [97]. The Nusselt number was determined by using measured values of cavity temperature, as a function of Re, Pr (Prandtl number), Fr, and L/D, again using empirically determined exponents for each parameter.

le 4-4 Constants and exponents for entrainment theory correlations [92]

el	Quantity	Constants $(C_1, C_2, C_3,$ or $C_4)$	L/D exponent	Re exponent	Fr exponent	Pr exponent
-caliber ogive	C_A	4.59	1.19			
	C_Q	0.424×10^{-2}	0.69	0.16	0.13	
	Nu	0.103×10^{-3}	−1.35	1.39	0.24	0.84
	ΔT_{max}	8.97	0.85	−1.23	−0.11	−0.84
ter-caliber ogive	C_A	2.06	1.18			
	C_Q	0.320×10^{-1}	0.74	0.46	0.26	
	Nu	0.546×10^{-2}	−0.74	1.00	0.34	0.49
	ΔT_{max}	0.284×10^{-2}	0.30	−0.54	−0.08	−0.49
ter-caliber ogive†	Nu	0.375×10^{-2}	−0.84	1.02	0.56	0.05
	ΔT	0.414×10^{-2}	0.40	−0.56	−0.30	−0.05

Correlation using ΔT data from Hord [88] as measured by the leading-edge thermocouple and C_Q and C_A from this study.

The entrainment theory described above can be related to the conventional B-factor approach according to the following relation:

$$B = \frac{C_Q}{C_A} \frac{\text{Pe}}{\text{Nu}} \qquad (4\text{-}39)$$

This relation indicates the truly complex nature of the B-factor, in that it depends upon four dimensionless coefficients as defined by the entrainment theory approach. This statement runs counter to Bonnin's earlier hypothesis [58–60] that two independent parameters would be sufficient to describe the thermodynamic effect.

A final relation for the calculation of ΔT follows [92]:

$$\Delta T_{max} = C_4 \left(\frac{L}{D}\right)^i \text{Re}^j \text{Fr}^k \text{Pr}^p \text{Pe} \frac{\rho_v}{\rho_L} \frac{\lambda}{C_P} \qquad (4\text{-}40)$$

The constants for this equation for two model geometries are given in Table 4-4.

4-2 BOILING AND CONDENSATION VERSUS CAVITATION

As discussed in Sec. 4-1-5, there are many cases normally considered under the heading of "cavitation" which involve many elements of "boiling" or, in the case of bubble collapse, more properly "condensation." Historically, cavitation bubble-dynamics analyses [1, 4] have for the most part concerned themselves only with cases where heat transfer, or thermodynamic effects, can be neglected and where mechanical effects such as inertia, viscosity, and surface tension predominate. On the other hand, boiling or condensation bubble-dynamics analyses neglected inertial and viscous effects, and concerned themselves primarily with heat-transfer and surface-tension restraints to bubble growth or collapse [98–100]. As indicated previously, there are many cases where both thermodynamic and mechanical effects are important. Physically, then, these are intermediate between classical cavitation and boiling models.

Bubble dynamics in highly subcooled boiling is then indistinguishable from cavitation cases where thermodynamic effects are important. In practice, subcooled boiling is distinguished from cavitation primarily by the way in which the bubbles are created. For any form of boiling, the predominant mechanism for bubble generation is heat transfer, whereas for cavitation it is pressure reduction due to fluid-dynamic effects or to acoustic radiation in "ultrasonic" cavitation. Bubble generation in subcooled boiling ordinarily occurs adjacent to a heated wall, under conditions where the bulk liquid temperature is below the saturation temperature for the existing pressure. Once generated, the bubbles are ejected from the wall by mechanisms which are beyond the scope of this discussion but are covered in detail in the boiling heat-transfer literature. Reference 98 provides a good summary. Once reaching the bulk stream beyond the thermal boundary layer, they are "cooled," so that their collapse is motivated by the fact that they are now in a region of relatively subcooled liquid. At this point, they are essentially indistinguishable from collapsing cavitation bubbles.

Damage, however, seldom results from boiling situations for several reasons. The bubbles are not adjacent to the wall when collapse starts, the driving suppression pressure differential is not large, and the liquid conditions are usually such that thermodynamic effects could be expected to severely restrict damaging capability, as discussed in the previous section on cavitation thermodynamic effects. In addition, boiling applications do not usually involve large enough bulk velocities to make cavitation damage a likely contingency.

Even though cavitation damage is not normally a problem, even in highly subcooled boiling applications, its possibility obviously cannot be absolutely excluded from consideration. However, aside from the damage possibility, there is also the noise associated with collapsing subcooled boiling bubbles. Thus the reliable acoustic detection of boiling becomes difficult, if not impossible, in such applications as liquid-cooled nuclear reactor cores, unless the possibility of cavitation in the system can be absolutely eliminated. In the present state of the art such an absolute exclusion of the possibility of cavitation is very difficult.

4-3 MISCELLANEOUS ENGINEERING EFFECTS

4-3-1 Degree and Intensity of Cavitation and Damage

"Degree" and "intensity" of cavitation are taken to describe respectively the extent of the cavitating region, i.e., the "cavitation condition," and the "intensity" of bubble collapse in that region, which is governed by the suppression pressure at the point of bubble collapse, but which cannot as yet be precisely defined. From a slightly different viewpoint, the degree of cavitation is primarily a function of cavitation number "sigma," σ. As a first approximation, constant sigma implies a fixed degree of cavitation, assuming that geometrical similarity is otherwise maintained. However, as discussed in detail in Chap. 3, very substantial cavitation "scale effects" are observed to exist in most cases, for reasons not yet entirely understood, so that the constant degree of cavitation in most practical cases can be maintained for changes in overall velocity, suppression pressure, or size only if precise empirical information is available for the particular case. Nevertheless, assuming classical scaling laws to apply, constant sigma implies constant degree of cavitation, which then implies fixed conditions for individual bubble-dynamics growth and collapse. Thus the effects of asymmetries, bubble-collapse volume ratios, etc., should be modeled in such a case to a first approximation.

For any degree of cavitation for a given geometry, various cavitation intensities are possible, depending upon absolute values of stream velocity or suppression pressure. The "intensity" reflects the actual collapse velocities, etc., which presumably scale as the root of suppression pressure as in any fluid-dynamic situation. Thus the probability and rate of damage increases in general for increased "intensities" at fixed degrees of cavitation. As a practical example, cavitating condensate pumps generally do not represent a damaging condition, whereas cavitating boiler feed pumps, in which mean velocities and NPSH are much higher, do represent an important cavitation-damaging situation, although S may be the same. Again, as with the degree of cavitation, there are large and important damage "scale effects," not at present entirely understood but discussed in detail in Sec. 5-4.

4-3-2 Thermodynamic Effects

The general thermodynamic effect, which provides a maximum damage temperature for all liquids so far tested with damage fall-off at either lower or higher temperature, has been discussed previously. This damage fall-off exists for lower as well as for higher temperatures, but it is much more substantial at the high-temperature end, so that cavitation damage is apparently not an important problem in many cases for high-temperature liquids. The reason for the low-temperature fall-off in damage rate is not well understood, but that at high temperature is presumably primarily the result of the classical thermodynamic effect. Of course, these results are as yet primarily from "vibratory" damage tests [75–80], so that their application to flowing systems is not as well documented. Nevertheless, there is no reason to believe that the trends are not approximately correct, and they are in fact consistent with some existing industrial experience [1, 55–57]. These trends are illustrated for water by Fig. 4-40 and for sodium tests [79, 101] by Figs. 4-46 to 4-48.

An example of pertinent industrial experience is the fact that cavitation damage has not been a problem for the coolant circulating pumps for water-cooled nuclear reactor systems such as PWR, BWR, or CANDU. This is pre-

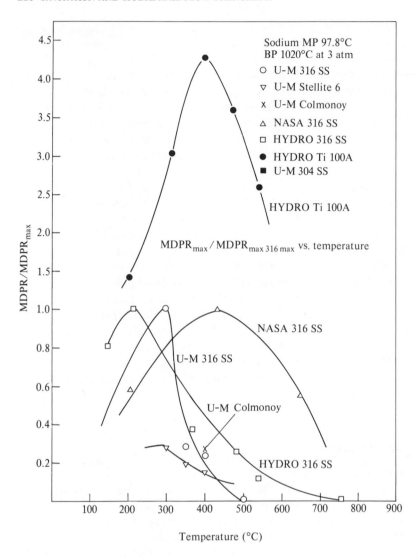

Figure 4-46 Comparison of temperature effects in sodium: U-M, NASA, HYDRONAUTICS [101]

sumably because the operating temperature is very far above the probable maximum damage temperature for water (Fig. 4-40), which is very unlikely to be above
~130°C, even at these elevated pressures. Unfortunately, this is probably not the
case for the sodium-cooled fast-breeder reactor (LMFBR), since the maximum
damage temperature appears (Fig. 4-46) to lie in the range 200 to 500°C. However,
insufficient field experience is as yet available to allow definite conclusions in
this regard.

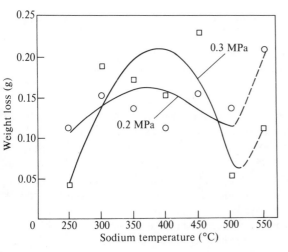

Figure 4-47 Weight loss vs. sodium temperature for the accelerated damage test at the University of Michigan [101].

4-3-3 Corrosion and Mechanical-Damage Effects

It has long been realized that corrosion and mechanical effects are often inextricably mixed in many apparent cavitation-damage cases (see also Chap. 5). This may partially be due to the effect of high-temperature gases such as oxygen within the collapsing bubble. A very important example of combined corrosion-cavitation is that of cavitation of marine propellors and other ship's components. In fact, "cathodic protection" has long been used to suppress the corrosion component of the overall erosion [102]. More recently it has been found that at least a portion of the beneficial effect of cathodic protection may be due to generation of electrolytic noncondensable gases along the material wall, which thus "cushions" the bubble collapse [103]. This mechanism is similar to some extent to that of the thermodynamic effect, where vapor cushioning is involved.

There is certainly a substantial experimentally demonstrable difference between the corrosion contributions to the overall erosion in corrosive liquids such as sea water and different materials, depending upon their corrodability, i.e., the additive corrosion effect is much greater for carbon steel than for austenitic stainless steels. This has been shown in recent tests at the author's laboratory [104] and elsewhere [105–107]. Presumably one of the major reasons for the unreliability of "accelerated" laboratory cavitation-damage tests such as, for example, the vibratory test, when applied to field-system predictions, is the fact that only the mechanical component is normally accelerated. Thus the comparison obtained between a hardened carbon steel and a stainless steel, for example, in a vibratory test where they may appear to be about equal, is not substantiated in most field tests, especially with sea water but also in some fresh water applications, where the stainless steel is normally far superior.

This problem was addressed by Plesset [106, 107] who utilized a "pulsed" vibratory test, where short pulses of cavitation were interspersed between much longer periods of "dead time." It was found that the ratio between cavitation and dead-time period durations affected drastically the relative damage resistance of various materials such as carbon steels and stainless steels when the tests were in salt water, but there was no effect for distilled water or an inert liquid such as toluene.

It is also generally recognized that the total damage resulting from combined cavitation and corrosive attack can be many times that which would result from the total of the two effects acting separately. This is presumably due to the interplay between these two forms of attack, i.e., cavitation provides a roughening and weakening of the surface resulting in increased mechanical cavitation attack, while mechanical cavitation removes the oxide or other film which would normally protect the material surface from rapid corrosive attack.

Figures 4-46 to 4-48 for vibratory facility damage data in sodium illustrate the interplay between mechanical and corrosive effects for that liquid on type 304 stainless steel (Fig. 4-47), and various other relatively incorrodible materials as well (Fig. 4-46). Figure 4-46 illustrates that for temperatures up to 500°C, the

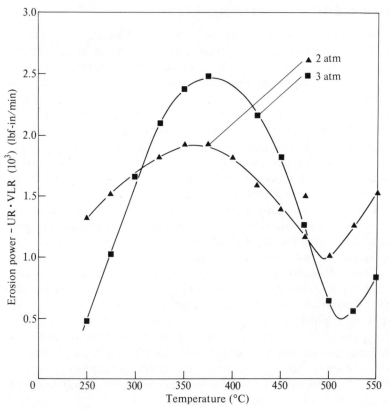

Figure 4-48 Ultimate resilience (UR) × volume loss rate (VLR) vs. test temperature [101].

damage rate continued to fall as the temperature was raised in the University of Michigan facility. Figure 4-47 shows that in later tests, for a further temperature increase to 550°C, the damage rate again rose. Presumably this high-temperature damage increase results both from reduced mechanical material properties at that temperature and increased corrosion. Figure 4-48 plots the product of volume-loss rate and ultimate resilience of the material, taking into account its decrease with increasing temperature, against test temperature. Ultimate resilience (discussed in Chap. 5) is taken as the best present parameter to represent failure energy per unit volume of the test material. Figure 4-48 thus essentially removes the effect of decreasing material properties at high temperatures. Since the damage rate at the maximum temperature (550°C) is still greater than that at 500°C, even in these terms, presumably the remaining effect is simply that of increased corrosivity at high temperatures.

4-3-4 Effects of Asymmetrical Collapse

The relatively recent studies indicating the importance of asymmetrical bubble collapse, already discussed, have shown the probable substantial importance in cavitation damage of liquid "microjet" impact, as well as pressure pulses from bubble rebound. This heightened understanding is particularly important in showing the close relationship between liquid-droplet-impact damage (where internal cavitation, also discussed in Chap. 6, may also occur) and cavitation damage. This has historically been assumed to be the case, primarily because of the close similarity between the different forms of damage, so that straight-forward impact tests were often used to evaluate the cavitation damage of materials [108].

Another important practical effect of the recently increased understanding of the asymmetrical collapse mode, as opposed to the classical Rayleigh spherical collapse model [4], is the effect of material surface rigidity upon the collapse model. As discussed in previous sections, a rigid surface "attracts" the bubble centroid during collapse and results in microjet generation toward the wall. Thus the possible damaging effect of bubbles not initially adjacent to the wall is increased. In addition, the complexity of the process is increased considerably, since both velocity and direction of the microjet are involved, as opposed to the Rayleigh model where spherical shock waves impinging upon the wall alone were supposed to be responsible for the damage, so that only initial bubble size and position were involved. Thus an additional "sorting mechanism" is created by the microjet model. This helps to explain the experimental observation [1, 2, 109, 110] that only one out of a very large number of bubbles ($\sim 10^4$ to $\sim 10^9$), observed to collapse near a damaged surface by high-speed photography, actually produces a detectable crater, even in relatively soft materials.

A further important effect of the asymmetrical collapse model and its dependence upon surface rigidity is the fact that a sufficiently flexible surface (or a free surface) will reverse the collapse direction, resulting in the "repulsion" of the bubble centroid from the surface and the generation of the microjet away from the surface [38–41]. Experiments at the author's laboratory with spark-generated bubbles collapsing near a thin-rubber diaphragm and sealing a small air pocket behind it, in a static water beaker, indicated precisely such a jet reversal and migration of the bubble centroid during collapse away from the diaphragm [39, 40]. Such a thin-rubber diaphragm arrangement could, of course, be considered either as an extremely flexible solid boundary or, alternatively, as a quasi-free surface. Hence it does not necessarily represent a practical construction for a flexible boundary. However, experiments by Gibson [38] at Cambridge University with soft rubber showed relatively similar results.

These relatively scanty experiments, and the confirming analyses, do indicate the possibility of designing a practical and sufficiently flexible surface to prevent or greatly reduce microjet impact, as well as shock waves originating from rebound upon a material wall adjacent to a cavitating liquid. Thus such a surface could be extremely resistant to cavitation damage and could represent a practical engineering solution for some cases. There is considerable past experimental data [41, 111] confirming the good cavitation-erosion resistance of certain rubber and elastomeric coatings in both vibratory and rotating-disc facilities. This is probably best explained in terms of the bubble-repulsion mechanism. In fact, the University of Michigan vibratory facility tests [41] showed that some such elastomerics exhibited cavitation-damage resistance superior to stainless steels in the same test facility. However, this may not be the case in some field applications where elastomeric coatings are proposed, since the bond between coating and base metal may fail due to high temperatures induced in the elastomeric material by mechanical strains under cavitation attack [111]. This thermal problem is augmented by the relatively poor thermal conductivity of elastomeric materials.

The droplet-impact resistance [41] of these same materials, however, was relatively very much worse, compared to the metals, than was their cavitation resistance. This is explicable by the fact that the existence of the impact, as opposed to the damaging cavitation bubble collapse, is in no way deterred by the flexibility of the surface. However, the resultant pressures upon the surface are somewhat reduced according to numerical studies at the author's laboratory [112, 113]. This problem is discussed in detail in Chap. 6.

In summary, it thus appears that while the best cavitation-damage resistance is usually provided by very hard materials, it is also true that in some cases very soft materials are also much more resistant than materials of intermediate hardness, such as many metallic alloys. Of possible structural metallic alloys, many aluminums, brasses and bronzes, cast iron, and others are much more vulnerable to cavitation damage than materials such as the 300-series stainless steels, for example. The best resistance for metals is provided by certain very hard alloys, and also by materials such as Stellite-6. These questions are discussed in much greater detail in Chap. 5.

REFERENCES

1. Knapp, R. T., J. W. Daily, and F. G. Hammitt: "Cavitation," McGraw-Hill, New York, 1970.
2. Knapp, R. T.: Recent Investigations of Cavitation and Cavitation Damage, *Trans. ASME*, vol. 77, pp. 1045–1054, 1955.
3. Besant, W. H.: "Hydrostatics and Hydrodynamics," art. 158, Cambridge University Press, London, 1859.
4. Lord Rayleigh (John William Strutt): On the Pressure Developed in a Liquid During the Collapse of a Spherical Cavity, *Phil. Mag.*, vol. 34, pp. 94–98, August, 1917.
5. Nieto, J., and W. Smith: "An Exact Solution of the Rayleigh-Besant Equation," University of Michigan Int. Rept. 18, ORA Project 03424, October, 1962.
6. Knapp, R. T., and A. Hollander: Laboratory Investigations of the Mechanism of Cavitation, *Trans. ASME*, vol. 70, pp. 419–435, 1948.
7. Poritsky, H.: The Collapse or Growth of a Spherical Bubble or Cavity in a Viscous Fluid, *Proc. First U.S. Nat'l Congr. Appl. Mech.*, 1952, pp. 813–821, ASME.
8. Goldstein, S.: "Modern Developments in Fluid Dynamics," Oxford University Press, London, 1938.
9. Ivany, R. D.: "Collapse of a Cavitation Bubble in Viscous Compressible Liquid—Numerical and Experimental Analysis," Ph.D. thesis, University of Michigan, 1965.
10. Ivany, R. D., and F. G. Hammitt: Cavitation Bubble Collapse in Viscous Compressible Liquids—Numerical Analysis, *Trans. ASME, J. Basic Engr.*, vol. 87, ser. D, pp. 977–985, 1965.
11. Cole, R. H.: "Underwater Explosions," Princeton University Press, 1948.
12. Kirkwood, J. G., and H. A. Bethe: "The Pressure Wave Produced by an Underwater Explosion, I," OSRD no. 588, 1942.
13. Gilmore, F. R.: The Growth or Collapse of a Spherical Bubble in a Viscous Compressible Liquid, *Proc. 1952 Heat Transfer and Fluid Mech. Inst.*, pp. 53–64, Stanford University Press.
14. Tait, P. G.: Report on Some of the Physical Properties of Fresh Water and Sea Water, *Phys. Chem.*, vol. 2, p. 71, 1888.
15. Flynn, H. G.: "Collapse of a Transient Cavity in a Compressible Liquid," Harvard University Tech. Memo no. 38, Acous. Res. Lab., Cambridge, Mass., 1957.
16. Mellen, R. H.: "An Experimental Study of the Collapse of a Spherical Cavity in Water," U.S. Navy Underwater Sound Lab. Res. Rept. 279, Ft. Trumbull, New London, Conn., 1956.
17. Mellen, R. H.: "Spherical Pressure Waves of Finite Amplitude from Collapsing Cavities," U.S. Navy Underwater Sound Lab. Res. Rept. 279, Ft. Trumbull, New London, Conn., 1956.
18. Schneider, A. J. R.: "Some Compressibility Effects in Cavitation Bubble Dynamics," Ph.D. thesis, California Institute of Technology, 1949.
19. Brand, R. S.: "The Collapse of a Spherical Cavity in a Compressible Liquid," Ph.D. thesis, Brown University, 1960.
20. Brand, R. S.: "The Shock Wave Produced by Collapse of a Spherical Cavity," Mech. Eng. Dept. Tech. Rept. 1, University of Connecticut, Storrs, Conn., 1962.
21. Hickling, R., and M. S. Plesset: Collapse and Rebound of a Spherical Bubble in Water, *Physics of Fluids*, vol. 7, pp. 7–14, 1964.
22. Hickling, R.: Effects of Thermal Conduction in Sonoluminescence, *J. Acoust. Soc. Amer.*, vol. 35, pp. 967–974, 1963.
23. Mitchell, T. M., and F. G. Hammitt: On the Effects of Heat Transfer Upon Collapsing Bubbles, *Nucl. Sci. and Engr.*, vol. 53, no. 3, pp. 263–276, March, 1974.
24. Plesset, M. S., and T. P. Mitchell: On the Stability of the Spherical Shape of a Vapor Cavity in a Liquid, *Quart. Appl. Math.*, vol. 13, pp. 419–430, 1956.
25. Welch, J. E., F. H. Harlow, J. P. Shannon, and B. J. Daly: "A Computing Technique for Solving Viscous, Incompressible, Transient Fluid-Flow Problems Involving Free Surfaces," Los Alamos Sci. Lab. Rept. LA-3425, 1966.
26. Harlow, F. H.: The Particle-in-Cell (PIC) Computing Method for Fluid Dynamics, *Methods in Computational Physics*, vol. 3, pp. 319–343, 1964.

27. Harlow, F. H., and J. E. Welch: Numerical Calculation of Time-Dependent Viscous Incompressible Flow of Fluid with Free Surface, *Phys. Fluids*, vol. 8, pp. 2182–2189, 1965.

28. Harlow, F. H., and A. A. Amsden: Numerical Calculation of Almost Incompressible Flow, *J. Comp. Phys.*, vol. 3, pp. 80–93, 1968.

29. Harlow, F. H.: "Numerical Methods for Fluid Dynamics, an Annotated Bibliography," Los Alamos Sci. Lab. Rept. LA-4281, 1969.

30. Mitchell, T. M., and F. G. Hammitt: Asymmetric Cavitation Bubble Collapse, *Trans. ASME, J. Fluids Engr.*, vol. 95, no. 1, pp. 29–37, March, 1973.

31. Mitchell, T. M., R. Cheesewright, and F. G. Hammitt: Numerical Studies of Asymmetric Bubble Collapse, *1968 ASME Cavitation Forum*, pp. 4–5, May, 1968.

32. Mitchell, T. M.: "Numerical Studies of Asymmetric and Thermodynamic Effects on Cavitation Bubble Collapse," Ph.D. thesis, Nuclear Engr. Dept., University of Michigan, 1970; see also University of Michigan ORA Rept. UMICH 03371-5-T, December, 1970.

33. Chapman, R. B.: "Nonspherical Vapor Bubble Collapse," Ph.D. thesis, California Institute of Technology, 1970.

34. Plesset, M. S., and R. B. Chapman: Collapse of an Initially Spherical Vapor Cavity in Neighborhood of a Solid Boundary, *J. Fluid Mech.*, vol. 47, no. 2, p. 283, May, 1971.

35. Benjamin, T. B., and A. T. Ellis: The Collapse of Cavitation Bubbles and the Pressures Thereby Produced Against Solid Boundaries, *Phil. Trans. Roy. Soc. (London)*, ser. A, vol. 260, pp. 221–240, 1966.

36. Chincholle, L.: Bubbles and the Rocket Effect, *J. Appl. Physics*, vol. 41, no. 11, pp. 4532–4538, October, 1970.

37. Chincholle, L.: L'Effet Fusée et l'Érosion Mecanique de Cavitation, *Bull. Tech. Suisse Romande*, vol. 94, no. 19, pp. 269–279, September, 1968.

38. Gibson, D. C.: Cavitation Adjacent to Plane Boundaries, *Proc. Conf. Inst. Eng., Australia*, April, 1969, pp. 210–214.

39. Timm, E. E., and F. G. Hammitt: Bubble Collapse Adjacent to a Rigid Wall, Flexible Wall and a Second Bubble, *1971 ASME Cavitation Forum*, pp. 18–20.

40. Timm, E. E.: "An Experimental Photographic Study of Vapor Bubble Collapse and Liquid Jet Impingement," Ph.D. thesis, Chem. Engr. Dept., University of Michigan, 1970; also available as University of Michigan ORA Proj. Rept. 01357-39-T, 1970.

41. Hammitt, F. G.: Cavitation and Droplet Impingement Damage of Aircraft Rain Erosion Materials in A. A. Fyall (ed.), *Proc. Third Int'l Rain Erosion Congr.*, August, 1970, pp. 907–932, Royal Aircraft Est., Farnborough, England.

42. Lauterborn, W., and H. Bolle: Experimental Investigations of Cavitation-Bubble Collapse in the Neighbourhood of a Solid Boundary, *J. Fluid Mech.*, vol. 72, no. 2, pp. 391–399, 1975.

43. Brunton, J. H.: Cavitation Damage, *Proc. Third Int'l Congr. on Rain Erosion*, August, 1970, Meersburg, Germany; see also Cavitation Erosion, *Proc. IUTAM Congr.*, 1971, Leningrad.

44. Kling, C. L.: "A High Speed Photographic Study of Cavitation Bubble Collapse," Ph.D. thesis, Nuclear Engr. Dept., University of Michigan, 1970; also available as University of Michigan ORA Rept. UMICH 03371-2-T, March, 1970.

45. Kling, C. L., and F. G. Hammitt: A Photographic Study of Spark Induced Cavitation Bubble Collapse, *Trans. ASME, J. Basic Engr.*, vol. 94, ser. D, no. 4, pp. 825–833, December, 1972.

46. Ivany, R. D., F. G. Hammitt, and T. M. Mitchell: Cavitation Bubble Collapse Observations in a Venturi, *Trans. ASME, J. Basic Engr.*, vol. 88, ser. D, pp. 649–657, 1966.

47. Robinson, M. J., and F. G. Hammitt: Detailed Damage Characteristics in a Cavitating Venturi, *Trans. ASME, J. Basic Engr.*, vol. 89, ser. D, no. 1, pp. 161–173, March, 1967.

48. Plesset, M. S.: Shock Waves from Cavitation Collapse, *Phil. Trans. Roy. Soc. (London)*, ser. A, vol. 260, pp. 241–244, 1966.

49. Yeh, H.-C., and W.-J. Yang: Dynamics of Bubbles Moving in Liquids with Pressure Gradient, *J. Appl. Phys.*, vol. 39, pp. 3156–3165, 1968.

50. Gibson, D. C.: "The Collapse of Vapour Cavities," Ph.D. thesis, Cambridge University, 1967.

51. Gibson, D. C.: The Kinetic and Thermal Expansion of Vapor Bubbles, *Trans. ASME, J. Basic Engr.*, vol. 94, pp. 89–96, March, 1972.

52. Chahine, G.: Interaction Between a Collapsing Bubble and a Free Stream, *Trans. ASME, J. Fluids Engr.*, pp. 709–716, December, 1977.
53. Bolle, H., and W. Lauterborn: "Nichtsymmetrischer Kollaps Lasererzeugter Kavitationsblasen," Drittes Physikalisches Institut der Universitat Göttingen, DAGA, pp. 645–648, 1975.
54. Kozyrev, S. P.: On Cumulative Collapse of Cavitation Cavities, *Trans. ASME, J. Basic Engr.*, vol. 90, ser. D, no. 1, pp. 116–124, March, 1968.
55. Stahl, H. A., and A. J. Stepanoff: Thermodynamic Aspects of Cavitation in Centrifugal Pumps, *Trans. ASME, J. Basic Engr.*, vol. 78, pp. 1691–1693, November, 1956.
56. Stepanoff, A. J.: Cavitation in Centrifugal Pumps with Liquids Other than Water, *Trans. ASME, J. Engr. for Power*, vol. 83, ser. A, pp. 79–90, 1961.
57. Stepanoff, A. J.: Cavitation Properties of Liquids, *Trans. ASME, J. Engr. for Power*, vol. 86, ser. A, pp. 195–200, 1964; see also A. J. Stepanoff and K. Kawaguchi, in F. Numachi (ed.), *Proc. 1962 IAHR Symp. on Cavitation and Hydraulic Machinery*, Sendai, Japan, pp. 71–85.
58. Bonnin, J. R.: Incipient Cavitation in Liquids Other Than Cold Water, *1971 ASME Cavitation Forum*, pp. 14–16.
59. Bonnin, J. R.: "Theoretical and Experimental Investigations on Incipient Cavitation in Different Liquids," ASME paper no. 72-WA/FE-31, 1972.
60. Bonnin, J. R.: Influence de la Temperature sur le Début de Cavitation dans l'Eau, Question 1, Rapport 1, *XII Journées de l'Hydraulique*, pp. 1–7, Soc. Hyd. de France, Paris, 1972.
61. Florscheutz, L. W., and B. T. Chao: On the Mechanics of Vapor Bubble Collapse—A Theoretical and Experimental Investigation, *Trans. ASME, J. Heat Transfer*, vol. 87, ser. C, pp. 209–220, 1965.
62. Wittke, D. D., and B. T. Chao: Collapse of Vapor Bubbles with Translatory Motion, *Trans. ASME, J. Heat Transfer*, vol. 89, pp. 17–24, 1967.
63. Hickling, R.: "I. Acoustic Radiation and Reflection from Spheres; II. Some Effects of Thermal Conduction and Compressibility in the Collapse of a Spherical Bubble in a Liquid," Ph.D. thesis, Div. of Engr., Inst. of Tech., Pasadena, Calif., 1962.
64. Hickling, R.: Some Physical Effects of Cavity Collapse in Liquids, *Trans. ASME, J. Basic Engr.*, vol. 88, ser. D, pp. 229–235, 1966.
65. Plesset, M. S., and S. A. Zwick: On the Dynamics of Small Vapor Bubbles in Liquids, *J. Math. and Phys.*, vol. 33, no. 4, pp. 308–330, January, 1955.
66. Plesset, M. S., and D. Y. Hsieh: Theory of Gas Bubble Dynamics in Oscillating Pressure Fields, *Physics of Fluids*, vol. 3, pp. 882–892, 1960.
67. Jarman, P. D.: Sonoluminescence: A Discussion, *J. Acoust. Soc. Amer.*, vol. 11, no. 11, pp. 1459–1462, November, 1960.
68. Jarman, P. D., and K. J. Taylor: Light Emission from Cavitating Water, *Brit. J. Appl. Phys.*, vol. 15, pp. 321–322, 1964.
69. Gavranek, V. V., D. N. Bol'shutkin, and V. I. Zel'dovich: The Thermal and Mechanical Action of a Cavitation Zone on the Surface of a Metal, *Fizika metallov i metallovedeniye*, vol. 10, no. 2, pp. 262–268, 1960.
70. Mitchell, T. M., and F. G. Hammitt: On the Effects of Heat Transfer upon Collapsing Bubbles, *Nucl. Sci. and Engr.*, vol. 53, no. 3, pp. 263–276, March, 1974.
71. Theofanous, T. G., L. Biasi, H. Isbin, and H. Fauske: Nonequilibrium Bubble Collapse—A Theoretical Study, AIChE Preprint 1, Presented at *Eleventh National Heat Transfer Conf.*, August, 1969, Minneapolis.
72. Theofanous, T. G., L. Biasi, H. S. Isbin, and H. K. Fauske: A Theoretical Study on Bubble Growth in Constant and Time Dependent Pressure Fields, *Chem. Eng. Sci.*, vol. 24, pp. 885–897, 1969.
73. Bornhorst, W. J., and G. N. Hatsopoulos: Bubble-Growth Calculation Without Neglect of Interfacial Discontinuities, *Trans. ASME, J. Appl. Mech.*, vol. 34, ser. E, pp. 847–853, 1967.
74. Bornhorst, W. J., and G. N. Hatspoulos: Analysis of a Liquid Vapor Phase Change by the Methods of Irreversible Thermodynamics, *Trans. ASME, J. Appl. Mech.*, vol. 34, ser. E, pp. 840–846, 1967.
75. Garcia, R.: "Comprehensive Cavitation Damage Data for Water and Various Liquid Metals

Including Correlations with Material and Fluid Properties," Ph.D. thesis, Nuclear Engr. Dept., University of Michigan, 1966; also available as University of Michigan ORA Rept. UMICH 05031-6-T, August, 1966.

76. Garcia, R., and F. G. Hammitt: Cavitation Damage and Correlations with Material and Fluid Properties, *Trans. ASME, J. Basic Engr.*, vol. 89, pp. 755–763, December, 1967.

77. Hammitt, F. G., and N. R. Bhatt: Cavitation Damage at Elevated Temperature and Pressure, *ASME 1972 Polyphase Flow and Cavitation Forum*, pp. 11–13.

78. Hammitt, F. G., and D. O. Rogers: Effects of Pressure and Temperature Variation in Vibratory Cavitation Damage Test, *J. Mech. Engr. Sci.*, vol. 12, no. 6, pp. 432–439, 1970.

79. Hammitt, F. G., and N. R. Bhatt: Sodium Cavitation Damage Tests in Vibratory Facility—Temperature and Pressure Effects, *ASME 1975 Cavitation and Polyphase Flow Forum*, pp. 22–23.

80. Devine, R., and M. S. Plesset: "Temperature Effects in Cavitation Damage," California Institute of Technology Rept. 85-27, Div. Engr. and Appl. Sci., 1964.

81. Knapp, R. T.: Recent Investigations of Cavitation and Cavitation Damage, *Trans. ASME*, vol. 77, pp. 1045–1054, 1955.

82. Gelder, T. F., R. S. Ruggeri, and R. D. Moore: "Cavitation Similarity Considerations Based on Measured Pressure and Temperature Depressions in Cavitated Regions of Freon-114," NASA Rept. TN D-3509, 1966.

83. Ruggeri, R. S., and R. D. Moore: "Method for Prediction of Pump Cavitation Performance for Various Liquids, Liquid Temperatures, and Rotative Speeds," NASA Rept. TN D-5292, 1969.

84. Ruggeri, R. S.: Experimental Studies on Thermodynamic Effects of Developed Cavitation, NASA Rept. SP-304, *Proc. Int'l Symp. on the Fluid Mech. and Design of Turbomachinery*, Aug. 30–Sept. 3, 1970, Penn. State University.

85. Hord, J.: "Cavitation in Liquid Cryogens—IV. Combined Correlations for Venturi, Hydrofoil, Ogives, and Pumps," NASA Rept. CR-2448, 1974.

86. Hord, J., L. M. Anderson, and W. J. Hall: "Cavitation in Liquid Cryogens, Volume I: Venturi," NASA Rept. CR-2054, May, 1972.

87. Hord, J.: "Cavitation in Liquid Cryogens, Volume II: Hydrofoil," NASA Rept. CR-2156, January, 1973.

88. Hord, J.: "Cavitation in Liquid Cryogens, Volume III: Ogives," NASA Rept. CR-2242, May, 1973.

89. Hord, J., and R. O. Voth: "Tabulated Values of Cavitation B-Factor for Helium, H_2, N_2, F_2, O_2, Refrigerant 114, and H_2O," Nat'l Bur. Stand., Tech. Note 397, February, 1971.

90. Holl, J. W., and A. L. Kornhauser: Thermodynamic Effects on Desinent Cavitation on Hemispherical Nosed Bodies in Water at Temperatures from 80 Deg F to 260 Deg F," *Trans. ASME, J. Basic Engr.*, vol. 92, March. 1970, pp. 44–58.

91. Kornhauser, A. L.: "Thermodynamic Effects on Desinent Cavitation in Water," M.Sc. thesis, Dept. of Aerospace Engr., Penn. State University, June, 1967.

92. Holl, J. W., M. L. Billet, and D. S. Weir: Thermodynamic Effects on Developed Cavitation, *Trans. ASME, J. Fluids Engr.*, vol. 97, pp. 507–514, December, 1975.

93. Holl, J. W., and G. F. Wislicenus: Scale Effects on Cavitation, *Trans. ASME, J. Basic Engr.*, vol. 83, ser. D, pp. 385–398, 1961.

94. Acosta, A. J., and B. R. Parkin: Discussion of Ref. 93 in *Trans. ASME, J. Fluids Engr.*

95. Billet, M. L., and D. S. Weir: "The Effect of Gas Diffusion and Vaporization on the Entrainment Coefficient for a Ventilated Cavity," Penn. State University Rept. TM 74-15, Appl. Research Lab., Jan. 24, 1974.

96. Billet, M. L., and D. S. Weir: The Effect of Gas Diffusion on the Flow Coefficient for a Ventilated Cavity, *Proc. Symp. on Cavity Flows, Joint Conf. of Fluids Engr. and Lubrication Div.*, May 5–7, 1975, ASME, Minneapolis, Minn.

97. Billet, M. L., J. W. Holl, and D. S. Weir: "Geometric Description of Developed Cavities on Zero and Quarter Caliber Ogive Bodies," Penn. State University Rept. TM 74-136, Appl. Research Lab., May 6, 1974.

98. Tong, L. S.: "Boiling Heat Transfer and Two-Phase Flow," John Wiley, New York, 1965.
99. Scriven, L. E.: On the Dynamics of Phase Growth, *Chem. Engr. Sci.*, vol. 10, pp. 1–13, 1959.
100. Plesset, M. S., and J. A. Zwick: The Growth of Vapor Bubbles in Superheated Liquids, *J. Appl. Phys.*, vol. 25, pp. 493–500, 1954.
101. Hammitt, F. G., et al.: Predictive Capability for Cavitation Damage from Bubble Collapse Pulse Count Spectra, *Proc. I. Mech. E. Conf. on Scaling for Performance Prediction in Rotodynamic Machines*, 6–8 Sept., 1977, University of Stirling, Scotland; also available as University of Michigan ORA Rept. UMICH 014456-3-I, November, 1976.
102. Petracchi, G.: Investigations of Cavitation Corrosion, *Metallurgia Italiana*, vol. 41, no. 1, pp. 1–6, 1949; see English Summary in *Engr. Digest*, vol. 10, no. 9, p. 314, 1949.
103. Plesset, M. S.: On Cathodic Protection in Cavitation Damage, *Trans. ASME, J. Basic Engr.*, vol. 82, no. 4, pp. 808–820, December, 1960.
104. Hammitt, F. G.: "Progress Rept. No. 3—Period: 1 September–1 December, 1976," University of Michigan ORA Rept. UMICH 014456-3-PR, November, 1976; see also F. G. Hammitt et al., University of Michigan ORA Rept. UMICH 014456-18-I, June, 1977.
105. Thiruvengadam, A.: Intensity of Cavitation Damage Encountered in Field Installations, *Trans. ASME, Cavitation in Fluid Machinery*, pp. 32–46, New York, 1965.
106. Plesset, M. S.: Pulsing Technique for Studying Cavitation Erosion of Metals, *Corrosion*, vol. 18, no. 5, pp. 181–188, 1962.
107. Plesset, M. S.: The Pulsation Method for Generating Cavitation Damage, ASME paper no. 62-WA-315, 1962, *Trans. ASME, J. Basic Engr.*, vol. 85, no. 3, pp. 360–364, September, 1963.
108. Ackeret, J., and P. DeHaller: Study of Corrosion Through the Impact of Water, *Schweiz. Bauztg.*, vol. 98, p. 309, 1931.
109. Robinson, M. J.: "On the Detailed Flow Structure and the Corresponding Damage to Test Specimens in a Cavitating Venturi," Ph.D. thesis, Engr. Dept., University of Michigan, 1965; also available as University of Michigan ORA Rept. UMICH 03424-16-T, August, 1965.
110. Robinson, M. J., and F. G. Hammitt: Detailed Damage Characteristics in a Cavitating Venturi, *Trans. ASME, J. Basic Engr.*, vol. 89, ser. D, no. 1, March, 1967.
111. Lichtman, J. Z., and D. H. Kallas: Erosion Resistance of Coatings—Methods for Evaluating Erosion (Cavitation) Damage, *Materials Protection*, vol. 6, no. 4, pp. 40–45, April, 1967.
112. Huang, Y. C.: "Numerical Studies of Unsteady, Two-Dimensional Liquid Impact Phenomena," Ph.D. thesis, Mech. Engr. Dept., University of Michigan, June, 1971; also available as ORA University of Michigan Rept. UMICH 03371-8-T.
113. Hwang, J.-B.: "The Impact Between a Liquid Drop and an Elastic Half-Space," Ph.D. thesis, Mech. Engr. Dept., University of Michigan, March, 1975; also available as University of Michigan ORA Rept. UMICH 012449-5-T.
114. Kuttruff, H., and U. Radek: Messungen des Druckverlaufs in Kavitationserzeugten Druckimpulsen, *Acustica*, vol. 21, no. 5, pp. 253–259, 1969.

CHAPTER

FIVE

CAVITATION AND JET IMPACT DAMAGE

5-1 LIQUID EROSIVE WEAR: THEORY AND TEST DEVICES

5-1-1 Introduction

Erosive wear of a solid surface can take place in a liquid or gaseous medium even without the presence of another phase in the fluid continuum. However, it can be greatly accelerated by the presence of additional phases. For the present purpose, we define "erosive wear" to be that provoked by fluid flow, but other than corrosion. It includes particularly, then, the phenomena of solid and liquid particle impact, where the particles may be carried by gas, vapor, or liquid, and liquid "cavitation," which is essentially a phenomenon involving vapor "particles," i.e., pockets or bubbles, in liquid. Cavitation is not entirely analogous, however, to the other droplet- or particle-impact phenomena, as will be discussed later. The main purpose of this section is to clarify "erosive wear" and its various facets. A more comprehensive review of the same subject is found elsewhere [1].

5-1-2 Mechanisms and Types of Erosive Wear

Erosive-wear mechanisms Since we have excluded chemical or corrosive effects from the category of phenomena here dubbed "erosive wear," it seems reasonable to suppose that material removal for this category of phenomenon must be due to the imposition on the surface of shear or normal stresses of sufficient magnitude to cause material failure, either through single blows or through fatigue-type

220

effects. Of course, in most real situations chemical effects are not completely absent, although there are certainly many cases where their effects are relatively negligible. Further, in most cases of "erosive wear" the existence of a potentially damaging level of stresses can be rationally justified, as explained in the following.

Erosive-wear phenomena

Single-phase flow—general   In the interest of a logical presentation, single-phase flow phenomena will be considered before multiphase, even though the most important erosive phenomena for reasonably strong materials appear to require the presence of more than one phase of the fluid. Of course, single-phase liquid flows are capable of river bank or beach erosion, for example, but such phenomena are not the subject of this section. It is concerned rather with the erosive wear of engineering materials such as structural metals, plastics, ceramics, etc. We are here considering such phenomena as liquid-droplet impact in wet steam, for example, to be a two-phase phenomenon; similarly, liquid or solid particles impact in an air or gas continuum. The jetting action of a fire-hose, on the other hand, we would consider here to be a single-phase phenomenon, provided it did not involve entrained solid particles, vapor bubbles to provide cavitation, or other auxiliary wear mechanisms.

In the nature of possibly erosive single-phase phenomena, there is the possibility of either very high velocity flows of liquid, vapor, or gas, the latter two being relatively similar in their damage capability. Damaging stresses to be provided on a surface can be included within the categories of shear and/or normal stress. Solid-surface shear should equal fluid shear at the fluid-solid interface, and would thus equal the product of viscosity and wall velocity gradient, i.e.,

$$\tau = \mu\left(\frac{\partial u}{\partial y}\right)_{\text{wall}} \tag{5-1}$$

where u is the velocity component parallel to the surface, y is the distance from it, and μ is the absolute viscosity.

Normal stress in the surface is approximately numerically equal to the fluid pressure at the surface. In the case of a steady-state impinging jet the maximum value this could attain would be the "stagnation pressure" in the fluid, i.e.,

$$\Delta p_{\text{stag}} = \frac{\rho V^2}{2g} \tag{5-2}$$

where V is the total fluid velocity and Δp_{stag} is measured above the ambient pressure.

In the case of a nonsteady liquid jet, the pressure at certain points can attain the approximate magnitude of the "water-hammer" pressure [1–5], i.e.,

$$\Delta p_{\text{WH}} = \frac{\rho V C}{g} \tag{5-3}$$

Table 5-1 Numerical example of fluid stresses at high velocity in cold water

	Flow of 500 ft/s ($\sim$150 m/s)	Flow of 1000 ft/s ($\sim$300 m/s)
Shear	1 lb/in²	2 lb/in²
Stagnation pressure	1,670 lb/in²	6,700 lb/in²
Water-hammer pressure	30,000 lb/in²	60,000 lb/in²

Note: 15 lb/in² $\cong$ 1 bar.

where C is velocity of sound in the liquid.† Of course, "water-hammer" type phenomena are important with gas or vapor flows only if the ambient pressure, i.e., density, is very large. For relatively low velocity liquid flows, water-hammer pressure can reach values sufficient to cause material surface failure, and hence erosion.

A numerical example at this point may be useful. Suppose a flow of cold water of 500 ft/s ($\sim$150 m/s), which is a very high velocity for water flow. For the case where this flow is assumed parallel to a wall, assume the viscous "boundary-layer thickness" to be 10^{-4} ft ($\sim$30 μm), which seems a likely minimum. Then the surface shear stress, stagnation pressure, and water-hammer pressure are approximately as listed in Table 5-1, along with values for 1000 ft/s ($\sim$300 m/s).

It is apparent from Table 5-1 that surface shear stress cannot be a damaging mechanism at velocities of interest unless the viscosity is extremely high. It also appears that in most cases the stagnation pressure is not sufficient to be damaging to most structural materials, so that high-velocity impacting liquid jets should not be damaging in most cases of engineering interest, unless non-steady-state behavior is involved, in which case fluid pressures could attain values of the general order of the water-hammer pressure. Another possibility exists if relatively large asperities exist on the surface. Stagnation pressure, rather than shear stress, could be exerted against these. The bending moment against an asperity of sufficient aspect ratio could cause a surface failure.

Examination of Table 5-1 and Eqs. (5-1) and (5-2) indicate that it is most unlikely that either shear stress or pressure induced by gas or vapor flows could be sufficient to damage such materials as normal structural metals, though gas velocities in some applications up to several $\sim$1000 m/s are possible. The wall shear is so low (Table 5-1) that even very high gas velocities would not raise it to damaging values. In addition, the viscosity of most gases is much less than that of the cold water used for calculation of Table 5-1. The stagnation pressure with gases or vapors is not likely to be damaging either, since they are proportional to density, which for gases or vapors is very much less than the cold water used for Table 5-1 (the factor is $\sim$10³ between atmospheric air and water).

---

† The derivation of the "water-hammer equation" is found in most elementary fluids texts.

Actual single-phase applications   Several single-phase flow applications in which erosive wear has sometimes occurred will be considered. For the most part these involve high-velocity water flows. However, even for these cases, a consideration of the possible normal and shear stresses induced by liquid flow seems to indicate that erosive wear is impossible in the absence of either corrosive effects or multi-phase phenomena such as cavitation, droplet impact, etc. Thus, in cases where erosion has in fact been observed in these applications, it is the author's opinion that one or more of these "extraneous" effects must have been involved.

From a slightly different viewpoint, for a given geometry, shear stress must increase roughly as (*a*) pU^2 (i.e., driving pressure for any fluid) and (*b*) kinematic viscosity (which is higher for air than for water).

1. *Pelton (hydraulic) turbine.* In this application water jets with velocity up to the order of 200 m/s impinge upon a rotating turbine wheel equipped with suitably designed "buckets," usually of hardened steel and often of the 400 series. If the design is correct, no significant erosion occurs, at least for thousands of hours of operation. In some cases, however, prohibitive erosion does occur quickly. This is presumably due to such factors as improper blade design leading to cavitation on the blade surfaces, or perhaps to entrained sand in the impinging water. In any case, it is not, in the writer's opinion, single-phase erosion.
2. *Boiler feed pump.* Modern boiler feed pumps also involve liquid velocities up to 200 m/s. With proper design, again, no substantial erosion occurs. However, there are many cases on record where large erosion has resulted (Fig. 5-1, for example) in such pumps. This is usually presumed to be due to cavitation,

Figure 5-1 Leading edge of a series 400 stainless-steel impeller for a boiler feed pump, exhibiting deep local damage caused by cavitation erosion.

even though it has occurred in some cases in the discharge casing, usually of the first stage. Cavitation in this region is possible, at least for off-design conditions. Again there is no plausible mechanism for erosion of these materials (probably 400-series steels) under the existing velocity conditions, except through cavitation and/or corrosion.

3. *Valve seats.* Very high velocities can exist across valve seats in some cases, at least of the magnitudes previously mentioned. Resultant erosion, sometimes called "wire-drawing," has been reported for both liquid and steam valves. The materials are often hardened steels. Again no plausible mechanism for erosion exists *unless* there is either substantial corrosion (unlikely for proper material choice) or multiphase phenomena are involved such as cavitation (for liquid-handling valves) and possible water-droplet impingement for steam valves, assuming that wet steam may be involved.

5-1-3 Multiphase Flow†

General As indicated in the foregoing, it seems most probable that in most engineering cases involving erosive wear, multiphase flow phenomena must be involved. These can involve a liquid gas or vapor continuum with solid or liquid particles (droplets), or a liquid continuum with entrained vapor (cavitation), or entrained gas. These cases will be discussed briefly in the following, with reference to the stress-raising mechanisms involved.

Solid-particle impingement High-velocity solid-particle impingement can certainly provoke undesired erosive wear in many well-known cases, e.g., dust erosion of helicopter blades, propellor blades, helicopter-drive gas-turbine compressor blades, etc. A case of useful solid-particle erosion is "sand-blasting." In all these cases, the phenomenon involved is that of the rapid motion of the eroded material through a continuum of gas with entrained solid particles. Other less clear-cut cases involving erosive solid-particle impingement are liquid or gaseous slurry flows. Applications are in the liquification of coal, coal or ore-bearing pipelines, etc. Tests have indicated that such flows are far more erosive than single-phase flows of the same velocity, but the precise mechanism of erosion at this point is not entirely clear.

The state of the art at this time does not provide methods for specifying the state of stresses on the eroded material surface (even to the extent possible for liquid impact) resulting from impact by particles of irregular shape, which is the usual case of interest. It is, of course, obvious that both shear and normal stresses of substantial magnitude will be provoked by such impacts, but no generalized governing relations are as yet available, to the author's knowledge.

The situation for slurry erosion is even more obscure than that for direct solid-particle impact with respect to being able to specify the material stresses or the detailed mechanisms causing the erosion. This is particularly true for

† See also Ref. 156.

liquid slurries, since the velocities are normally relatively low, so that stresses from direct impact should not be of damaging magnitude. An extremely damaging situation, nevertheless, is that provided by cavitating slurries which have sometimes occurred in pumps (e.g., dredging pumps) or in solids-bearing transport pipelines (e.g., ore-bearing).

Liquid-droplet impingement   Liquid-droplet impingement erosive-wear applications usually involve the rapid motion of the eroded material through a gaseous or vapor continuum with entrained liquid droplets. Important examples are the motion of high-speed aircraft or missiles, or propellor or helicopter blades, through air, or the motion of steam-turbine blades through a vapor continuum including relatively large water droplets. A somewhat similar situation can occur for aircraft gas-turbine compressor blades under atmospheric rain conditions. Inverse cases, where the droplets are projected against relatively stationary target materials, are not usual because it is not generally possible to accelerate liquid droplets of potentially damaging size to damaging velocities without droplet disintegration, i.e., a critical Weber number from the viewpoint of droplet stability is involved.

The stress regimes applying for liquid-droplet impact erosion can be estimated much more closely than those applying for the solid particle or slurry cases discussed above. In general, the order of magnitude of normal stresses can be obtained from the water-hammer relation [Eq. (5-3)]. Some improvement can be made if the result is corrected for the nonrigidity of the target material and for the effects of liquid compression on liquid sonic velocity and density. Droplet shape also affects the stress regime, as shown by various recent numerical and experimental studies, some from the author's laboratory [2-4]. The last of these [4] is a numerical study where the target material was assumed elastic rather than rigid, as in the earlier cases, so that realistic target material stresses could be computed. However, the precise state of the analysis of liquid-droplet impact is beyond the scope of the present chapter, and will not be discussed further here. Suffice it to say that the general level of stress magnitudes computable (and measured) for this case is sufficient to explain and justify the erosion observed.

Another interesting point which can be made with regard to droplet-impact erosion is that radial velocities along the impacted surface are generated by droplet impact which can be several times the original impact velocity. It has been suggested that the shear stress caused by this high-velocity flow parallel to the surface might contribute importantly to the damage. However, this hypothesis seems unlikely considering the numerical results in Table 5-1 and the form of Eq. (5-1). Even though the radial velocity in an extreme case might be 10 times that used for the example of Table 5-1 (500 ft/s or ∼150 m/s), the shear stress induced by this flow would still be very small, since it is proportional only to velocity to the first power. However, the impingement of this high-velocity radial flow against a small asperity raised from the surface could create failure stresses. This process, as well as that of the droplet impact in general, is well illustrated in Fig. 5-2 (from Ref. 5). This handbook article, coauthored by the present writer, summarizes the droplet impact and cavitation processes discussed here.

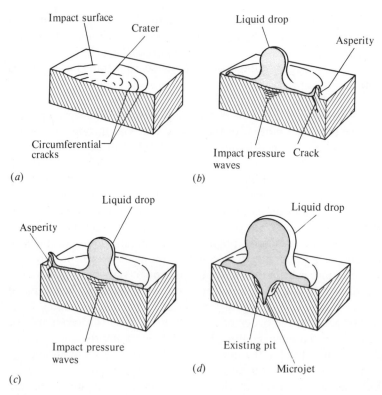

Figure 5-2 Damaging mechanisms for liquid-impingement erosion [5]. (*a*) Solid surface showing initial impact of a drop of liquid that produces circumferential cracks in the area or impact, or produces shallow craters in very ductile material. (*b*) High-velocity radial flow of liquid away from the impact area by a nearby surface asperity, which cracks at its base. (*c*) Subsequent impact by another drop of liquid breaks the asperity. (*d*) Direct hit on a deep pit results in accelerated damage, because shock waves bouncing off the sides of the pit cause the formation of a high-energy microjet within the pit [5].

Cavitation Whereas droplet and solid-particle impact involve liquid and solid particles respectively in a gas or vapor continuum, cavitation involves vapor (with some gas content) "particles" in a liquid continuum. However, since these "particles" involve only relatively low density material with little mass, their "impact" with target material is not in general a likely cause of erosion; rather it is a case of the highly specialized phenomenon of bubble collapse, discussed in Chap. 4. Of course, this statement may not apply to the combined phenomenon case of a cavitating slurry (mentioned earlier).

Though particle impact per se is not the presumed cause of cavitation erosion, a combination of shock waves in the liquid and liquid "microjet" impact upon the eroded surface seems to represent, at this time, the most likely detailed mechanism for cavitation erosion. This problem is thoroughly discussed in Chap. 4 as well as in Ref. 6, and is summarized in Ref. 5, as well as in many research

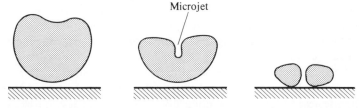

Figure 5-3 Schematic representation of successive stages of nonsymmetrical cavity collapse with micro-jet impingement against a metallic surface [5].

articles too numerous to mention here. Bubble collapse adjacent to a surface with development of liquid microjet is shown in Fig. 5-3 (from Ref. 5). The shock waves emitted during the bubble "rebound" [7] which often follows original collapse (Fig. 5-4) are believed to provide in many cases important assistance to the damaging process originating from the microjet impact. At least, the liquid pressures upon a neighboring wall during bubble collapse appear to be considerably less than those during rebound [8, 9], and appear to be of sufficient magnitude in fact to contribute to damage for most materials.

Actual calculation of the stress regime applied to an eroded surface by cavitation is not possible in the present state of the art. This is not surprising when one considers the complex mix of processes which are involved (not to mention the important contribution of corrosion in many cases). The problem of stress calculation appears even more difficult, as compared with that of droplet impingement, when it is realized that the size and position of the collapsing bubbles, to which the damage is presumably due, is not fixed or well known in most cases. Natural cavitation fields include bubbles covering a large range of

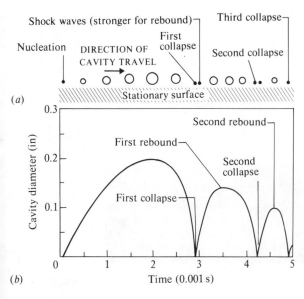

Figure 5-4 (a) Schematic representation of successive stages of growth, collapse, and rebound of a single traveling cavity. (b) Graph of cavity diameter as a function of time for the cavity in (a) [6, 7].

diameters. In the usual engineering case, neither range of diameter, distribution over this range, nor the number of bubbles involved is known to any degree of precision. Hence, for the general engineering case, estimation of the stress regimes to which a cavitated surface will be exposed by a given flow regime is essentially impossible, whereas for the droplet-impact case quite reasonable estimates can be made as previously discussed. For laboratory cavitation-erosion test devices, the situation is only slightly less obscure (depending upon the type of test device), as will be discussed later. While from numerical analyses [6, 8–10] it can be shown that the potentialities for sufficient stress magnitudes to account for the observed damage exist, it is still true that the best evidence of the stress regimes to which cavitated surfaces have been exposed can be obtained either from examination of the damaged surfaces themselves or (more difficult to achieve) damage debris. Since the damaged surfaces from cavitation and droplet impact often have a very similar appearance, it can be presumed that the two processes are quite similar in their effects upon surfaces. Of course, in most cases the attack by cavitation is on a smaller and finer scale, so that individual-blow craters from cavitation have a diameter typically of only a few mils [11] and it appears that the microjet diameter is typically only a few microns [12]. Typical individual-blow cavitation craters on stainless steel are shown in Fig. 5-5. In a typical case such craters

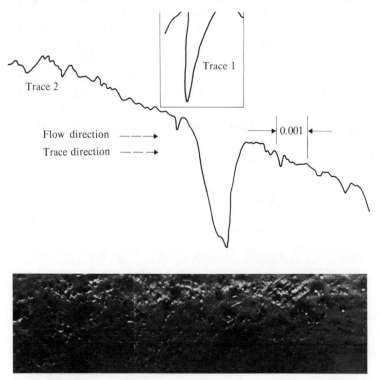

Figure 5-5 Individual-blow cavitation craters on stainless steel [11, 101] University of Michigan venturi.

presumably cover the entire surface by an essentially "random" bombardment until large-scale fatigue failure eventually occurs, producing eventual large-scale failure (Fig. 5-1).

5-1-4 Erosive-Wear Testing Devices

Applications   The applications for erosive-wear testing devices can be subdivided in the following manner. This division is not entirely parallel to that based on erosion phenomena previously discussed, since the test devices attempt in general, for practical reasons, to model one primary factor of the application involved, rather than the phenomenon itself.

High fluid velocity devices   In these, only "single-phase" erosion is to be evaluated. If cavitation of droplet or particle impact occurs it is unintentional, but it may be instrumental in the results. Such devices are intended for the study of erosion in steam or liquid ("noncavitating") valves, i.e., "wire-drawing," boiler feed-pump casings, etc.

Solid-particle or droplet-impact devices   In these, the material to be eroded rapidly traverses a field of essentially stationary particles or droplets. In most cases the target material is whirled through a field of falling particles or droplets, but in some cases translational motion is used. In some cases liquid jets rather than droplets are impacted. Impacting liquid devices can sometimes generate secondary cavitation, which may contribute importantly to the damage, but usually this is not intentional.

Flowing cavitation devices   In these, cavitation is caused by converting pressure "head" into kinetic "head." Numerous geometries have been used for this purpose, as will be discussed later. In general these could be characterized under the terms "venturi," "rotating disc," and "miscellaneous." These devices are meant to cause cavitation erosion under flow conditions as realistic as possible, since damage-modeling laws are highly uncertain.

Vibratory cavitation devices   In these, cavitation is provoked in an essentially static fluid, as opposed to the flowing cavitation devices discussed above. Such a device, sometimes called "magnetostriction" or "ultrasonic" tester, usually relies on the rapidly reciprocating motion of a submerged test plate at relatively high frequency to provoke cavitation by pressure oscillation in an essentially static liquid. The necessary pressure oscillation is due to the very high acceleration imposed upon the liquid by the vibration. This type of device is used for the study of cavitation damage, since it is the most economical, both for purchase and operation, of the possible cavitation-damage test devices. It is also a strongly "accelerated" device, in that it can provide substantial damage on even the most resistant of materials within relatively short test periods. However, its major disadvantage is that it does not, by its nature, relate cavitation damage to flowing system parameters such as velocity and pressure, so that the conversion of "vibratory" results to projected performance in field devices is extremely uncertain, if not impossible.

Actual test devices

High-velocity single-phase erosion-wear test devices Various tests have been made at various times to evaluate high-velocity single-phase erosion in cases where this has occurred in field machines, so that laboratory tests seemed warranted. However, no relatively standardized machine of this type appears to exist. A case in point was the work at Detroit Edison in the 1940s to evaluate erosion in boiler feed-pump casings and regulating valves [13, 14] which were exposed to relatively high velocity but, supposedly, not cavitation. Some corrosive contribution no doubt was also included with some of the materials used (carbon steels, etc., but also including the 400 and 300 series used later in this application). High velocities (~ 60 m/s) were attained by accelerating the pressurized water through a small slit formed by the materials to be tested. Backpressure was limited by the equipment available for the test, so that although the absence of cavitation was one of the test objectives it is nevertheless quite likely, in the author's opinion, that it contributed importantly to the results, which included considerable erosion of most materials tested. As previously discussed, without cavitation (or corrosive attack, probably not important for the stainless steels tested) there is no plausible mechanism to explain the erosion observed.

Other partially pertinent cases in point are the "rotating-wheel" devices developed originally in the 1930s, probably first by Ackeret and de Haller [15]. This device is shown schematically in Fig. 5-6 (from Ref. 6), and consists of a rotating "wheel" to the periphery of which the specimens to be eroded are attached. These are rotated through a relatively low-velocity water jet with direction parallel

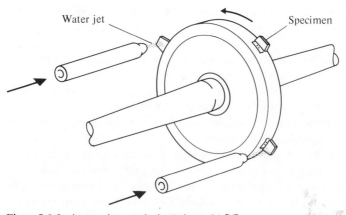

Figure 5-6 Jet-impact damage device (schematic) [6].

to the wheel axis. Since the impact velocity for these devices is typically no more than 100 m/s, it is difficult to explain the rapid erosion of some of the hardened materials tested without the contribution of local cavitation as well as liquid impact. According to Table 5-1, the water-hammer pressure for this device at 100 m/s would be $\sim 29,000$ lb/in² (2000 bar), but even materials such as stellite are rapidly eroded. These devices were originally developed to study erosion of impulse hydraulic turbines such as "Pelton wheels." It was assumed then that this erosion was of a similar nature to that encountered in large steam turbines (which is now clearly known to be a case of liquid-droplet impact). Actually, the Pelton-wheel erosion is probably mainly due to cavitation, but involves a very high liquid velocity parallel to the blading surfaces, as previously discussed. Thus, since this rotating-wheel test device was developed to study Pelton-wheel erosion, which at first glance appears to be a case of high-velocity single-phase erosion, its introduction in this article at this point is pertinent.

Solid-particle or droplet-impact erosive-wear test device Various devices of this type have been developed and used over the years, including the relatively low-velocity "rotating-wheel" device discussed above (Fig. 5-6). In recent years, solid-wheel devices for rotating speeds up to perhaps 500 m/s have been built in various laboratories throughout the world, particularly for the study of the droplet-impact problem existing in the low-pressure end of large steam turbines. These more modern wheels are generally enclosed within a strong steel casing, both for protection in case of failure and to allow operation under vacuum, and both to model more closely the steam turbine problem and to reduce drive power for the device. Relatively low-velocity liquid droplets or jets are caused to impact the rotating test specimens. Various test facilities of this type, existing in England, are well described in Ref. 16, for example. Somewhat comparable facilities also exist in this country and in Russia, but little descriptive data have yet been published.

In addition to the wheel devices described above, designed particularly for the steam-turbine application where the materials to be tested are generally of a highly resistant nature, such as stellites, hardened steels, etc., another group of facilities has been developed in recent years for droplet-impact erosion testing, both in this country and Europe, of aircraft- and missile-component materials where the application is "rain erosion," i.e., the erosion encountered when such components are flown through rainstorms. For applications where the flight velocity exceeds Mach 1 (~ 350 m/s), particularly, erosion can occur very rapidly, since the materials involved are not optimum for erosion resistance but are rather chosen for other prerequisites particular to the application, i.e., radomes, propellor or helicopter blades, etc. For this application, rotating arms rather than disks are normally used. Relatively large diameters, and hence low rpm are usually required for such a test device, since very large "g" loads must not be imposed upon the test materials. This requirement is obviously not of such great importance for the very strong metallic alloys to be tested in the turbine application. Also, required test times for the aircraft-type device are obviously much shorter. The device of this kind with the largest and highest speed (~ 900 m/s) is that at Bell Aerospace

[17]. The diameter of the rotating element is 7 m. The rotating arm at Michigan is shown in Chap. 8, where test facilities are further discussed (Fig. 8.3).

Another type of device for the study of aircraft- and missile-component rain-erosion resistance at very high velocity is the rocket sled, where test materials can be driven through an artificial rain field. The largest such device, to the author's knowledge, is that at Holloman Air Force Base [18, 19], where test velocities up to ~Mach 5 (~1700 m/s) have been utilized. This type of device allows higher velocities than do rotating-arm devices, which are limited by centrifugal stresses in the arm. The rocket sled has the advantage of allowing the test of many material specimens in a single run; hence they are tested under closely identical conditions. However, the test is relatively expensive and has the disadvantage that intermediate observation of the progress of erosion is not practical.

Many of these aircraft-component test devices have also been used for dust-erosion tests, which is an important present-day problem for such applications as helicopter blades.

Flowing cavitation devices

1. *General and miscellaneous.* Flowing cavitation-erosion test devices include machines involving both rotating elements and translatory flows. In general, these are well described in Ref. 6, from which some of the figures used here were taken. No really standard device has yet devolved in this field, and a variety of devices have been used. These can be considered under the main headings of "venturi" and "rotating-disc" devices. However, there exist several miscellaneous devices such as test specimens submerged in large water tunnels (used by Knapp, Ref. 7) and a vibrating reed in a flowing stream [20]. However, since these and other miscellaneous devices are not of major importance to present-day cavitation-damage evaluations, they will not be discussed further here.

2. *Venturi devices.* Venturi devices are here taken to include all those flow devices employing a flow restriction to convert pressure into kinetic head, creating a cavitating region when the static pressure falls to the level of the vapor pressure. For damage studies, relatively standard venturis (Fig. 5-7, from the University of Michigan, for example), as well as several quite special designs, have been used. Of these the earliest is probably that of Boetcher [21] reported in 1936 (Fig. 5-8 from Ref. 6; see also Ref. 21). As will be noted from Fig. 5-8, the arrangement is such that the cavitating jet impinges upon the test specimen. Such a venturi geometry does in fact provide a very intense damaging regime, as compared, for example, to the University of Michigan design (see Fig. 5-7 and Ref. 11) which, however, does model more closely the usual flow conditions found in hydraulic machines (Table 2 from Ref. 22).

   Another special venturi-damage design which has been used fairly broadly in various countries since its introduction in 1955 [23] is that of Shal'nev

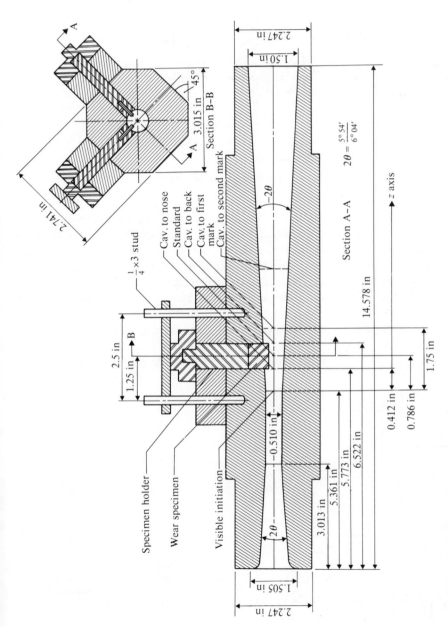

Figure 5-7 U-M damage venturi schematic [11, 101].

233

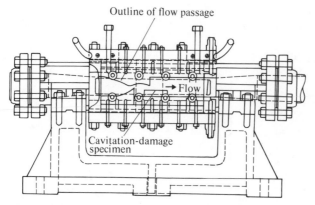

Figure 5-8 Holtwood laboratory cavitation-damage test section [6, 21].

(Fig. 5-9) in Moscow. The flow geometry consists of a rectangular throat of constant flow area, across which a small cylindrical pin is placed. Cavitation occurs in the wake of this pin, and the damage specimens are located flush with the wall and downstream of the pin (Fig. 5-9).† The damaging intensity induced by this geometry is also much higher than the University of Michigan design (Fig. 5-7). However, the flow regime is that of separated vortices, which may model a relatively special type of cavitation quite closely, but is not particularly similar to the more usual flow regimes encountered in flow machinery.

3. *Cavitating disk devices.* A "rotating-disk" device developed for the study of

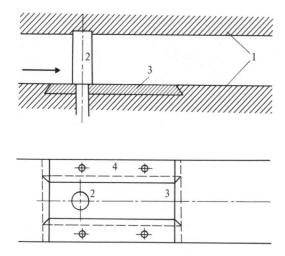

Figure 5-9 Shal'nev-type venturi, cavitation behind a circular profile: 1. Walls of the experimental chamber. 2. Model. 3. Test piece. 4. Test piece holding device [23].

† Test at University of Michigan used a similar pin in a venturi diffuser, and found very rapid damage rates.

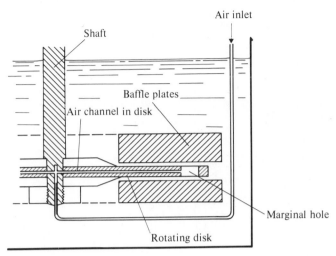

Figure 5-10 Rasmussen-type rotating disk for cavitation damage with holes for air supply [24].

cavitation damage was reported [24] in 1955 by Rasmussen (Fig. 5-10). The flow geometry consists of a flat disk, fitted with pins or through-holes at various radial locations. The disk is caused to rotate in the test liquid which is contained within a circular casing. The casing is fitted with radial baffles to prevent gross rotation of the overall fluid. The traverse of the disk pins or holes through the relatively stationary surrounding liquid causes cavitation clouds which follow the rotating disk and collapse upon test specimens fitted flush with the disk surface. Figure 5-11 (from Ref. 6) is a schematic of a more recent rotating-disk facility built by Pratt and Whitney Aircraft for eventual use with liquid metals [25]. Eroded specimens of refractory metal are also shown in Fig. 5-11.

This type of facility also produces damage very rapidly—more so than the Boetcher and Shal'nev types of venturi (Table 5-2). In all these cases, however, the flow regimes involved are really quite different. By its very nature, the regime provided by the rotating disk resembles closely that involved for regions of separated flow in turbomachines.

Another valid comparison between these flowing damage tests is the expense of the facilities involved. The venturis obviously require a loop facility with driving pump and much other instrumentation and controls. The rotating disk, however, is not a simple or cheap facility in itself, as can be seen from Fig. 5-11 which includes the cross-section drawing for the Pratt and Whitney device [25]. An accurate statement comparing the cost of the rotating-disk and venturi-damage facilities is not possible at this time, since too many unknown and complicating factors are involved. However, it is certainly true that the vibratory type of damage devices, to be discussed next, are considerably

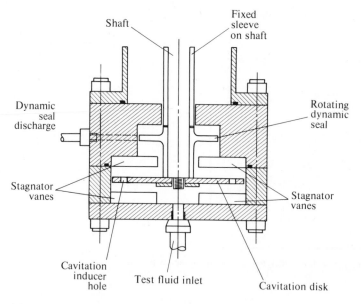

(a)

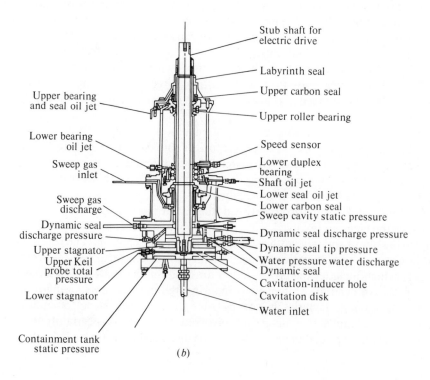

(b)

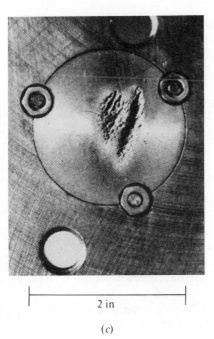

├───────────────────┤
2 in

(c)

Figure 5-11 Water rotating-disk cavitation-damage test device at Pratt and Whitney Aircraft, CANEL (Wood et al). (a) Cross section of the PWA rotating disk device. (b) Disk installed in housing (schematic). (c) Damaged specimen (Cb-1Zr alloy) after 30 hr. cavitation is generated from the hole above the specimen [6, 25].

more economical, certainly in the first cost. Operating costs, primarily that of operator salary, are probably similar. Length of test required for a given material, i.e., damage intensity (Table 5-2), for the vibratory device covers the same general order as the others.

Table 5-2 Comparative damage intensities for different types of facilities [22]

Type of facilities		Intensity, $W/cm^2 \times 10^7$
Magnetostriction†		
Devices 1–7		0.004–2.5
Venturis		
8–9	Boetcher type	$0.1–0.1 \times 10^{-2}$
10	Shal'nev type	0.1
11	Shal'nev type	0.03
12	Shal'nev type	0.1
13	U-M	0.3×10^{-4} [11, 101]
Rotating disk		
14		4
15		0.34
16		1.0

† Vibratory, ultrasonic, or piezoelectric devices are included.

Vibratory cavitation devices   The vibratory type of cavitation-damage test, already mentioned, is certainly the simplest, cheapest, and most common of all presently known cavitation-damage test devices. It is also capable of providing erosion rates of the same general order as the flowing systems already discussed. It is also the only one for which an ASTM Standard Method has been promulgated [26]. Figure 5-12 (from Ref. 6) is the schematic of the University of Michigan device of this type, which is designed for a variety of liquids, temperatures, and pressures. This unit is somewhat more complex than the standard ASTM device [26], since the open beaker is replaced by a sealed tank. This type of test device is most useful for the comparison of material resistances and the evaluation of effects of different fluids, temperatures, and pressures, but it is not suitable for evaluation of probable cavitation erosion in the usual fluid-handling machine, since the very important flow parameter of velocity is not modeled. In the present state of the art it is thus not possible to predict damage in a flowing situation from vibratory test results. Recent work in the author's laboratory and elsewhere [27–30] to correlate the damage rate from such a device with bubble-collapse pulse counts measured by an acoustic probe show some quite good correlations (discussed later).

Variations of the vibratory device have been used (but not yet standardized) wherein a cavitation field is provided by the vibrating horn, but the specimen to be tested is held stationary in the cavitating field rather than attached to the end of the vibratory horn, as in the standard arrangement. This arrangement is useful for the testing of materials which cannot be vibrated by the horn without deleterious extraneous effects. Since the stationary specimen is usually located with

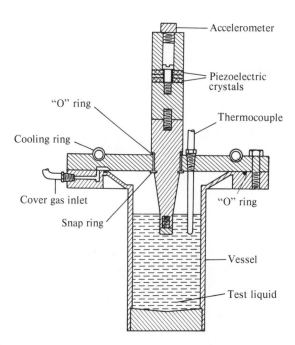

Figure 5-12 Vibratory cavitation-damage facility at the University of Michigan. (Patented seal arrangement.)

only a small clearance from the vibrating horn, this stationary-specimen test geometry is particularly useful for the testing of materials for bearings, since the bearing geometry is well modeled even though the effects of velocity are absent.

5-1-5 Conclusions

Fluid-induced erosion, both single- and multiphase, has been considered according to the various phenomena from which it may be generated. These include both simple high-velocity single-phase flows and also liquid- and solid-particle impact, as well as cavitation. These latter phenomena are considered as multiphase in nature. It is concluded that in cases of engineering interest there is no plausible mechanism for single-phase erosion of relatively strong materials unless essentially multiphase phenomena as droplet or particle impact and/or cavitation are present. Finally, the various types of erosion testing devices are considered and described as to their range of utility, limitations, and relative merits. Further discussions of some of these test devices are provided in Chap. 8.

5-2 BASIC CAVITATION-DAMAGE MECHANISMS AND STATE OF THE ART

5-2-1 Introduction

It is generally recognized today that the flow phenomenon called "cavitation," involving a generally heterogeneous mixture of vapor and gas pockets or "voids," some of which can be approximately described as bubbles, frequently causes a rapid erosion of adjacent material structure. This is often much more rapid than would be expected for single-phase flow erosion or corrosion for the same conditions of velocity, temperature, and turbulence. While rapid erosion or damage is only one observable result of this flow condition, it is the one with which this chapter is primarily concerned.

There is at present an enormous body of research literature concerned with the processes of bubble collapse and cavitation damage. This has accumulated at an increasing rate since the pioneering work of Rayleigh [31] in 1917. It is the purpose of the present section to briefly summarize the significant results of the very considerable research which has been concentrated on this problem over the years, both from the viewpoint of basic understanding of the phenomenon and that of practical information of use to the designer of fluid machinery.

In spite of more than half a century of research there is still a very incomplete understanding of the mechanisms by which a "cavitation field" causes rapid damage to adjacent solid material. Hence, it seems logical to start this discussion by listing those points which seem to be clearly and directly based upon observations or experimental measurements, and thus appear at present to be essentially indisputable. These basic experimental observations lead to simple theoretical concepts from which apparently clear and incontrovertible ideas are generated. From this point of common agreement, more-complex theory and less-definite experimental observations lead to more speculative concepts regarding the detailed mechanisms of cavitation damage, until an area is reached where no general agreement exists. For previous partial summarizations of similar material see Refs. 6 and 32.

The foregoing relates primarily to the basic understanding of the phenomenon of cavitation damage, considered primarily from the viewpoint of fluid mechanics rather than material reaction. So far, this basic approach has produced only scattered practical information of utility to the fluid machinery designer. Hence, as in many fields of engineering, it has been necessary to formulate semiempirical relationships and general rules for the use of the designer, using the specialized test results which have become available. Generally these rules attempt to predict, with varying degrees of precision, the effect of various independent variables upon cavitation damage in real situations. In a fully developed form such relations must involve both fluid and material behavior, since significant coupling between these often exists in the real situation. This body of semiempirical knowledge will be very briefly reviewed and avenues for useful future research in these areas discussed.

5-2-2 Commonly Agreed Basic Principles Related to Cavitation Damage

Primary experimental facts and conclusions   It is first necessary to review various well-known indisputable experimental facts upon which the consideration of cavitation damage must be based. These are primarily the following.

General observations
1. Rapid pitting and erosion often occur in flows where cavitation is observed to exist. Its existence can be determined audibly by acoustic instrumentation, visually if windows in the containment systems are provided, by machine vibrations, or through decrease or other change in performance from the single-phase flow condition. For example, a measurable decrease in head is produced from a cavitating centrifugal pump for a given flow and rotating speed. If the cavitation region in a fluid machine is observed visually, it appears as a "frothy" region. If optical instrumentation of suitable time and space resolution is used, it is found that the "frothy" region is actually composed of a heterogeneous mixture of odd-shaped "voids," many of which are roughly spherical bubbles. In some cases a relatively clear cavity attached to the structure is found, but this is then often surrounded and followed by a "frothy" region of traveling "voids." In a cavitating flow the rate of attack can be many times that due to erosion and corrosion alone in the absence of cavitation.
2. Cavitation can damage, under certain conditions, even the strongest of materials such as stellites, tool steels, and any other known structural materials. This damage can occur rapidly, even in cases where chemical corrosion in single-

phase flow with the same liquid-material combination would not be significant, e.g., cavitation in petroleum products on metals or glass.

3. Cavitation pitting shows the characteristics of mechanical attack. Such well-known mechanical manifestations as, for example, slip lines in metals have frequently been observed. The single craters which are formed in the early stages of the attack appear under a low-power microscope as "moon craters," i.e., more or less symmetrical craters often with a raised rim, as if formed by single impact rather than corrosion. In fact, damage to materials from liquid-impact tests closely resembles cavitation damage both qualitatively and quantitatively.

4. Mechanical cavitation attack and corrosion can supplement each other through obvious mechanisms resulting in a damage-rate increase, in cases where both are important, to many times the sum of damage rates from corrosion and cavitation acting separately and independently.

5. An important theoretical contribution to the development of the concept of the mechanical cavitation-damage mechanism was given by Rayleigh [31]. He showed (see also Chap. 4) that a collapsing spherical vapor bubble has the potential for generating extremely high pressures and velocities in the fluid near the point of collapse. The original analysis, based entirely on ideal-fluid concepts including that of spherical symmetry, shows that these quantities, i.e., pressure and velocity, become infinite. Thus, while more realistic assumptions are required to evaluate pressures and velocities quantitatively, it is apparent that the possibility exists for values large enough to be damaging even to very strong materials.

Certain obvious and important conclusions can be drawn from the general observations noted above. They primarily apply to cases where mechanical effects predominate, although, as has already been mentioned, it is evident that if corrosion effects are significant they too can add greatly to the overall damage rate.

Conclusions from general observations

1. Since observed cavitation fields usually contain large numbers of essentially spherical bubbles of various diameters, and since as Rayleigh showed [31] the collapse of such bubbles could create pressures and velocities large enough to be damaging, it is likely that the surface of a material exposed to cavitation will experience a multiplicity of impulse impositions of widely varying intensities and with locally random spatial distribution. The Rayleigh theory [31] shows that the duration of imposition of such impulses due to individual bubble collapses is extremely short. Furthermore, the impulse magnitudes and collapse times are greater for larger bubbles for a given collapsing pressure differential. Since individual symmetrical craters are observed, it is apparent that some of these impulses are sufficient to cause permanent material deformations. Since the spectrum of impulses varies widely, it is to be expected that individual craters with diameters covering a given range will be formed (as has been observed [33–35]) and that in fact many "blows" (i.e., impulses) may be of

insufficient strength to cause permanent material deformation. A large number of these weaker blows, however, may be sufficient to also contribute to eventual fatigue failure. Thus it is to be expected that cavitation damage will often eventually take the form of fatigue failures, and this is in fact observed. The concept of a spectrum of blows resulting from a spectrum of bubble sizes and locations is well summarized by Fig. 5-13, previously published by the author to describe cavitation damage in a venturi [34].

2. As the surface roughness increases due to accumulated cavitation (or corrosion) damage, the flow pattern near the surface will frequently be importantly altered. In addition the substantial cold-working of the material surface may affect its ability to resist further damage (increased strength and hardness will tend to increase its damage resistance while increased brittleness will have the opposite effect). Thus it is to be expected that the rate of cavitation damage in a given situation will not be constant with time. Often an "incubation period" is observed before substantial material loss occurs, presumably while fatiguing processes proceed to a point necessary to cause failure. The damage rate then

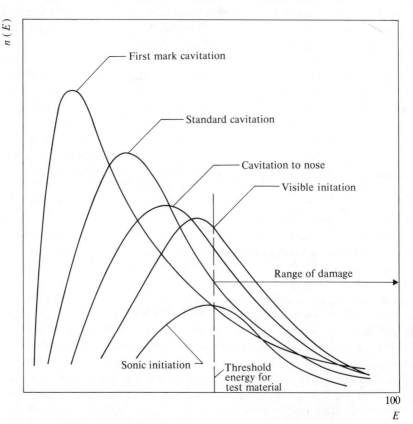

Figure 5-13 Hypothesized bubble energy spectra for various cavitation conditions at a constant velocity, for a given material in U-M venturi [34].

often increases to a maximum, after which it decreases. Later secondary and tertiary, etc., maxima may occur. This behavior probably depends primarily upon the interplay of flow-pattern alteration by virtue of accumulated roughness and material surface-property changes which are themselves due to the accumulated permanent deformations and stressings. This general situation is well described in Refs. 5 and 156.

More speculative observations and conclusions It will be noted that the foregoing generally agreed concepts relating cavitation damage do not attempt to specify the detailed mechanism whereby bubbles collapsing in the liquid continuum can damage materials submerged in or containing the liquid. It has been stated already that a simple ideal-fluid analysis [31] shows that very high pressures and velocities may exist for a very short time period over a very small space around the center of collapse of the bubble, if this collapse proceeds with spherical symmetry. The detailed mechanisms whereby the material surface of even very strong materials can be stressed sufficiently to cause damage was left ambiguous, since this is an area where at present agreement does not exist. However, certain additional items of experimental and theoretical evidence can be presented to throw some light on the probable mechanisms involved. Some of these additional items are discussed below.

1. In typical situations only one out of perhaps ten thousand bubbles seen to collapse close to a surface actually causes a crater [35, 36], though such craters, judging from their symmetry and unchanging contours with additional exposure [34], result from individual bubble collapses. However, in very carefully controlled laboratory conditions, even in a flowing system such as a venturi, a one-to-one correspondence between bubble collapses and observed craters can be attained if all parameters are adjusted precisely and correctly [37, 38]. These facts seem to indicate that some very selective mechanism is involved in delineating damaging from nondamaging bubble collapses. The model of symmetrical bubble collapse provides only bubble diameter and wall distance as "sorting" parameters, while an additional sorting parameter exists, i.e., orientation, if the collapse is nonsymmetrical. Intuitively, this latter factor seems more consistent with the very large ratios of damaging to nondamaging collapses which are in fact observed.
2. The collapse of bubbles with approximate spherical symmetry through a radius ratio, i.e., change in radius, sufficient to generate damaging pressures according to the Rayleigh analysis [31], close enough to a material surface so that damage might occur, and in real flow situations involving pressure and velocity gradients, turbulence, etc., seems unlikely. This statement is based upon excellent photographs which have been obtained quite recently [37–42] as well as theoretical analyses [8, 9, 43–48]. Wall effects, pressure gradients, gravity, and initial motion are all sufficient to radically change the mode of collapse from one of approximate spherical symmetry to an approximately toroidal collapse wherein the bubble is apparently pierced by a small microjet of liquid

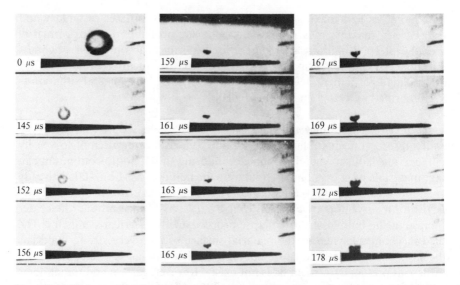

Figure 5-14 High-speed motion pictures. Selected frames from a sequence taken at 550,000 frames/s of a spark-generated cavitation bubble near a splitter in a venturi, exposure 1.8 μs per frame, flow right to left, magnification ×5 [37].

before the bubble volume has been reduced by more than a factor of 10 to 100. This type of collapse is shown clearly in recent pictures (Fig. 5-14) obtained in the author's laboratory [37, 38] for bubbles collapsing in a venturi adjacent to a knife-edge which is aligned parallel to the flow.

3. Theoretical analyses of bubble collapse assuming spherical symmetry as well as real-fluid parameters such as viscosity, surface tension, and compressibility [8, 9] indicate that the pressures around a collapsing bubble at a minimum distance of the initial bubble radius from the center of collapse, i.e., the minimum possible distance from the collapse center to the wall to be damaged if the collapse center is stationary, are not sufficient to explain the damage observed on most materials. However, if the bubble rebounds from a minimum-volume condition, as it theoretically would if it contained some noncondensed gas or vapor, the calculated pressures are greater in the surrounding liquid, so that damage from this mechanism becomes more likely [8, 9]. Also, theory indicates that the center of a collapsing bubble will tend to move toward an adjacent wall (or away from a free surface) during collapse. No realistic analysis has yet been made to show how important such movement might be in the consideration of damage mechanisms, though an analysis assuming the bubble to remain spherical during the motion [49] shows a significant effect upon the pressures exerted on the wall.

Motion picture sequences of collapsing bubbles often show rebounds, i.e., growth of the vapor mass after passing through the minimum volume condition, though not usually as spherical bubbles [37]. A rebound would be theoretically expected if the bubble contained trapped gas, or vapor which behaves as gas during the very short critical portion of the collapse, which may last only a few microseconds. Also, in some cases the pressure distribution around a bubble moving at considerable velocity relative to the fluid [44] may have the same effect. However, a rebound is even more likely if the bubble collapses in a toroidal mode, since the centrifugal pressures around the vortex ring provide a restoring mechanism in much the same fashion as trapped gas.

Individual craters caused by cavitation are often very similar to craters generated by the impact of a high-velocity liquid. This similarity is especially striking if one compares the craters formed by actual liquid-droplet impact on Plexiglas (Fig. 5-15) with those from cavitation in a venturi (Fig. 5-16) obtained in the author's laboratory, since the damage pattern for droplet impact on Plexiglas is rather unique, having an undamaged center area surrounded by an annular failure area. Another crater configuration illustrative in this

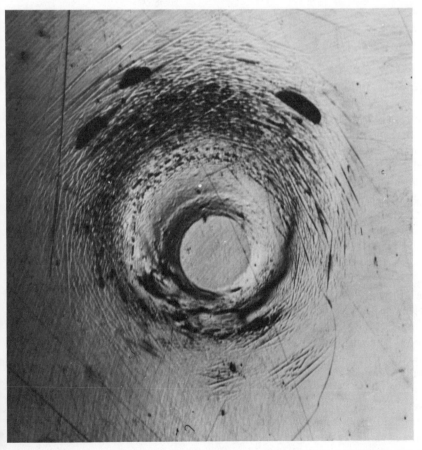

Figure 5-15 Water-droplet impact crater on Plexiglas, magnification × 100 (after A. Fyall, RAE, Farnborough, England).

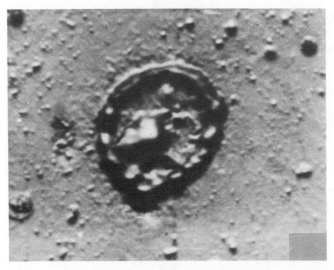

Figure 5-16 Crater produced by cavitating water in University of Michigan venturi on Plexiglas, magnification × 4,000.

respect was obtained in the author's laboratory from venturi cavitation upon stainless steel plated with a very thin coating of cadmium (Fig. 5-17). The cadmium is completely removed in the center so that the underlying stainless steel is exposed. This region is surrounded by an annular area where the cadmium is partially removed, suggesting the impact of a liquid jet which then accelerates radially after impact (a common observation for the impact of actual liquid jets [50]) and "washes" away the thin cadmium plate. It is difficult to imagine a similar result caused by the imposition of a spherical shock front which would merely press the cadmium plate deeper into the surface, leaving a crater within which the cadmium plate would remain on the surface. This was confirmed at the author's laboratory in an experiment wherein hard steel balls were impacted upon the same surface at high velocity (~ 100 m/s).

The foregoing points all indicate the strong probability that in most engineering situations the mechanical portion of cavitation damage is due more to the impact of a high-velocity microjet upon the damaged surface than to the imposition of shock waves in the liquid emanating from the center of collapse of a spherically collapsing bubble, as Rayleigh [31] assumed. However, even if the bubble does not collapse spherically, the liquid pressures will rise substantially around the bubble as the volume is significantly reduced from the initial value. Intuitively, the pressure rise will be less, the greater the departure from spherical symmetry, and the collapse will generally be slower, although local pressure rises due to the asymmetry might be greater in the nonsymmetrical case. Also, a given diminution of volume in an ideal fluid will produce the same total kinetic energy in the fluid for both symmetrical and nonsymmetrical cases, since this is equal to

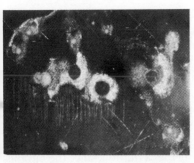

Coating thickness: 2.5×10^{-5} in.

Figure 5-17 Craters produced by cavitating water on 0.6 μm cadmium-plated stainless steel, magnification × 180, University of Michigan venturi.

$(p_\infty - p_v) \Delta V$† for the bubble [51]. Unfortunately, little quantitative information pertinent to this highly complex problem yet exists. Nevertheless, in real cases, if liquid microjet damage appears to be the most important mechanical damage component, it is also likely that the pressure rise around the collapsing bubble caused by the variations of bubble volume according to the Rayleigh mechanism may also be an important damaging mechanism. It must be admitted that these two schools of thought on the relative importance of likely damage mechanisms exist at present.

An indication that very high pressures (and hence temperatures) do sometimes exist within collapsing bubbles is afforded by the observation of "sonoluminescence." An indication that thermal effects are of substantial importance in this phenomenon has been provided [52, 53]. However, there is no indication [54, 55] that bubble collapse through the large radius ratio required to compress trapped gas to the extent necessary to cause luminescence can occur near enough to material surfaces to cause damage, since such a complete collapse requires a highly symmetrical environment. Schlieren and interferometric pictures have succeeded in showing strong density gradients or shock waves in the liquid around collapsing bubbles [54, 55].

Another manifestation of cavitation damage which is sometimes observed that is difficult to explain either in terms of shock-wave effects or microjets is that of "worm-hole" pits, i.e., very deep curving pits of a large length-diameter ratio. While chemical effects are probably predominant in this phenomenon, they may also be the result of (a) a wave-guide effect [46] which tends to amplify shock-wave pressures generated in the liquid or (b) a microjet mechanism wherein the jet is repeatedly generated across a liquid-gas interface at the bottom of the pit where a vapor-gas mixture may be trapped. A concave surface, conducive to the generation of such jets, would be formed if the liquid wets the walls of the "worm-hole." The jet could conceivably be triggered by the imposition of pressure loading either from shock waves or jets at the outside end of the worm-hole [56].

The foregoing items indicate that many points relating to the actual

---

† p_∞ is upstream pressure and p_v is vapor pressure.

mechanism, even of purely mechanical damage, remain unresolved and that there is no general agreement, even on the type of event which is occurring. However, certain apparently relatively firm conclusions can be drawn.

1. It seems obvious that mechanical cavitation damage is the result of the highly transient imposition of very intense and highly local forces on the surface. Since these are associated with bubble collapse rather than bubble nucleation, it is apparent that the damage occurs in the collapse region. Hence, modification of the flow geometry in a damaged region of a fluid machine to prevent further damage will be usually ineffective in eliminating the cavitation, since the bubbles initiate at some point upstream and only collapse in the damaged region. Obvious as this sounds today [6], it has at times been disputed in past literature.

   The fact that cavitation loading on a surface is very transient and local is important in the selection of protective coatings or surface treatment. It is also important in the attempt to correlate cavitation-damage rates with material properties, since the mechanical properties of materials measured in the ordinary fashion do not reflect accurately the resistance of the same materials to the highly transient loading encountered in cavitation.

2. According to present evidence, the cavitation-induced loading on a material surface results from a combination of liquid shock-wave effects generated primarily during the rebound† of a bubble, with the impact of a high-velocity liquid microjet directly on the surface to be damaged. Such a microjet is generated when the bubble collapse becomes substantially nonsymmetrical, and in such cases the magnitude of shock waves during collapse (as opposed to rebound) is probably reduced. Photographic evidence shows that approximately symmetrical collapses are the exception (or perhaps even an impossibility) in the vicinity of solid surfaces and/or in regions of strong pressure and velocity gradients. Considering the jet impact mechanism in further detail, it is clear that shock waves within the jet itself are important in the impact phenomenon, and local cavitation around the point of jet impact may also be a damage mechanism. This appears to be so for large jets which damage relatively strong materials at surprisingly low impact velocities.‡ This damage may be at least partially the result of local cavitation as the jet is deflected around the target specimen.

Additional research for basic understanding of damage mechanisms   It is apparent from consideration of the foregoing that many years of additional research may be required to fully delineate the presently rather sketchy picture of the cavitation-damage mechanisms. Such basic studies could well consider the following areas in which more precise information is required.

---

† Volume increase after initial collapse.

‡ Type-316 stainless steel was quickly eroded by a 5-mm water jet impacting at only about 100 m/s [57] in tests conducted for an ASTM study.

Detailed bubble-collapse behavior   Powerful tools are becoming increasingly available today which are useful in this respect, such as ultra-high-speed photographic equipment and other sophisticated optical techniques. Since the critical part of bubble collapse requires only a few microseconds and involves an object only a small fraction of a millimeter in diameter, it is clear that extremely sophisticated photographic equipment is required. Another possible method for obtaining new and useful information is holographic photography with a nanosecond light pulse. This technique can be combined with Schlieren photography or differential interferometry, hopefully to show density gradients or shock waves in the liquid. Such shock waves have already been shown by Ellis [54] and Lauterborn and Bolle [55], for example.

Other important measurements which could be made around a collapsing bubble include local temperature distributions, using a microthermocouple probe [58], and acoustic output from bubble collapse, discussed in more detail later. Several measurements of this general type have been made, but more precise information and correlation with other conditions of the experiment would be useful. Another good possibility which has been employed to some extent is the direct measurement of the peak pressures exerted upon a specimen by an adjacent bubble collapse. Some excellent measurements of this last type have been reported [59, 60], but additional information would be desirable.

The effect of fluid properties, flow-field parameters, and wall effects   These include deflection under bubble-collapse loading (discussed later) on bubble kinetics. It would be very desirable to know the effects of pressure and velocity gradients, boundary-layer parameters, etc., on the very complex chain of events apparently necessary to produce a damaging bubble collapse. If more detailed information of this type could be achieved, it might become possible to modify the design of fluid-handling machines in such a way that cavitation damage would be largely avoided or reduced. Along this same line, it might eventually become possible to measure the size and number distribution of the gas nuclei upstream of a cavitating region and, knowing the flow patterns approaching the region, predict the cavitation bubble distribution within the region. If the damage mechanisms were understood to the extent necessary to predict the required size, location, and orientation of damaging bubbles, it would then be possible to predict the rate of damage to be incurred from a given flow situation. If this were possible, it might then be only a small additional step to modify the flow-path design in such a way that damage would be grossly reduced or avoided entirely. A few studies have appeared which attempt to predict the trajectory and distribution of bubbles from an initial distribution and given flow regime [61, 62].

5-2-3 Semiempirical Results of Utility to Design Engineers

Various groups of information have been accrued through the many years of cavitation experimentation which can be summarized in a form useful to the design engineer. Some of these are discussed below.

Basic information

Effects of flow velocity and pressure   In many flowing devices such as rotating discs [25], jet impact devices (which have been used to study "cavitation," since it was observed that damage produced by such impact and true cavitation damage were very similar), tunnel devices using separated flow past a pin such as that pioneered by Shal'nev [63], or flow over an ogive as used by Knapp [6, 36], it has been observed that damage rates are proportional to a relatively high power of the velocity. The sixth power was suggested by Knapp [36], and this seems fairly representative for the damage obtained with these types of devices. Later tests have shown that the exponent varies with many factors such as accumulated damage [25, 33, 34]. It appears that no general rule is possible, since there is an interrelation involved between velocity and pressure in the collapse region that depends upon the actual flow regime. It seems clear theoretically that pressure is actually the primary variable. If pressure in the collapse region increases rapidly with velocity so that collapse intensity is strongly increased, then damage rates should be very sensitive to flow velocity. On the other hand, if the pressure in the collapse region is not affected by velocity, as in a conventional venturi, then damage may not increase strongly with velocity. Such a condition was observed in the author's laboratory [34]. Nevertheless, from the designer's viewpoint it must be recognized that damage rates may increase very rapidly with velocity, so that a small increase in velocity may convert an otherwise nondamaging, but nevertheless cavitating flow into one of substantial damaging capability. In cases where there is little or no cavitation, a very small increase in velocity may cause a cavitation field to form, with subsequent significant damage. In such a case, the "velocity exponent" could be very large. Although cavitation damage is primarily pressure dependent, it is the velocity fields which are directly under the control of the designer.

Effects of gas content in the fluid   There appear to be two opposing effects. If total gas content is increased it is likely that entrained gas, generally thought to be most important (as compared to dissolved gas) for bubble nucleation, will also increase. There should then be more cavitation bubbles produced for the same pressure, temperature, and velocity, i.e., same "cavitation sigma," conditions. Thus damage should increase. On the other hand, if the cavitation bubbles actually contain a higher quantity of noncondensable gas, the bubble collapses are restrained and reversed at a larger radius than otherwise, so that the resultant pressure waves in the liquid are reduced in amplitude. The analogous effect on the microjet collapse mechanism is less clear, and intuitively appears to be less important. Still, for either mechanism, damage would be reduced. The interplay of these opposing trends is uncertain in the general case, but experience appears to indicate that large quantities of injected gas do indeed substantially reduce cavitation damage. These trends are discussed in further detail in Ref. 64 and also in Chap. 3.

Effect of cavitation number   The effect of cavitation number $[\sigma = 2(p_\infty - p_v)/\rho V_\infty^2]$ upon cavitation-damage rates is similar to that of gas content, in that much the same opposing trends are evident. If cavitation number is increased for a given flow situation (e.g., by raising the pressure and maintaining constant velocity), the number and mean diameter of bubbles will be decreased but their collapsing pressure differential will be increased. Thus collapse violence will be increased, although the number of bubbles will be reduced. Hence, it is conceivable that a slight rise in the cavitation number, if accomplished by a rise in the pressure at constant velocity, could cause an increase in damage. This has in fact been reported for various vibratory test facilities [65–68]. It is, of course, clear that a sufficiently large pressure increase will cause a reduction in damage, since cavitation will cease entirely if the pressure is raised sufficiently.

If the cavitation number is increased in a given situation by reducing the velocity and maintaining constant pressure, the general evidence related to a velocity effect, already discussed, indicates that the damage will probably be decreased.

Effect of fluid temperature at constant cavitation number   If fluid temperature is raised at a constant cavitation number, the density of vapor within the bubbles is increased. Due to thermal restraints, bubble growth and collapse are inhibited. Thus the mean diameter of bubbles may be reduced, tending to reduce cavitation damage. Due to the thermal restraints upon collapse velocities, damage is also reduced. The overall mechanism, described as the "thermodynamic effect" [69–72], involves a thermal restraint upon bubble collapse and growth due to a potential alteration of the vapor temperature in the bubble. This will become actual, and affect the pressure in the bubble, if heat transfer in the vicinity of the bubble is not adequate to transfer quickly enough the latent-heat component involved in either bubble growth or collapse. The resultant vapor-pressure change within the bubble is in such a direction as to inhibit either growth or collapse.

The effect of increased fluid temperature in reducing cavitation damage has been adequately demonstrated in vibratory cavitation-damage tests [65–68, 72–73], where it is shown that damage decreases rapidly for temperatures larger than the approximate average of melting and freezing temperatures for the liquid. Figure 5-18a demonstrates the effect from tests in the author's laboratory [72], and somewhat similar data were also reported earlier [73]. The decrease of damage rate at temperatures approaching the boiling temperature in such a test is, of course, also partially a result of the decrease of the collapsing pressure differential as the fluid vapor pressure is increased appreciably, though the same effect is observed when static NPSH is maintained constant [65–68] (Fig. 5-18b). In all cases, the decrease in damage rate becomes significant at temperatures below those for which the increase in vapor pressure seems significant.

A study carried out at the author's laboratory of this effect, using a variety of fluids including water and high-temperature liquid metals but conducted at constant pressure with a vibratory cavitation-damage device, showed that damage was not affected by thermal restraints as long as a modified "thermodynamic

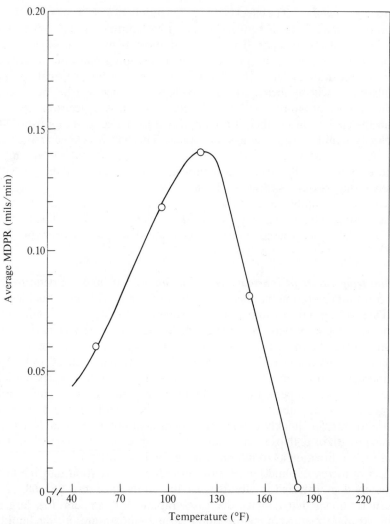

Figure 5-18 (*a*) Effect of temperature on average damage rate for 304 stainless steel cavitated in water at one atmosphere pressure. University of Michigan tests, Garcia [72, 81].

parameter" was in a range typical of low vapor pressure fluids, but that damage became very strongly reduced when this same parameter reached values typical of fluids such as hot water [72]. The correlation presented in Ref. 72 is not entirely satisfactory, and somewhat more-complex relations have since been presented by Bonnin [70, 71] using two "thermodynamic parameters."

Material-property effects—general Cavitation-damage rates are, of course, very strongly affected by material properties, but no generally applicable relations appear to exist. The subject is extremely complex and cannot be treated in adequate detail here, although additional information is given in Sec. 5-3. While any body

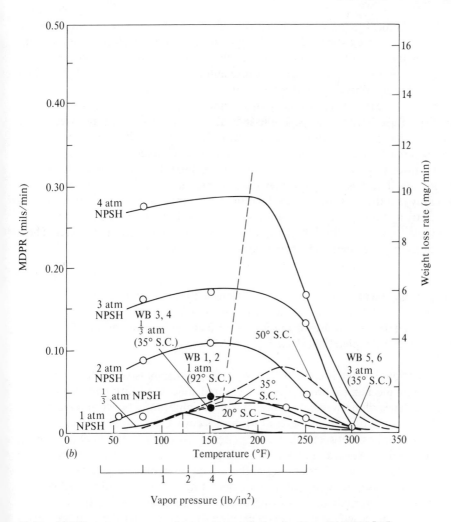

(b) Max MDPR vs. temperature and vapor pressure for bearing brass (SAE 660) [67].

of damage data on various materials can be adequately fitted by a sufficiently complex expression combining various material properties, there is little likelihood that such an expression will fit new data unless it is based on adequate physical reasoning. In addition, such complicated empirical expressions are likely to be too complex for general utility. On the other hand, relatively simple expressions do not adequately fit the data unless a probable error factor in the order of 3 times is considered satisfactory. This general problem has been explored recently in the author's laboratory [72, 74] and elsewhere [75–77] using very comprehensive sets of data generated both in vibratory and venturi facilities. As a result of these studies and others, it appears that the statistically best-overall correlation with a single mechanical property for metals is obtained in terms of "ultimate resilience" = (tensile strength)^2/elastic modulus, as originally suggested by Hobbs [68], or with other relatively similar combinations of these terms. Ultimate resilience is the energy per unit volume necessary to cause failure, if the failure were of the brittle type, so that ductility apparently does not play a very effective role.

Since the earliest days of cavitation-damage investigations, it has been the practice to use hardness as a simple indicator of probable cavitation resistance of a material. For brittle materials, particularly, hardness still seems perhaps the most suitable index. Furthermore, it appears to be relatively generally applicable within groups of materials of the same general type. It is further recommended by the fact that it is comparatively easy to measure for a given material. It is probably still the most widely used parameter for this purpose, and, as shown by the relatively complex statistical studies mentioned above, its predicting value is not substantially worse than that of any other mechanical-property parameter. These questions are discussed further in Sec. 5-3.

Complicating factors   Further complications in the attempt to predict cavitation-damage resistance of a new material in terms of easily measurable mechanical properties exist as a result of the following facts, even if the effect of corrosive influences are completely neglected.

1. Cavitation damage is created by a very transient impulsive loading, while the standard mechanical properties are measured under semistatic conditions. Time dependence of mechanical properties differs significantly between materials.
2. The response of a given material to pressure waves or liquid impact involves some coupling between fluid and material properties. A suitable coupling parameter is presently not known, and perhaps must await the development of a more precise understanding of the cavitation-damage mechanisms.
3. The material to be damaged in most operating machines is under substantial stress in its normal mode of operation. The cavitation-induced stresses are thus superimposed on the already existing stresses, so that the resultant stress regime depends upon its initial state of stress. Thus damage rates in operating components may not be entirely predictable from tests on unstressed specimens. Some initial work on this problem has been done in the author's laboratory [78, 79] and also by Shal'nev [80].

In addition to the above, and perhaps even more important, are the possible effects of corrosion on the otherwise mechanical damage problem. It is primarily this effect which often renders laboratory cavitation-damage tests inapplicable to prototype field conditions, since scaling of the correct "mix" of mechanical and corrosive effects in accelerated laboratory tests is almost impossible to achieve. The problem of prediction of erosion resistance from laboratory tests will be discussed in Sec. 5-5.

Useful additional research   It is clear from the foregoing discussion that there are many avenues for additional research which would be useful in assisting the equipment designer. However, it is difficult to list a limited number of especially desirable studies, as was done in the case of basic research, since the results to be expected from applied research of the type involved here are generally highly specialized to the actual geometry, etc., which was used. While the gap between the basic understanding of the cavitation-damage process remains as large as it is at present, so that the practicality of predicting engineering results from basic principles remains small, it is clear that many applied investigations will continue to be necessary.

5-3 MATERIAL-PARAMETER EFFECTS

5-3-1 Introduction

Over the past half century very much experimental data upon cavitation and liquid-droplet impingement erosion rates for numerous materials and with various liquids over a range of temperatures and pressures has been accumulated. This is reported in articles too numerous to list here, but is summarized up to the mid-1960s in Ref. 6. However, in general it has not been possible over this extended period to provide good and usable correlations between measurable material and/or fluid properties and measured erosion rates. It has also not been possible to predict with reasonable engineering precision erosion rates in field devices from laboratory tests. While this is still generally the case, some meaningful progress has been made, and this will be reported in the present section. A portion of the pertinent research was done in this laboratory [5, 32, 78, 81].

One of the major objectives of much past and present erosion research, either cavitation or impingement, must be to establish a mathematical model with fluid flow and material parameters as input data, which would allow the engineering prediction of erosion rates for given, as yet untested, materials. A precise model of this sort has so far, however, eluded investigators. This appears to be inevitable in view of the highly complex and varied nature of the erosion processes, even though produced by droplet impingement or cavitation, for example, alone. Nevertheless, it is desirable, using a large and diverse group of data, to attempt to determine optimum correlation relationships and also to determine roughly what degree of precision can be expected from correlation models using easily measured standard engineering parameters as input data. Fairly comprehensive sets of data, including both impingement and cavitation results, have been used for the study reported here. This combination of data seems reasonable due to the presumed basic similarity of the erosion phenomena from impingement and cavitation, as discussed elsewhere in this book. The model chosen for discussion here has been made dimensionally consistent and as simple as possible, in the hopes of obtaining maximum generality and applicability. This objective is also enhanced by the use of a diverse data set including items generated in different impact and cavitation-type tests.

5-3-2 Mathematical Model [74, 82]

The relatively poor fit achieved so far in correlating measured erosion data with material properties indicates the desirability of a predicting model based closely on the details of the physical processes involved. However, such a desirable resolution has not yet been attained.

The best hope of achieving a relationship of the generality necessary to allow possible applicability over a broad range of materials requires a relation which is directly related to a physical model of the erosion process, is dimensionally consistent, and is as simple as possible. While it will be possible often to achieve a better fit for a given data set with more complex mathematical expressions, its utility in fitting other data sets has always been poor. Following this line of reasoning, the basic energy flux model suggested by Hoff, Langbein, and Rieger [82] may be useful. This model assumes simply that the product of the rate of volume loss per unit of exposed area (MDPR) times the exposed area (A_e) is proportional to the product of the impacting kinetic energy per unit of projected area and the projected area. The constant of proportionality is the quotient of the efficiency of energy transfer between the impacting-drop and material-damage processes (η) and a material parameter (ε) describing the energy per unit material volume absorbed in the material in such a way as to cause damage. This relation is expressed as

$$\text{MDPR} = \frac{\eta}{\varepsilon} \frac{A_p}{A_e} \frac{\rho_{\text{eff}}}{2} V^3 \tag{5-4}$$

This equation attempts to compartmentalize and rationalize the effects involved. It also shows the real complexity of the phenomenon and the extreme difficulty of a reasonably rational analysis. The evaluation of the material parameter ε in terms of mechanical material properties, as well as of the effect of the energy-transfer efficiency term η, which depends upon material and fluid parameters, are areas in which some progress may be possible.

Incidentally, for implementation of Eq. (5-4) it is necessary to use only data wherein the total MDPR versus exposure curve is available so that only comparable portions of this curve will be compared. Precise definitions of the pertinent parameters here required are given in Ref. 83. It is by now well recognized, as perhaps first emphasized by Thiruvengadam [84], that damage is *not* linear with time. For the present purpose, data from various types of facilities, both impact and cavitation, have been compiled together and used to evaluate and generate "best-fit" erosion-predicting equations.

The efficiency η of energy transfer is influenced by several factors, and may perhaps be considered as a product of several separate terms reflecting each of these mechanisms. Considering the details of the jet impact process (assumed applicable to both cavitation and droplet impact) between a liquid drop and a material surface, η should be a function of (*a*) material and liquid properties, perhaps as reflected by the acoustic impedance ratio [85] between material and liquid, (*b*) geometrical factors involved in the collision, i.e., shape of impacting drop, angle of impact, surface roughness, etc., and (*c*) velocity of impact which will affect the pressure applied to the surface, and hence the degree of surface deformation and the departure from the concept of an elastic material. Since material and liquid properties involve no other parameters of the collision, we have lumped their consideration into that of the energy parameter ε, assuming as a first approximation that this portion of the efficiency term may be some function of the acoustic impedance ratio. No comprehensive attempt has yet been made to evaluate the remaining portions of η, but some possibilities are discussed next.

5-3-3 Evaluation of Energy Parameter ε

General remarks From the foregoing, it is desired to find a material mechanical property with units of energy per unit volume having the characteristic that for a given test (impingement or cavitation) with fixed test parameters (velocity, fluid conditions, geometry, etc.), the product of MDPR $\cdot$ ε will be as nearly constant as possible over a broad range of test materials. The material property must appear only to the first power. Thus a polynomial expression for the energy term will not suffice, since this would destroy the dimensional consistency of Eq. (5-4). Also, to be of use in a predicting equation, the energy term must be measurable in a simple mechanical test. Hopefully it would already be available in the literature for most standard materials.

Among the parameters meeting these conditions are those energy terms which can be computed from the standard stress-strain curve. Our own previous work [81] and, more importantly, that of Hobbs [68] and Rao et al. [77] (all, incidentally, for cavitation tests) and of Heymann [76, 86] (for a combined data set) suggest that the best single-parameter correlation is to be found with "ultimate resilience" = (tensile strength)^2/2E, i.e., the area under the elastic portion of a stress-strain curve if elastic† strain were continued up to the full tensile strength (TS). Thiruvengadam [84], on the other hand, has reported that the best fit is in terms of "strain energy" (SE, the area under the complete stress-strain curve of a material). This latter parameter can be evaluated either as the "engineering strain energy" (ESE), i.e., the area under the conventional stress-strain curve where tensile strength is computed from the observed breaking load without consideration of reduced area, or "true strain energy" (TSE), where actual breaking stress is used. Approximations of both have been used for the present purpose.

For ultimate resilience (UR), for simplicity, the observed breaking load only has been used, since for many materials reduction of area data is not available. Also, our own previous work [81, 85] indicated this to be preferable.

† E = elastic (Young's) modulus.

Another parameter which can be considered approximately as an energy per unit volume, and which is easily available for most materials, is hardness. This also has been considered (Table 5-3) as a possible correlating parameter in the form of Brinell hardness, but is difficult to use in a "rational" damage equation.

Selection of data for evaluation [74]   To achieve maximum applicability and generality, as broad a group of data as possible has been used for evaluation of the "material coefficient" of ε/η from Eq. (5-4), including some data from cavitation tests and also from impingement tests. It was also necessary that accurate stress-strain curves for the materials used be available. In addition, the damage data must be such that the entire MDPR versus time curve be available so that a comparable portion of this curve could [72, 81] be used in all cases. Consistent with our own previous practice and that of Hobbs [68], as well as following generally adopted procedure today [87], maximum MDPR has been used as the characteristic value for the material. The largest single portion of our present data set is that generated by our own vibratory cavitation facility in water following the present ASTM standard conditions [26], i.e., 24°C, 20 kHz, 2 mil double amplitude. Other data have been incorporated into the analysis only when tests were available for at least one common material, preferably from identical material specimens from the same bar stock (if at all possible).

A ratio between maximum MDPR for the common material in the differing tests or facilities and the additional materials tested in the other facility (or test condition), normalized to the common material in our own vibratory facility, was established. Thus values of these comparative amplitude constants are relevant to our particular vibratory facility. In this manner, it is possible to incorporate data from various types of tests, since the efficiency factors involving test geometry and velocity are thus removed from consideration. Data from the following sources, in addition to our own vibratory cavitation data, have been used:

1. Impact tests by King of the Royal Aircraft Establishment [88] on Dornier rotating-arm facility.
2. Impact tests by Electricité de France [57, 87] on rotating wheel.
3. Venturi tests by Rao et al. [77].
4. Vibratory cavitation tests in the author's laboratory [79] using a stationary specimen positioned in close proximity to a vibrating horn (same horn is also used in standard set-up).

The materials used and their mechanical properties are listed in Table 5-3.

Best-fit results attained

Predominant mechanical property ε   Previous work here and elsewhere led to the conclusion that the most likely form for an energy parameter ε would be a combination of ultimate resilience and strain energy so arranged that the resultant

Material	YS	TS	Y	EL	HARD	MDPR	UR	SE	NUR	NSE
BS1433 copper	0.300E 05	0.360E 05	0.180E 08	0.180E 00	0.900E 02	0.647E 01	0.360E 02	0.648E 04	1.000	1.000
Stainless steel 316	0.310E 05	0.813E 05	0.260E 08	0.690E 00	0.748E 02	0.301E 00	0.127E 03	0.561E 05	3.531	8.657
Nickle 270	0.800E 04	0.488E 05	0.277E 08	0.610E 00	0.249E 02	0.128E 01	0.430E 02	0.298E 05	1.194	4.594
Al 6061	0.407E 05	0.475E 05	0.910E 07	0.220E 00	0.600E 02	0.436E 01	0.124E 03	0.104E 05	3.444	1.613
Stainless steel 304	0.647E 05	0.945E 05	0.290E 08	0.638E 00	0.237E 03	0.330E 00	0.154E 03	0.603E 05	4.277	9.304
Bronze no. 1	0.243E 05	0.452E 05	0.128E 08	0.230E 00	0.189E 03	0.189E 01	0.798E 02	0.104E 05	2.217	1.604
Bronze no. 2	0.790E 05	0.112E 06	0.147E 08	0.205E 00	0.304E 03	0.163E 00	0.426E 03	0.229E 05	11.834	3.537
Bronze no. 3	0.880E 05	0.119E 06	0.172E 08	0.150E 00	0.225E 03	0.220E 00	0.411E 03	0.178E 05	11.410	2.752
Bronze no. 4	0.190E 05	0.282E 05	0.121E 08	0.600E-01	0.152E 03	0.176E 01	0.329E 02	0.169E 04	0.913	0.261
Bronze no. 5	0.105E 05	0.189E 05	0.558E 07	0.130E 00	0.974E 02	0.330E 01	0.320E 02	0.246E 04	0.889	0.379
Bronze no. 6	0.162E 05	0.193E 05	0.711E 07	0.300E-01	0.152E 03	0.257E 01	0.262E 02	0.579E 03	0.728	0.089
Stainless steel no. 1	0.115E 06	0.157E 06	0.263E 08	0.220E 00	0.290E 03	0.252E 00	0.470E 03	0.346E 05	13.050	5.337
Stainless steel no. 2	0.186E 06	0.188E 06	0.257E 08	0.750E-01	0.418E 03	0.270E 00	0.691E 03	0.141E 05	19.189	2.179
Stainless steel no. 3	0.104E 06	0.126E 06	0.251E 08	0.195E 00	0.264E 03	0.430E 00	0.319E 03	0.247E 05	8.865	3.807
Copper	0.282E 05	0.333E 05	0.160E 08	0.543E 00	0.968E 02	0.671E 01	0.347E 02	0.181E 05	0.963	2.790
Brass (65-35)	0.489E 05	0.605E 05	0.157E 08	0.393E 00	0.146E 03	0.170E 01	0.117E 03	0.238E 05	3.238	3.669
Mild steel 1020	0.897E 05	0.965E 05	0.300E 08	0.259E 00	0.227E 03	0.808E 00	0.155E 03	0.250E 05	4.311	3.857
Stainless steel 304	0.410E 05	0.994E 05	0.290E 08	0.168E 00	0.315E 03	0.332E 00	0.170E 03	0.167E 05	4.732	2.577
ASTM										
B144 (SAE660)	0.175E 05	0.225E 05	0.140E 08	0.173E 00	0.174E 03	0.147E 01	0.181E 02	0.389E 04	0.502	0.601
Magnesium	0.241E 05	0.392E 05	0.650E 07	0.255E 00	0.885E 02	0.434E 01	0.118E 03	0.100E 05	3.283	1.543
Aluminum 3003-0	0.680E 04	0.159E 05	0.900E 07	0.541E 00	0.512E 02	0.304E 02	0.140E 02	0.860E 04	0.390	1.327
Copper	0.300E 05	0.360E 05	0.180E 08	0.180E 00	0.900E 02	0.647E 01	0.360E 02	0.648E 04	1.000	1.000
CR-130 steel	0.290E 05	0.780E 05	0.290E 08	0.280E 00	0.255E 03	0.465E 01	0.105E 03	0.218E 05	2.914	3.370
Al alloy	0.450E 05	0.560E 05	0.100E 08	0.100E 00	0.114E 03	0.802E 01	0.157E 03	0.560E 04	4.356	0.864
Aluminum	0.150E 05	0.160E 05	0.900E 07	0.500E-01	0.270E 02	0.255E 02	0.142E 02	0.800E 03	0.395	0.123
Copper	0.142E 05	0.310E 05	0.170E 08	0.500E 00	0.600E 02	0.824E 01	0.283E 02	0.155E 05	0.785	2.392
Phosphor bronze	0.394E 05	0.416E 05	0.150E 08	0.110E 00	0.950E 02	0.440E 01	0.577E 02	0.458E 04	1.602	0.706
Brass	0.157E 05	0.260E 05	0.160E 08	0.530E 00	0.150E 03	0.200E 01	0.211E 02	0.138E 05	0.587	2.127
Mild steel	0.484E 05	0.650E 05	0.280E 08	0.600E-01	0.950E 02	0.236E 01	0.754E 02	0.390E 04	2.096	0.602
Stainless steel	0.354E 05	0.930E 05	0.280E 08	0.570E 00	0.170E 03	0.653E 00	0.154E 03	0.530E 05	4.290	8.181
Stainless steel 316	0.310E 05	0.813E 05	0.260E 08	0.690E 00	0.748E 02	0.713E 00	0.127E 03	0.561E 05	3.531	8.657
Nickle 270	0.800E 04	0.488E 05	0.277E 08	0.610E 00	0.249E 02	0.126E 01	0.430E 02	0.298E 05	1.194	4.594
Al 6061	0.407E 05	0.475E 05	0.910E 07	0.220E 00	0.600E 02	0.436E 01	0.124E 03	0.104E 05	3.444	1.613
Stellite 6-B	0.710E 05	0.138E 06	0.304E 08	0.210E 00	0.322E 03	0.180E 01	0.313E 03	0.290E 05	8.728	4.475
Tool steel no. 1	0.540E 05	0.110E 06	0.275E 08	0.175E-01	0.235E 03	0.730E 01	0.220E 02	0.193E 04	6.111	0.298

YS = yield strength, lb/in^2; TS = tensile strength, lb/in^2; Y = elastic modulus, lb/in^2; EL = elongation, %; HARD = Brinell hardness; MDPR = maximum mean depth of penetration rate, mils/h (all values are corrected to U-M vibratory facility); UR = ultimate resilience = TS^2/2E, lb/in^2; SE = strain energy to failure; NUR = ultimate resilience normalized to BS 1433 copper; NSE = strain energy normalized to BS 1433 copper.

term would have the units of energy/volume. To attain reasonable flexibility within this limitation, the following relation was assumed:

$$\varepsilon = C_1\left(\frac{\text{UR}}{\text{ESE}}\right)^a \text{UR} + C_2\left(\frac{\text{UR}}{\text{ESE}}\right)^b \text{SE} \tag{5-5}$$

where C_1, C_2, a, and b are constants to be computed by a least-square-fit regression analysis of the data. Investigation of this relation showed that the best values for a and b were close to zero, so that a simpler relation than Eq. (5-5) was indicated. An additive constant C_0 was used since this improved the data fit. The physical interpretation of C_0 is that of a threshold energy necessary to cause measurable damage, i.e., a concept analogous to that of threshold velocity:

$$\varepsilon = C_0 + C_1\text{UR} + C_2\text{SE} \tag{5-6}$$

Using a least-mean-square-fit analysis with Eq. (5-6) or the following special case versions of it:

$$\varepsilon = C_0 + C_1\text{UR} \tag{5-7a}$$

$$\varepsilon = C_0 + C_1\text{SE} \tag{5-7b}$$

$$\varepsilon = C_0 + C_1\text{TSE} \tag{5-7c}$$

$$\varepsilon = C_1\text{UR} \tag{5-7d}$$

it was found that the best correlation coefficient and the smallest percentage of standard error of estimate resulted from Eq. (5-6) itself. However, Eq. (5-7a) was in all cases nearly as good, indicating that ultimate resilience was the material parameter of major importance. This was further verified by the dominance of the second term over the third in Eq. (5-6). The statistics of the correlation with either Eq. (5-7b) or Eq. (5-7c) were relatively very poor with TSE worse than SE. Hence SE is used in Eq. (5-6). This data is summarized in Tables 5-4 and 5-5. While the correlation with Eq. (5-6) is better than that with Eq. (5-7a), it is only slightly so. Hence for simplicity it is reasonable to use Eq. (5-7a) in preference to Eq. (5-6), so that the only mechanical property involved in the correlation becomes ultimate resilience. Since the best value of C_0 in Eq. (5-7a) is relatively very small, it is justifiable to use Eq. (5-7d) where this threshold energy term is neglected. Values of C_0, C_1, and C_2 are listed in Table 5-4.

The standard error of estimate [74] is computed in such a way that it is always approximately proportional to mean magnitude, to give equal weight to both weak and strong materials in the correlation, and to allow the reasonably accurate prediction of MDPR for materials of low ε.

Determination of efficiency factor η It seems reasonable that $\eta = \eta_a\eta_b...\eta_i$ where η_i represents each different portion of an overall efficiency η. Then one factor of the overall energy-transfer efficiency term in the basic equation (5-4) becomes η_a, which to a first approximation can be considered as a function of the acoustic impedance ratio (AI) between liquid and material (AI $= \rho_L C_L/\rho_S C_S$).

Material	Epsilon	$C_1 = 0.914$ $C_2 = 1.897$ $C_1 + C_2 \times UR$	$C_1 = 2.875$ $C_2 = 1.824$ $C_1 + C_2 \times SE$	$C_1 = 6.487$ $C_2 = 0.445$ $C_1 + C_2 \times TSE$	$C_1 = 1.633$ $C_2 = 0.889$ $C_1 \times UR + C_2 \times SE$	$C_0 = -1.773$ $C_1 = 1.735$ $C_2 = 1.139$ $C_0 + C_1 \times UR + C_2 \times SE$	Sources
BS1433 copper	1.000	2.811	4.699	6.931	2.522	1.102	
Stainless steel 316	21.482	7.611	18.664	12.474	13.464	14.218	
Nickle 270	5.044	3.179	11.254	14.510	6.035	5.534	
Al 6061	1.482	7.446	5.817	7.103	7.057	6.040	
Stainless steel 304	19.594	9.026	19.845	13.565	15.258	16.250	U-M vibratory cavitation facility
Bronze no. 1	3.421	5.119	5.801	6.924	5.046	3.902	
Bronze no. 2	39.669	23.359	9.326	7.612	22.468	22.789	
Bronze no. 3	29.391	22.556	7.895	7.278	21.079	21.161	
Bronze no. 4	3.674	2.645	3.352	6.553	1.723	0.109	
Bronze no. 5	1.959	2.601	3.567	6.581	1.789	0.202	
Bronze no. 6	2.516	2.295	3.038	6.511	1.268	-0.408	
Stainless steel no. 1	25.659	25.666	12.609	8.452	26.055	26.951	
Stainless steel no. 2	23.948	37.310	6.850	7.230	33.271	34.005	U-M vibratory cavitation facility with stationary specimen
Stainless steel no. 3	15.037	17.729	9.818	8.706	17.861	17.946	
Copper	0.963	2.740	7.965	7.240	4.053	3.077	
Brass (65-35)	3.801	7.056	9.567	7.455	8.550	8.026	
Mild steel 1020	8.002	9.091	9.910	7.571	10.470	10.102	
Stainless steel 304	19.476	9.889	7.575	7.017	10.018	9.374	
ASTM B 144 (SAE660)	4.384	1.867	3.971	6.643	1.354	-0.217	
Magnesium	1.490	7.142	5.689	6.849	6.733	5.682	
Aluminum 3003-0	0.213	1.654	5.296	6.764	1.818	0.417	
Copper	1.000	2.811	4.699	6.931	2.522	1.102	
CR-130 steel	1.391	6.441	9.022	7.654	7.755	7.123	RAE-Dornier rotating-arm facility
Al alloy	0.806	9.175	4.451	6.740	7.880	6.769	
Aluminum	0.254	1.663	3.100	6.523	0.755	-0.946	
Copper	0.785	2.403	7.238	7.320	3.409	2.315	
Phosphor bronze	1.470	3.953	4.163	6.787	3.244	1.812	
Brass	3.228	2.027	6.754	7.323	2.849	1.669	U-M venturi facility
Mild steel	2.739	4.889	3.973	6.724	3.957	2.549	
Stainless steel	9.902	9.051	17.795	9.383	14.280	14.992	
Stainless steel 316	9.069	7.611	18.664	12.474	13.464	14.218	
Nickle 270	5.111	3.179	11.254	14.510	6.035	5.534	Rotating-wheel impact facility
Al 6061	1.482	7.446	5.817	7.103	7.057	6.040	
Correlation coefficient		0.808	0.466	0.236	0.854	0.856	
Standard error of estimate		5.914	8.874	9.744	5.502	5.191	

Table 5-5 Recommended correlating equations [74] for Eq. (5-12)

1/MDPR (h/mil)

Material, normalized	Epsilon	$C_0 = 0.463$ $C_1 = +0.899$ $C_0 + C_1 \times UR$	$C_0 + C_2 \times UR$	$C_1 = 2.330$ $C_1 \times UR$	$C_0 + C_2 \times UR$	Sources
BS1433 copper	1.000	2.462	1.981	2.330	2.007	
Stainless steel 316	21.482	7.520	6.993	8.225	7.086	
Nickle 270	5.044	2.850	2.365	2.782	2.396	
Al 6061	1.482	7.346	6.820	8.022	6.911	U-M vibratory cavitation facility
Stainless steel 304	19.594	9.011	8.471	9.964	8.584	
Bronze no. 1	3.421	4.894	4.391	5.164	4.449	
Bronze no. 2	39.669	24.115	23.438	27.568	23.750	
Bronze no. 3	29.391	23.269	22.600	26.582	22.900	
Bronze no. 4	3.674	2.288	1.808	2.127	1.832	
Bronze no. 5	1.959	2.240	1.761	2.071	1.784	
Bronze no. 6	2.516	1.918	1.442	1.696	1.461	
Stainless steel no. 1	25.659	26.547	25.847	30.402	26.191	
Stainless steel no. 2	23.948	38.817	38.007	44.704	38.513	U-M vibratory cavitation facility with stationary specimen
Stainless steel no. 3	15.037	18.182	17.559	20.653	17.792	
Copper	0.963	2.387	1.906	2.242	1.932	
Brass (65-35)	3.801	6.935	6.413	7.543	6.499	
Mild steel 1020	8.002	9.080	8.539	10.043	8.653	
Stainless steel 304	19.476	9.921	9.372	11.024	9.497	
ASTM B144 (SAE660)	4.384	1.467	0.995	1.170	1.008	
Magnesium	1.490	7.026	6.503	7.649	6.590	
Aluminum 3003-0	0.213	1.243	0.773	0.909	0.783	
Copper	1.000	2.462	1.981	2.330	2.007	
CR-130 steel	1.391	6.287	5.771	6.788	5.848	RAE-Dornier rotating-arm facility
Al alloy	0.806	9.169	8.627	10.147	8.742	
Aluminum	0.254	1.253	0.782	0.920	0.793	
Copper	0.785	2.032	1.555	1.829	1.576	
Phosphor bronze	1.470	3.666	3.174	3.733	3.216	U-M venturi facility
Brass	3.228	1.636	1.162	1.367	1.178	
Mild steel	2.739	4.652	4.151	4.882	4.206	
Stainless steel	9.902	9.038	8.497	9.994	8.610	
Stainless steel 316	9.069	7.520	6.993	8.225	7.086	Rotating-wheel impact facility
Nickle 270	5.111	2.850	2.365	2.782	2.396	
Al 6061	1.482	7.346	6.820	8.022	6.911	

Correlation coefficient 0.898 0.898

The "water-hammer equation" for droplet impact, discussed in detail in Chap. 6, is usually assumed for materials of finite elasticity to give a reasonable approximation for the pressure applied to the material surface under droplet impact [3]. Examination of this equation indicates the importance of AI in determining this pressure, and in fact suggests a functional form of AI, $f(\text{AI})$, which might be used:

$$\Delta p = \frac{\rho_i C V}{\text{AI} + 1} \tag{5-8}$$

so

$$f(\text{AI}) = \text{AI} + 1$$

$f(\text{AI})$ is taken as a direct factor in the relation describing the pressure generated at the point of impact. Since pressure has units of energy per volume and $f(\text{AI})$ is nondimensional, its use here is dimensionally consistent with the general model assumed. Another possible form of $f(\text{AI})$ is the "transmission coefficient," giving the ratio of absorbed to reflected energy for the case of a shock wave impinging upon a solid surface in a continuous medium (which is not identical to the present case). Then

$$f(\text{AI}) = \frac{(\text{AI} + 1)^2}{4\text{AI}} \tag{5-9}$$

The best-fit correlations have been investigated for both forms. It was assumed that

$$\eta_a = f(\text{AI})^n \quad \text{where } n = \pm 1, \pm 2, \pm 3 \tag{5-10}$$

Table 5-6 summarizes the results. It appears that there is no substantial improvement in the correlation to be attained by the use of $f(\text{AI})$ in any of these forms. This is surprising in the light of Heymann's result [76, 86] that the fit with UR was improved by using $\text{UR} \cdot E$, since $E^2 \cong \rho E = \rho^2 C^2$ for the metals used. As also suggested by Heymann [76], it seems necessary that η_a differ substantially between materials, since the ratio between the extremes of material erosion resistance is orders of magnitude greater than that between the corresponding material energy properties. Nevertheless, in light of the present results, η_a must be assumed unity and omitted from subsequent relations. Further discussion of efficiency factors is provided in Sec. 5-5.

Nonlinear parameter fits

1. *Polynomial energy parameter fit.* The postulated basic equation (5-4) requires a first-power energy term for dimensional consistency. In order to verify that the assumption of such a linear relationship with energy is reasonable, polynomial data fits of the type

$$\varepsilon = C_0 + C_1\text{UR} + C_2(\text{UR})^2 + C_3(\text{UR})^3 \tag{5-11}$$

were investigated (Table 5-7). While there was some improvement over the

Table 5-6 Acoustic impedance correction

| $f(AI)^n$ | AI + 1 | | $(AI + 1)^2/4AI$ | |
| | Correlation coefficient | Standard error of estimate | Correlation coefficient | Standard error of estimate |
n				
0	0.808	2.007	0.808	2.007
1	0.807	2.005	0.807	2.101
2	0.807	2.003	0.782	2.324
3	0.806	2.001	0.743	2.668
−1	0.808	2.009	0.781	2.070
−2	0.808	2.011	0.721	2.431
−3	0.809	2.014	0.582	3.745

linear fit in "correlation coefficient," the "standard error of estimate" becomes worse. Thus it appears that the linear relationship, being physically reasonable, is most suited for the present purpose, where the maintenance of dimensional consistency is important.

2. *Fit with* $UR \cdot E^2$. Heymann's study [76, 86] showed some improvement by using $UR \cdot E^2$ rather than $UR \cdot E$. However, the statistical fit for either of these terms according to the present data set is not as good as that with UR alone, and of course is dimensionally inconsistent with the assumed model (Table 5-7).

Recommended relations The following relations seem most useful [in "English engineering units" of pound-force (lbf), feet (ft), and seconds (s)] for common metals and alloys:

$$\varepsilon = C_0 + C_1 UR \qquad (5\text{-}12a)$$

$$\varepsilon = C_1 UR \qquad (5\text{-}12b)$$

Table 5-7 Equations using nonlinear parameters

Equation	Correlation coefficient	Standard error of estimate
$\varepsilon = 2.330\ UR$	0.808	2.007
$\varepsilon = -2.681 + 3.343\ UR - 0.087\ UR^2$	0.870	5.616
$\varepsilon = 0.266\ UR + 0.412\ UR^2 - 0.019\ UR^3$	0.919	4.459
$\varepsilon = 3.685\ UR \times E^2$	0.678	5.714
$\varepsilon = 1.147 + 1.444\ UR \times E^2$	0.678	4.271

Table 5-8 Statistical correlation coefficients

Correlating relation	n (where applicable)	Sample correlation coefficient	95% confidence limits for population correlation coefficients	Factorial standard error of estimate
$\dfrac{1}{\text{MDPR}} = C(\text{UR})^n$	0.998	0.811	0.64–0.91	2.52
$\dfrac{1}{\text{MDPR}} = C(\text{UR})$	—	0.811	0.64–0.91	2.52
$\dfrac{1}{\text{MDPR}} = C(\text{UR} \times \text{BHN})^n$	0.720	0.798	0.62–0.89	2.25
$\dfrac{1}{\text{MDPR}} = C(\text{UR} \times E^2)^n$	0.659	0.744	0.52–0.86	2.35
$\dfrac{1}{\text{MDPR}} = C(\text{BHN})$	—	0.742	0.52–0.86	2.75
$\dfrac{1}{\text{MDPR}} = C(\text{BHN})^n$	1.788	0.734	0.52–0.85	2.38
$\dfrac{1}{\text{MDPR}} = C(\text{UR} \times \text{BHN})$	—	0.716	0.49–0.84	2.57
$\dfrac{1}{\text{MDPR}} = C(\text{UR} \times E^2)$	—	0.684	0.44–0.82	2.86
$\dfrac{1}{\text{MDPR}} = C(\text{SE})^n$	0.738	0.517	0.21–0.73	3.24
$\dfrac{1}{\text{MDPR}} = C(\text{SE})$	—	0.498	0.17–0.72	3.30

		(5-12a)	(5-12b)
where	$C_0 =$	0.463	—
	$C_1 =$	1.999	2.330
Coefficient of correlation =		0.808	0.808
Standard error of estimate = (multiplicative factor)		1.981	2.007

Table 5-9 Classification of 22 alloys or alloy groups according to their normalized erosion resistance relative to 18Cr-8Ni austenitic stainless steel having a hardness of 170 dph [5]

Material	Hardness, BHN or dph	Normalized erosion resistance
Carbon steel	110 to 190	
Ausformed 12% Cr tool steel (nonstandard)	450 to 620	
Maraging steel	500 to 650	
Gray iron	140 to 230	
Tool steels (H26, T1, T2, and T3)	600 to 900	
Austenitic stainless steel (series 300)	140 to 230	
Type 410 stainless steel	200 to 400	
Types 639 and 637 stainless steel	320 to 460	
Stellite 6	380 to 450	
Stellite 6B	380 to 500	
Stelline 12 (cast iron)	480	
Aluminum	20 to 90	
Aluminum alloys	100 to 200	
Copper alloys 260.268.280	60 to 200	
Copper alloys 614 and 953	150 to 180	
Copper alloys 628, 630, and 955	140 to 220	
Copper alloys 675, 662, 863, and 865	120 to 230	
Copper alloys 713 and 719	70 to 200	
Copper alloy 903	60 to 100	
Nickel	—	
Inconel	150 to 380	
Monel	120 to 360	

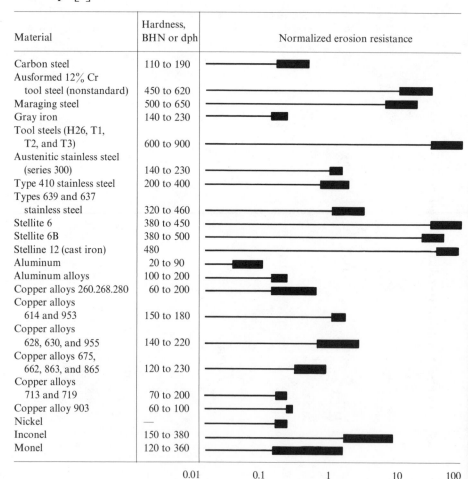

0.01   0.1   1   10   100

Since the improvement due to the inclusion of the threshold energy C_0 is small, the simplest form, Eq. (5-12b), is perhaps most useful.

Tables 5-4 and 5-5 list the full data set used along with measured and predicted values of ε (which is equivalent to MDPR for data normalized in the fashion used here), according to Eq. (5-12).

The predicted and measured values are tabulated for both Eq. (5-12a, b), along with the deviations for each material. Figure 5-19 presents the same information graphically for the recommended Eq. (5-12b), where the "triangular" standard error of estimate band is shown. Tables 5-7 and 5-8 summarize statistically various best-fit equations including those which are nonlinear [74]. Table 5-9 summarizes the resistance of a different group of materials as a function of hardness. A best-fit hardness dependence for 1/MDPR is 1/MDPR $\sim$ (BHN)$^{1.8}$ per Heymann [5].†

Figure 5-20 shows information very similar to Fig. 5-19 from a different but also very comprehensive data set, provided by Heymann [75] in a discussion

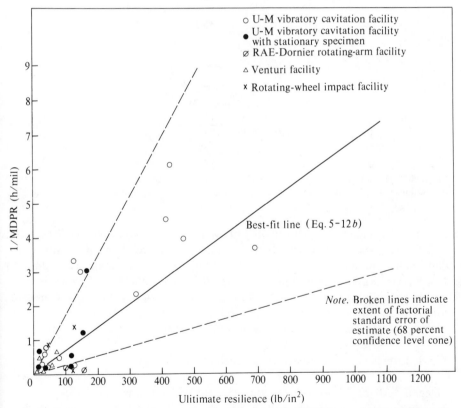

Figure 5-19 Best-fit correlation and standard deviation cone for 1/MDPR vs. ultimate resilience for thirty-three materials.

† Since BHN $\tilde{\alpha}$ TS for many materials, a BHN^2 dependence is $\sim$ equivalent to a linear VR fit for materials of same E.

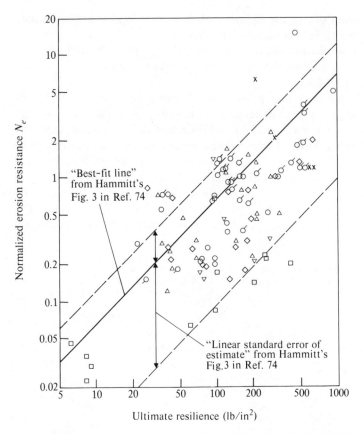

Figure 5-20 Erosion resistance versus ultimate resilience: comparison of Hammitt's and Heymann's correlations [74, 75]. Data is from Heymann [75], and best-fit and standard error from Hammitt [74].

to our earlier paper [74]. Heymann plotted his and our data together (Fig. 5-20), and the very close agreement between these two correlations based upon independent, but large, data sets is heartening. Heymann's analysis leading to Fig. 5-20 is further discussed in his article [76] at the same ASTM symposium.

The amplitude constants apply to the U-M vibratory facility only. Constants for other facilities are found by multiplying the given constants by the ratio between maximum damage rates in the other facility and the U-M facility.

5-3-4 Interplay of Corrosion and Mechanical Effects

General   All the previous discussion, as well as the mathematical model proposed, has considered only the mechanical aspects of cavitation erosion of materials; i.e., possible corrosive effects have not been considered. However, it is obvious that with certain liquid-material combinations corrosive effects will also be significant, and of course corrosion-induced erosion is a major topic in its own right. Corrosion per se is not within the scope of this book; it is treated in detail in various other works. However, there are many important applications which include not only cavitation effects but also substantial corrosive attack. The most obvious case may be marine propellors fabricated of corrodible materials, or other machines operated in sea water. Other cases are hydraulic turbines (in some cases even in fresh water), machines operating with other corrosive liquids, and numerous similar applications. An important case, usually involving only very mild cavitation combined with strong corrosive effects, are diesel engine coolant "liners." In this case the cavitation is apparently of vibratory rather than flow origin.

Mechanism and results of combined cavitation-corrosion erosion   Numerous experimental observations clearly indicate that the combined erosion due to a combination of cavitation and corrosive effects is often very much greater than the summation which would be obtained if the two effects were to act singly. This multiplied combined effect is usually attributed to the following interactive mechanisms.

1. Mechanical cavitation removes the protective layer which normally inhibits rapid corrosion with most materials, thus continually exposing fresh metal for renewed corrosive attack.
2. Corrosion roughens the surface causing increased local cavitation beyond that which would be incurred in the absence of such surface roughening. The corrosion rate may also be increased in some cases by the presence of high-temperature gases (such as oxygen) within the collapsing bubbles.

   One of the earliest attempts to separate the effects of corrosion and mechanical attack using a vibratory facility was that by Wheeler, as reported in detail in Ref. 6.

Cathodic and anodic voltages and other effects

Cathodic and anodic voltages   "Cathodic protection," i.e., application of a negative electric potential, to a material under corrosive attack in water, particularly sea water, has been used for many years to reduce or prevent ordinary corrosive attack. Experiments have also been made to evaluate the effect of cathodic protection upon cavitation damage [89, 90]. It has been found that cavitation damage is thus reduced for various metallic materials [89]. However, it was later stated [90] that this was primarily a result of electrolytic gas evolution at the eroded surface, which then "cushioned" cavitation bubble collapse, rather than of the expected inhibition of the corrosive component of the overall attack. Presumably each of these mechanisms is oversimplified in general; both, as well perhaps as other presently unknown mechanisms, are involved in many real applications.

   Recent experiments have also been made to include the effects of both positive and negative voltages applied to cavitated metallic specimens [91–93]. Generally it seems to be indicated that for vibratory horn cavitation-damage tests there is a continuous progression toward increased damage as the applied voltage is increased from strongly negative, through zero, into the positive range. Much greater damage rates are observed for positive (versus zero) voltage. This observation does not appear consistent with the concept of reduced damage due to "gas cushioning" [90], since approximately similar rates of gas evolution are observed at either positive or negative applied voltages. It thus appears at this time that the effects of applied voltages are not well understood and that additional careful research is required.

Gas effects As discussed in the foregoing, "gas cushioning" has been proposed in the case of "cathodic protection" as a possible mechanism for reduced cavitation damage. This mechanism also presumably explains the reduction of damage observed for large gas contents with various liquids, as discussed in more detail in Sec. 5-4. Also, air injection is used in some field applications for the purposeful reduction of cavitation damage. However, in some cases there may be a partially countering effect due to the increase in corrosion which would be expected with higher oxygen content. Again this is a subject for which very little quantitative information exists at this time.

Stress corrosion The phenomenon of "stress corrosion" represents an obvious interplay between applied mechanical stress and corrosion. In those cases where liquid-handling components encounter "stress corrosion," there is also a possible interplay between cavitation and corrosion effects, if cavitation is in fact present. Again this appears to be a field wherein very little quantitative information is available.

Cavitating slurries Cavitating slurries involve erosion which may result from both corrosive and particle-impact effects, in addition to ordinary cavitation mechanical attack. While some field experience shows that very high rates of erosion may result from cavitating slurries, there is virtually no qualitative information available at this point.

5-3-5 Summarization

The most likely form for an equation relating material, liquid, and test parameters with cavitation or impingement erosion rates, with good hope for general applicability, is one which is based on a simple physical model with dimensional consistency. For the evaluation of impingement erosion rates, consistent with the previous suggestion of Hoff, Langbein, and Rieger [82], the equation

$$\text{MDPR} = \frac{\eta}{\varepsilon} \frac{A_p}{A_e} \frac{\rho_{\text{eff}}}{2} V^3 \tag{5-4}$$

where $\eta = \eta_a \eta_b \dots \eta_i$ has been chosen.

A statistical evaluation of ε, which must have units of energy per volume, has shown the best fit with a comprehensive data set generated in various laboratories with various types of test devices and various test materials, including both impingement and cavitation data, in the form

$$\frac{1}{\text{MDPR}} \propto \varepsilon = C_1 \text{UR} \qquad (5\text{-}12b)$$

Neither higher-power terms in UR nor terms in SE (terms previously defined) improved the statistics of the fit substantially, and the fit in terms only of SE was relatively very poor. It is thus concluded that for the large group of metals used at the author's laboratory the best linear energy per volume mechanical-property correlation for the volume-loss rate under droplet impingement or cavitation attack is the expression [Eq. (5-12b)] in ultimate resilience alone.

Traditionally, hardness has been used as the indicator of cavitation or impingement liquid-erosion resistance. This parameter at least has the merit of extremely easy laboratory determination, where values are not already available. In terms of a predicting equation, BHN (Brinell hardness) requires an exponent of ~1.8 (Table 5-8), as opposed to ultimate resilience where the best exponent is essentially unity [Eq. (5-12b)]. Another very useful predicting equation is then

$$\frac{1}{\text{MDPR}} \propto \varepsilon = C(\text{BHN})^{1.8} \qquad (5\text{-}13)\dagger$$

The statistical fit of Eq. (5-13) is not quite as good as that of Eq. (5-12b), but it is not very much worse (Table 5-8), and in most cases BHN is much more easily determinable than is UR.

For these, or any correlating equations, there are certain materials (e.g., Stellite) which lie relatively far from the predicting line. In any case, a standard error of the order of $\times 3$ must be expected. Since there is a range of $\sim \times 10^4$ between the resistances of weak materials such as, for example, soft aluminum and strong ones such as Stellite 6-B (see Table 5-9), this apparently large standard error may not be too surprising.

Liquid-impact erosion, in particular, is discussed in further detail in Chap. 6.

5-4 CAVITATION-DAMAGE SCALE EFFECTS

5-4-1 Introduction

A review of the state of the art for cavitation-damage "scale effects" is made here. Nucleation scale effects were discussed in Chap. 3. The most prominent of such damage scale effects (for constant sigma) occur with variation of velocity, pressure, or size, but no precise prediction of the magnitude of these effects is as yet possible. A relatively comprehensive bibliography of pertinent literature is included, but this is by no means complete.

"Cavitation-damage scale effects" are here taken to include all those phenomena which result in a change in cavitation-damage rates in a given flow regime

† See footnote page 267.

occurring as a result of changes in operating conditions such as velocity, pressure, machine size, fluid and fluid conditions (e.g., temperature), all at constant sigma, with geometric and nominal similarity maintained. In most cases the operative mechanisms, and the fact that such changes in damage rates exist at constant sigma, are fairly clear. However, it is in general not possible in the present state of the art to estimate the changes to be expected other than empirically. In most cases, reliable empirical information is also not available, although in many cases it is at least possible to know the direction of the trends to be expected. Hopefully, within the next decade or two, these matters will be substantially clarified as a result of continuing research. For the present purposes, it is best to divide the overall subject into several portions according to the type of effect to be considered, although in many applications several of the mechanisms operate simultaneously. The various separate effects will be considered in the following sections. The operative mechanisms will be described as they are presently understood, and important literature sources will be cited. More detail on this subject is found in the survey report [94] prepared by the Working Group No. 1 of the Hydraulic Machinery Section of IAHR† and a subsequent summary article [95].

5-4-2 Velocity and Pressure Effects

The most prominent and well-known cavitation-damage scale effects are probably those due to variations of velocity or suppression pressure. Assuming, however, that "scale effects" presuppose the maintenance of constant sigma, it is obvious that velocity and suppression pressure cannot be varied arbitrarily and independently if sigma is to be maintained constant. It is also obvious that tests at varying pressure and/or velocity, but with varying sigma, will result in large changes in cavitation-damage rates. However, these are not scale effects.

Pseudo scale effects Considering cavitation-damage tests wherein suppression pressure is varied but velocity is held constant, sigma varies as determined by the variation of suppression pressure. In such cases it is obvious that, as suppression pressure (and sigma) is reduced from a very high value at which cavitation is completely suppressed, the cavitation-damage rate will increase from zero toward a maximum. As suppression pressure is further reduced, experience and theory show that damage rate will decrease again toward zero. This general behavior is clearly indicated in tests using a vibratory-type facility [96–100] and in flowing tests such as in venturi-type devices [11, 101–105]. However, these results are not, strictly speaking, damage "scale" effects, since sigma is varied.

Other related tests in which the effect of velocity variation has been examined at constant suppression pressure are those using liquid-jet or droplet-impact devices. While these are not basically cavitation tests (though local cavitation may be involved), it has been assumed for many years, and demonstrated experimentally to some extent, that cavitation and liquid-impact erosion are very similar

---

† International Association for Hydraulic Research.

phenomena, perhaps governed by the same parameters. In these liquid-impact tests, damage exponents vary over the range ~ 5 to ~ 10 [106–109]. The same is approximately true for cavitation tests using rotating-disc devices in which the nominal pressure is held constant but the velocity varied [24, 25, 110]. This, and other somewhat inconsistent results, may be due to the highly complex three-dimensional nature of these flows.

5-4-3 "True" Damage Scale Effects

True damage scale effects are those encountered at constant sigma with variation in velocity (or pressure), while the other parameter is also varied in such a way as to maintain constant sigma. Under these conditions, to a first approximation, i.e., neglecting ordinary "scale effects," if the size is also constant the number and size of bubbles in the collapse region should be very approximately constant, as should the flow regime in general, although the very nonlinear nature of bubble growth and collapse equations makes even this approximation hazardous. This is not a good approximation for cavitating wakes, where periodic cavity shedding is involved [111]. Here the number of bubbles and the cavity-shedding frequency may increase with velocity. However, the energy involved in the collapse of a bubble of a given size would then be directly proportional (to a first approximation) to suppression pressure, and hence its damage capability would be greater for greater suppression pressures. Of course, it cannot be assumed as a consequence that the damage rate will also be proportional to the suppression pressure. This general situation, using many idealized assumptions, has been examined by Thiruvengadam [112–115].

"True" damage scale effects are then hereby defined as those for which sigma and flow geometry are constant. As pointed out in the foregoing, many tests involving the effects of velocity change upon cavitation-damage rates do not meet these conditions, and of course vibratory tests are inapplicable, since velocity is not a pertinent parameter for such tests. However, tests of this type reported in the literature will be discussed in the following.

California Institute of Technology (CIT) tests In water-tunnel tests at CIT in which velocity was varied but sigma was held constant, Knapp [6, 116] deduced the now well-known velocity-damage "exponent law"† and assigned the exponent a value of 6. His observations were based on counting pits in a soft aluminum specimen of ogival shape immersed in the working section of the large water tunnel at CIT. No weight losses were involved—only pit counts. Velocity was varied, but minimum pressure in the working section remained p_v. The variation of pitting rate with velocity on soft aluminum was later confirmed in a test on a water turbine in the field [6, 117]. A somewhat similar program involving damage comparisons between vibratory, rotating-disk, and hydraulic-turbine devices with

---

† Damage rate $\propto V^n$. It is sometimes assumed that damage rate $\propto (V - V_0)^m$, where V_0 is a "threshold velocity" below which no substantial damage occurs. Exponent m is then less than exponent n. Damage rate = volume loss rate or MDPR.

different test materials has been reported at the Institute of Fluid Flow Machines, Polish Academy of Science, Gdansk, Poland [118–120].

Industrial experience It is generally recognized in the industry that pumps, turbines, propellors, or other machines designed for an acceptable sigma from the viewpoint of head drop, loss of efficiency, etc., will be satisfactory from the viewpoint of cavitation damage if the velocity is sufficiently low. Thus, in most cases, sigma is fixed from the viewpoint of fluid-dynamic performance, and damage will be negligible (or at least acceptable) if the pressure and velocity are sufficiently low. For fixed sigma, a low velocity, of course, implies a low pressure, and vice versa. A case in point is the comparison between condensate pumps and boiler feed pumps. It is common practice to design condensate pumps for operation in the cavitating regime, which has been used sometimes as a method of flow control. In this type of pump, operating with very little suppression pressure, the velocities are necessarily low to provide the desired sigma. On the other hand, for boiler feed pumps (operating after the condensate and booster pumps in a power-plant system, and thus provided with much higher suppression pressure) a small amount of cavitation can lead quickly to prohibitive damage. Other similarly related industrial experiences exist, but do not in general lead to quantitative predicting relations regarding the damage scale effect. Examples of related industrial experience are reported from Voith by Thuss [121] and from Electricité de France by Giraud [122].

5-4-4 Venturi-Damage Tests at Constant Sigma

Tests at the University of Michigan (U-M) in a cylindrical-conical venturi, wherein damage specimens were inserted through the wall into the region of bubble collapse, were conducted more recently [11, 101]. Figure 5-7 shows the venturi geometry used. Many series of tests have been performed on various materials at essentially constant sigma. However, the geometry of the cavitating region as viewed through the transparent wall was kept constant so that sigma actually varied slightly. In general, of course, there should be an unambiguous relation between sigma and cavity length.

Test fluids were water and mercury. These tests showed that the "velocity exponent" depended upon the "degree of cavitation," i.e., the extent of the cavitating region into the venturi diffuser, with the exponent being greater for more limited extent of the cavity region. In all cases the exponent was less than had been expected from previous tests such as those of Knapp [6, 116, 117]. It ranged for water from ~ 1.7 to 4.9 for a cavity length extending into the diffuser up to the position of the test specimen. In the mercury tests, the increase of damage with velocity was generally small, and in some cases a decrease was found [11, 102–103]. In these particular tests, a reduced cavitation-damage exponent for increased cavity length was generally observed, explainable by the fact that the pressure in the collapse region for this particular geometry (Fig. 5-7) is insensitive to velocity for the cases of well-developed cavitation, since the damage test specimen is then completely immersed within the cavity region, where the pressure is essentially vapor pressure at all velocities.

On the other hand, for a degree of cavitation near initiation, the pressure in the bubble-collapse region increases approximately with the square of the velocity, thus strongly increasing the collapse "intensity" for increased velocity. These results emphasize the fact that, in general, in order to estimate the effect of velocity change on cavitation-damage rates, it is first necessary to estimate the effect of velocity change on suppression pressure, both in the initiation and collapse regions.

Other recent damage tests in a venturi-type geometry† (Fig. 5-21) are those at the Indian Institute of Science at Bangalore [104]. While the velocity was varied over a considerable range, tests at constant sigma and variable velocity were not made. In any case, the velocity exponent increased from ∼6 to 17 as velocity was increased, but then became negative, i.e., damage decreased with a further increase of velocity, with the exponent eventually reaching a negative value of ∼74. A large scatter of velocity-exponent data is also consistent with the U-M tests described above. Again, it appears that the key lies in a knowledge of the behavior of the pressure regime and cavity extent in a particular system.

A summarization of the sigma versus damage-rate effects from the U-M venturi tests is shown in Fig. 5-22. Here sigma is based upon the minimum wall pressure measured for the various flow conditions. Figure 5-7 shows the venturi flow path in which these tests were made and Fig. 5-23 the typical measured wall-pressure profiles from which the sigma values used in Fig. 5-22 were calculated.

The general shape of the sigma versus damage-rate curve (Fig. 5-22) is as would be expected, i.e., very low or very high sigma produces a low damage rate (MDPR = mean depth of penetration rate), whereas an intermediate sigma produces a maximum damage rate.‡ However, the value of the maximum damage sigma and the detailed shape of the curve depends on other related parameters such as test liquid and material, velocity, temperature, and details of the flow geometry (which differs slightly for the two curves shown in Fig. 5-22). It can thus be concluded that cavitation sigma is a very important parameter for developing an overall correlation between externally measurable parameters and damage rate, but that it is not alone a sufficient correlating parameter. While it is perhaps the most logical externally measurable parameter with which to start in the formulation of damage-rate correlations, it must be considered along with the other related parameters discussed above.

The reasoning leading to the expected shape of the damage-sigma curve is simply that, for sufficiently high sigma, cavitation is entirely suppressed; for very low sigma, driving pressure to collapse the bubbles and generate damage is lacking, i.e., "supercavitation" develops. The case of saturated boiling is obviously very similar. Incidentally, the same pressure-damage relationship is exhibited by the vibratory test, but facility limitations so far limit this to the low-pressure part of the overall regime [100, 123].

† A Shal'nev-type was also used.
‡ Only the high sigma end of the curve was tested (Fig. 5-22) because of equipment limitations.

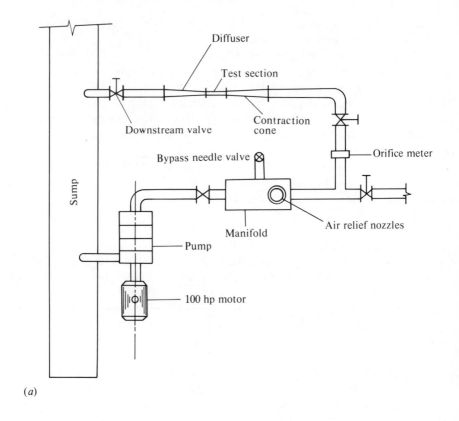

(a)

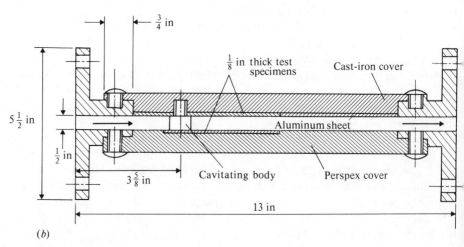

(b)

Figure 5-21 Cavitating venturi facility, Indian Institute of Science, Bangalore [104], showing Shal'nev-type venturi.

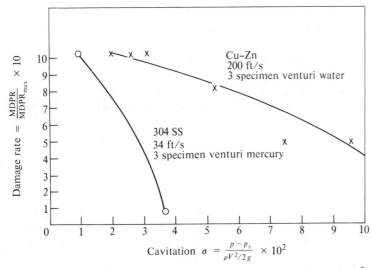

Figure 5-22 Damage rate vs. cavitation sigma in venturi (water and mercury) [11, 101], University of Michigan.

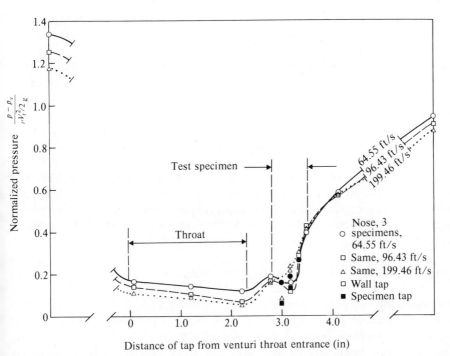

Figure 5-23 Normalized pressure profile for "cavitation to nose," three specimens, in water, at various velocities. University of Michigan venturi [11, 101].

A somewhat similar and related investigation, producing relatively similar results, is that by Hutton and LoboGuerrero at the University of Southampton in England [124].

Another problem to be considered, aside from prediction of damage rates, is the prediction of the pressure and velocity conditions which correspond to the inception of cavitation damage in a given situation. This approach has been followed by Ball, Tullis, and Stripling [125] in a study of cavitation downstream of sudden enlargements. They present empirical equations from which the velocity for inception of cavitation damage can be estimated.

5-4-5 Other Parameter Effects

Size effects The problem of cavitation-damage scale effects with change in machine size (as from model to prototype) at constant sigma (or constant "degree of cavitation" from the viewpoint of head loss, efficiency loss, etc.) is highly complex. In the first place, due to ordinary performance scale effects, constant sigma and constant degree of cavitation are not synonymous. Neglecting that rather secondary difficulty for the moment, and assuming that damage scale effects imply constant degree of cavitation (or cavitating regime) in a given machine, the area to be damaged will be proportional to the square of any characteristic dimension. The average bubble size attained, however, will be greater for increased machine size, since the time of exposure of the bubble to the same underpressures will be greater, and hence the energy of bubble collapse will be greater. Due to the non-linearity of the bubble-growth and -collapse equations, the total bubble-collapse energy (considering all bubbles in the field) will not in general be directly proportional to size scale, but (intuitively) will probably increase with an exponent considerably greater than unity.† A numerical study of this situation by Canavelis [126] predicted an increase of total volume loss on this basis proportional to D^5. Hence it appears that the volume-loss exponent with size lies between 2 (based on the simple area effect) and at least 5. Experimental studies in a venturi by Shal'nev, Varga, and Sebestyen [111] indicate a diameter exponent of 3† and a velocity exponent of 5. More recent studies by Meier and Grein [127] and Schiele and Mollenkopf [128] also indicate a D^3 and V^6 dependence. Malyshev and Pylaev [129] confirm the D^3 dependence. General experience in the hydraulic-turbine industry appears to be that a diameter exponent of about 3 is realistic. Tests by Thuss at Voith [121] indicate a velocity exponent of ~ 4.

Temperature ("thermodynamic") effects The effect of fluid temperature variation on cavitation damage under conditions of constant sigma and similar geometry can reasonably be included in the family of damage scale effects, although it is perhaps not usually considered as such. This effect has been explored in most detail in vibratory horn cavitation-damage tests [65–67, 72, 73, 96–100, 123, 130]. To a first approximation it has been shown that, for a variety of fluids (including

---

† In general, of course, collapse energy is proportional to $p \, dv \propto p D^3$.

liquid metals [72, 97–98, 123, 130]), there is a very strong diminution of damage rate as temperature is increased beyond a certain intermediate value for which the damage rate is a maximum. The foregoing assumes no change in material mechanical properties due to temperature variation; i.e., the diminution of damage at increased temperature has been observed even though the material were at the same time weakened substantially by the high temperature.

In the vibratory tests there is also a relatively mild decrease of damage rate as the temperature is reduced below the temperature for the maximum damage rate. The reason for this decrease in damage at low temperatures is not fully known, but probably involves increased viscosity and other changes in the liquid properties with temperature, including gas solubility. The decrease in damage at high temperatures, which is much more pronounced, is presumably primarily the result of "thermodynamic effects," also discussed in Chaps. 2–4. The overall result of these effects is the confirmed existence of a maximum damage temperature for all liquids tested.

This general behavior, common for all fluids so far tested, covers the temperature range between the freezing and boiling temperatures corresponding to the static pressure under which the test is conducted. The maximum damage temperature is in general about midway between these limits. There is no reason to doubt that the same trends exist in flowing situations, but there are as yet no pertinent precise data. It is thus likely that cavitation damage is not a serious problem for liquid conditions for which the thermodynamic effects are predominant. This is probably true of water at temperatures in the order of 150°C [99, 100] or sodium at high temperatures (for example, ~ 320°C [123, 130]), and also for cryogenic liquids such as hydrogen or oxygen, but at much lower temperatures.

These trends have long been recognized industrially, the name "thermodynamic effects" (which is really a misnomer) originating probably from the pioneering paper on this subject by Stahl and Stepanoff [69]. The primary mechanism involved is the fact that heat transfer from the vapor in the collapsing bubble to the surrounding liquid becomes a limiting mechanism as the temperature is increased, because of the rapid increase of vapor density with temperature. While very high liquid velocities can be achieved if the collapse is inertia controlled, this is not the case for thermally controlled collapses, so that cavitation-damage rates are greatly reduced under these conditions. While there have been a great many theoretical and experimental studies of this problem, particularly in recent years [72, 123, 130–135], it is still not possible to predict a priori the degree of reduction of damage to be expected from these effects in a given situation with any fluid.

Fluid property effects Fluid property effects, other than those due to temperature, may also not normally be considered as true scale effects, even though they involve tests conducted at constant sigma and similar geometry. However, as with temperature, they are very important, particularly from the viewpoint of damage.

Aside from temperature, probably the most important fluid property, from the viewpoint of damage, is liquid density. Constant sigma for fixed velocity implies constant suppression "head."† Thus, to a first approximation, bubble-collapse velocities would also be fixed. The pressure then generated by contact between the collapsing bubble wall and the structure, i.e., the microjet impact or shock-wave imposition, would then be directly proportional to liquid density, if other pertinent properties (such as, for example, the velocity of propagation of sound in the liquid) remained constant. The situation is obviously highly complex, but it is certain that most fluids with higher density, under constant sigma and velocity conditions, will produce higher pressures upon neighboring structural walls and thus be more damaging. However, as yet no directly applicable experimental data appear to exist.

Another fluid property which has in the past been discussed in relation to damage [136] is surface tension. While numerical studies show that it is not likely to affect the collapse process [9], in a given situation it can importantly affect nucleation, and thus the existence of cavitation itself. If cavitation does in fact exist, however, the number and size of bubbles in the collapse region will be affected, thus having an important indirect effect on damage. The foregoing argument leads to the conclusion that high surface tension will reduce cavitation damage by reducing the number and size of bubbles. No direct evidence is available to the author's knowledge, since any experiment in which different values of surface tension are involved also involves changes in numerous other related parameters.

Other fluid properties of probably less importance to damage (except for extreme variations) are viscosity, sonic velocity, and bulk modulus. Numerical calculations of bubble collapse [9] show that even relatively large viscosities do not have much effect upon collapse velocity. Presumably, increased viscosity can only reduce damage, but no pertinent tests where only viscosity is varied are available, to the author's knowledge. The same is true of sonic velocity and bulk modulus, although in both cases reduced values should in general reduce cavitation damage.

A final and most important fluid property to be considered is corrosivity. This is certainly not in the nature of a "scale effect," but it can obviously strongly affect damage rates under conditions of constant sigma and similar geometry. However, its increase certainly leads to increased damage (except for "completely" noncorrodible materials). In the same category the solids content of the liquid might be considered, in cases where this is a factor (as with slurry flows). Little information on this effect appears to exist at the moment, although it can be important in certain solids-transporting pipeline applications (coal or ore-bearing pipelines).

5-4-6 Conclusions

One general conclusion is that damage rate can increase very strongly with increased velocity, pressure, or size, when sigma is maintained constant. Damage

† Here defined as energy/mass, i.e., ft-lbf/lbm in English "engineering units."

rates are in general more sensitive to these parameters than to any others. The velocity damage exponent is likely to lie in the range from 4 to 6 and the diameter exponent in the range from 2 to 4. Considerable further systematic experimentation is required before these effects can be delineated more closely. Uncertainty in erosion scale effects is also due to the lack of any universally accepted criteria in the measure of damage parameters. Only damage intensities from rather identical eroding environments can usefully be compared at this time.

5-5 DAMAGE-PREDICTING CAPABILITY AND POSSIBILITIES

5-5-1 Introduction

One of the major problems at present concerning liquid cavitation and impact erosion is our relative inability to predict erosion rates, or even the likely existence of important erosion, in field devices, from available laboratory tests or theoretical models. This problem exists for all liquids of interest ranging from cryogenics to water, and also including liquid metals such as sodium, used as coolant for nuclear fast-breeder reactors. Even the relative erosion resistance of materials measured in the same type of facility, but with minor differences in operating parameters, are often not the same, even to a useful engineering approximation. In addition, there is no presently known technique for applying this type of test result to field devices. This was illustrated forcefully in a recent "round-robin" conducted by the ASTM Committee G-2 [87] for the simplest and most readily standardizable type of cavitation test device, i.e., the vibratory horn, described also in Ref. 6 and numerous other papers and articles.†

A similar situation exists for liquid-impact erosion, for which a second ASTM "round-robin" has also been held [137]. Liquid impact is even more difficult to standardize than cavitation because of the present lack of a relatively conventional test device, a situation which is even worse for the case of solid-particle impact which is not within the primary scope of this book. This section attempts to summarize the status of liquid-erosion laboratory testing and its relation to erosion in field machines. In addition, some of the pertinent experimental and theoretical work to improve the present state of the art is summarized.

One of the major difficulties involved in the prediction of field-machine erosion from laboratory tests is due to the fact that liquid erosion involves both mechanical and corrosive effects. In long-term field exposure, the corrosive effects can be relatively much more important than in the laboratory devices, which in general are intended to accelerate the mechanical effects and thus provide tests of short-enough duration to be feasible from the viewpoint of laboratory tests. To provide realistic modeling for specific field applications, it is necessary either to accelerate both modes of attack in suitable proportion or to provide methods of measuring them separately, i.e., identify somehow the mechanical and corrosive

† See ANSI/ASTM Standard Method G 32-77.

effects in laboratory tests, so that a more meaningful application to field devices can be made. Work to achieve these goals is described.

One objective then is to measure, with greater precision than has been previously possible, the characteristics of the cavitation or liquid-impact regime in both laboratory and field devices, and thus make more meaningful predictions than previously possible of probable prototype field results from laboratory tests. To accomplish this goal, one possibility is to measure and count actual individual pressure pulses from bubble collapse or droplet impact originating from the flow regime in both laboratory and field devices. This would have the advantage of increasing basic understanding of the cavitation or impact flow regimes involved, and also of providing a comparison of their "intensity" in various field and laboratory devices.

Liquid impingement is now supposed [5, 6] to be a major contributory mechanism in cavitation erosion through the asymmetric collapse of bubbles and the generation of a liquid "microjet," which impinges upon the material surface. The author's laboratory and others have, therefore, developed various numerical models of individual droplet or jet impact, with the goal of providing the capability for estimating actual stresses and strains in a material so impacted. These models of course apply to direct droplet or jet impact as well as to cavitation. Hopefully, it may eventually become possible to then compute expected rates of erosion if size, velocity, shape, and number of impacting droplets is known. The likelihood of a realistic calculating capability of this type for droplet impact appears to be much greater than for cavitation erosion, because in most cases the details of the attacking flow regime, i.e. "microjet," pressure pulses, etc., are known much less precisely. It is for this reason that many present efforts on cavitation are primarily concerned with providing a technique for measuring and counting bubble-collapse pulses. These spectra can hopefully be used as a measure of the "intensity" of the cavitation-regime attack. Brief details of the mathematical modeling of liquid impact and bubble collapse will be provided here, though the question is also discussed elsewhere in this book.

5-5-2 Computer Modeling of Bubble Collapse and Droplet and Jet Impact

General background The phenomenon of liquid-solid impact has technological importance in various engineering applications, including steam turbines, rain erosion of aircraft or missile components, and also cavitation through the liquid "microjet" mechanism previously discussed. Its detailed study using sophisticated computer models is thus worthwhile in attempting to promote the ability to predict damage. It has thus been pursued vigorously in the author's laboratory over the past several years, as well as elsewhere. For simplicity, primarily our own work will be discussed as illustrative [2–4, 138–140]. Previously we had also conducted several detailed studies of bubble collapse utilizing both high-speed photography [12, 141, 142] and computer models [9, 53, 143, 144]. Some of these studies [9, 53] considered spherically symmetric bubble collapse, but included all pertinent real-fluid effects, including "thermodynamic" restraints [53]. Our other studies considered the effects of nonsymmetrical influences upon bubble collapse [12, 141–144] such as the presence of adjacent walls, pressure gradients, relative ("slip") velocity around the collapsing bubble, and multibubble effects [142], but neglected real-fluid effects other than viscosity. Photographic studies here [12, 141–142] and elsewhere [39–41] had shown the controlling importance in many cases of these nonsymmetric effects. Unfortunately, it does not appear feasible at this point to include together asymmetry and real-fluid effects, or the effects of multiple bubbles, in a completely general computer solution.

In summary [5, 6] the results of the above and other recent studies indicate the probable predominance of the liquid microjet mechanism in cavitation damage. It is possible from these studies to estimate the probable velocity (~ 300 m/s) and diameter (~ 1 to 10 μm) of such microjets in typical cases. Hence, the microjet velocity magnitude is roughly typical of other droplet or jet impact applications, such as the steam-turbine droplet-erosion problem, but its diameter is orders of magnitude less, since droplet diameters in the order of 1 mm are typical in conventional droplet-erosion problems. Hence it may be assumed that cavitation and conventional droplet erosion are very similar in mechanism as well as appearance, although the cavitation erosion may be generally of finer texture, and this has in fact been the general observation for many years. Full details of these and other computer studies are given in the appropriate sections of Chaps. 4 and 6.

5-5-3 Cavitation and Impact Erosion-Prediction Techniques

General The previous section has discussed computer modeling of the basic phenomena and high-speed photography as basic tools for developing the ability to predict cavitation or liquid-impact erosion rates. It may eventually become possible using this approach to estimate realistically the stress-time regimes existing in the material as a result of liquid impact or cavitation. This capability is closer now for impact than for cavitation, since the cavitation phenomenon is basically considerably more complex. However, even if this capability should be attained, there would still remain the apparently even more difficult problem of accurately modeling the material removal process, given the pertinent time-dependent stress-strain values. Computer modeling of the erosion process would obviously depend upon the type of material and the relative importance of corrosive effects, as well as upon many other complicating factors, so that a general solution of this portion of the overall liquid-erosion process seems even more remote than that of the fluid-flow phenomena, previously discussed. Nevertheless, continued, though relatively gradual, progress from the side of computer modeling of the basic phenomena is certainly useful in increasing basic understanding of the phenomena involved and gradually improving overall predicting capabilities.

Noise and bubble-collapse pulse spectra Since the capability for predicting erosion rates a priori from relatively basic principles does not presently exist, another possibility is to measure some easily and quickly measurable aspect of the cavitation or impact phenomenon and then correlate this measurement with measured damage. The measurement of "noise" provides such a possibility, which appears to be of potential importance.

The use of noise for the detection of cavitation has been an accepted practice for years, and several recent attempts have been made to correlate overall noise with erosion [145–147]. Some success has been obtained in specific experiments with individual pumps or venturis, showing that damage rate and noise curves roughly correlate, maximizing, for example, near the pump head fall-off point (knee of curve). However, it appears that useful general correlations cannot be attained in this way. Hence, we have here attempted to develop a technique whereby bubble-collapse pressure pulse "spectra" are measured. If it is assumed that bubble-collapse pressure pulse durations are roughly uniform (or at least uniquely related to the pressure magnitude), then the area under such a spectrum curve provides a measure of the total impulse (or energy) delivered to the damaged surface or to a probe presumably located at a position equivalent to that of the surface to be damaged from the viewpoint of the cavitation field. Such a technique is sensitive to not only the number of "blows" delivered to the surface but also to their strength. It thus appears to provide vital information beyond that provided either by a simple noise measurement or a noise frequency spectrum. Some relatively limited but successful experimentation with methods similar to those suggested here has been reported [27–29, 147, 148], using primarily vibratory cavitation-damage facilities, as we have also in our work to date [30, 149–151]. Similar work using a venturi facility is now in progress at U-M [152].

Experiments here have included cavitation-damage rate measurements in both water and molten sodium over a range of temperatures and suppression pressures in a 2 mil† vibratory horn cavitation-damage facility (20 kHz). Detailed results are found elsewhere [30, 149–151], but Fig. 5-24 illustrates the type of correlation obtained between measured bubble-collapse pressure spectrum area and measured damage rate (MDPR = mean depth of penetration rate, i.e., volume loss rate per unit of exposed area). The best-fit relation is described by

$$\text{MDPR} = C(\text{spectrum area})^{1/n} \tag{5-14}$$

where C is a constant obtained empirically. The average value obtained for the exponent n is ~ 5 for high-intensity tests and ~ 1 for lower-intensity results.

The pulse-count energy-spectrum area measurement will also allow computation of the "efficiency" η_{cav} of the delivery of energy to the damaged surface from the region of bubble collapse, as compared with the energy operative in the erosion process. This latter quantity, according to much previous work, can best be expressed in the form of an ultimate resilience, UR (see Sec. 5-3), i.e., "strain energy to failure," if failures are of a brittle (rather than a ductile) nature, as generally appears to be the case for cavitation. Such a relationship would then be of the form:

$$\text{Spectrum area} = C \cdot \eta_{cav} \cdot \text{UR} \tag{5-15}$$

† 2 mil $\cong$ 51 μm.

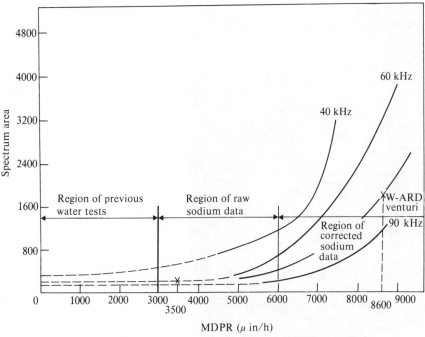

Figure 5-24 Correlation of pulse pressure spectrum area and MDPR [149–151].

This equation considers the erosion-rate model discussed in Sec. 5-3, where erosion energy efficiencies of this type are discussed. However, the measurement of spectrum area would allow the computation of the otherwise unknown "efficiency factor." In Eq. (5-15), C is a constant involving the calibration constant for the bubble-collapse pulse-measuring pressure probe, and η is the cavitation-erosion "efficiency" sought. Preliminary results from our cavitation vibratory facility (Fig. 5-25) [150, 151] indicate $\eta_{cav} \cong 10^{-8}$. Such a very small value is intuitively not surprising.†

If found appropriate after additional research, material failure energy properties other than UR could be used in Eq. (5-14). If η_{cav} were known for a class of cavitation regimes, then erosion rate (MDPR) could be calculated from the pulse-count measurement.

Figure 5-26 shows typical individual bubble-collapse pulses for water in the U-M venturi measured by a Kistler and a U-M designed microprobe. Figure 5-13 shows schematic pulse-count spectra postulated in an earlier (1963) paper by the author [34] for cavitation in our venturi. The standard percentage of error for the MDPR correlation [Eq. (5-14), Fig. 5-24] was only ~20 percent. This seems surprisingly small for experiments of this kind. On-going work [152] continues this type of experiment in a venturi facility with water; eventually in field devices. No field studies have been reported by the time of writing.

---

† It is largely a result of the area ratio between probe and microjet.

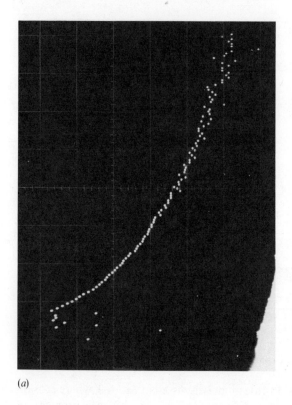

(a)

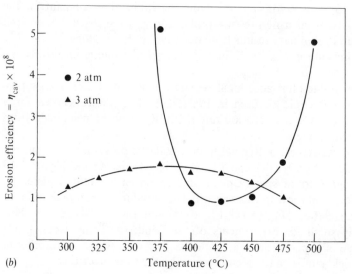

(b)

Figure 5-25 (a) Spectrum of pressure pulses from venturi [30]. (b) Temperature vs. cavitation-erosion efficiency for 80 kHz, vibratory facility [150, 151].

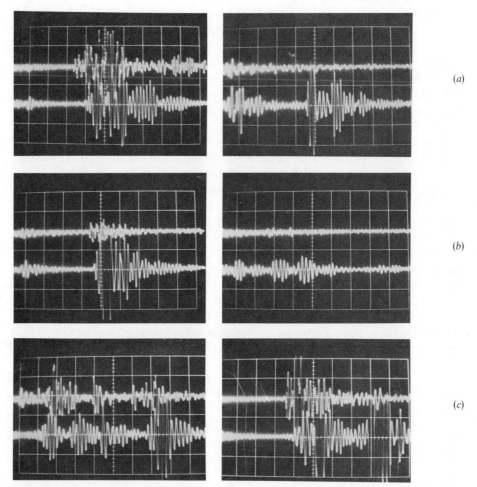

Figure 5-26 Response of U-M and Kistler microprobes (positioned on a plane of symmetry in a venturi) to pressure waves emanated by collapsing cavities [150–152].

Though the evidence yet available is relatively sparse, it appears that an a priori erosion rate predicting technique may be developed in the relatively near future for application to many field devices. Such a capability, within even very rough limits of possible engineering utility, is unfortunately virtually nonexistent today.

While the pressure-pulse spectrum technique has so far [149–152] been applied only to cavitation erosion, it appears applicable also to impact-erosion cases, depending of course on geometrical considerations. Its eventual application to rotating components also seems feasible, but obviously involves geometric considerations pertinent to specific applications.

Figure 5-26 shows typical bubble-collapse pulses measured in the U-M cavitating venturi. The rapid oscillations accompanying the initial pulse are believed to result from the probe "ringing" frequencies of ∼0.1 mHz.

Erosion "acoustic emission noise" "Acoustic emission" from material deformation (microcrack formation, etc.) is a well-known phenomenon, so that it is conceivable that this could be used to detect and measure erosion directly from an acoustic probe attached to the eroding surface. However, at the present time this appears to be improbable of success, since it appears that the level of acoustic emission noise is orders of magnitude less than that of cavitation, so that the cavitation noise itself would mask the acoustic emission noise, making it impossible to detect. However, since little or no pertinent research appears to have as yet been done, it is possible that this expectation may prove to be in error.

Somewhat pertinent to this possibility is an observation from a high-speed motion picture sequence taken here of an apparent ejected particle from soft aluminum. It had an ejection velocity of ∼100 m/s normal to the surface, and was of elongated shape (∼1 mm length). If such high ejection energy is typical, the associated noise could be sufficient to detect, considering the good acoustic transmission to be expected from a probe attached to the surface as compared to that from bubbles collapsing in the liquid.

Erosion particle size Pertinent to the possibility of detecting erosion by its associated "emission noise," as discussed in the foregoing, is knowledge of eroded

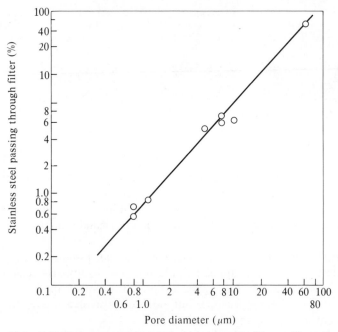

Figure 5-27 Percentage stainless steel which permeates filter vs. filter rating.

particle size and mode of removal from the eroded surface. Some information concerning the "ejection velocity" of such a particle was given in the last section. Very few studies exist as yet concerning the size distribution of cavitation debris. However, one such study was conducted some years ago here [153] where neutron-irradiated stainless-steel test specimens were used in a water venturi system. The irradiated eroded particles could thus be isolated from other miscellaneous particles in the system and sieved. It was found that particle diameters ranged up to ~0.1 mm, with the greatest number of particles being much smaller. Typical results are shown in Fig. 5-27. It can thus be concluded from this study, and whatever other fragmentary information exists, that the debris particle sizes are certainly not uniform, that they no doubt depend strongly upon the material and flow parameters, that no minimum size can be specified, and that the maximum in some cases could be in the order of a millimeter.

Suppression pressure (or NPSH) effects on damage rates Figure 5-28 shows schematically the presumed effect of suppression pressure or NPSH upon damage rate (e.g., MDPR). Figure 5-18b shows actual measurements from a vibratory damage facility, indicating the surprising result (at first glance) of a strong increase in MDPR for an increase in NPSH. The trend with suppression pressure would, of course, be the same. Figure 5-28 indicates that this is only one portion

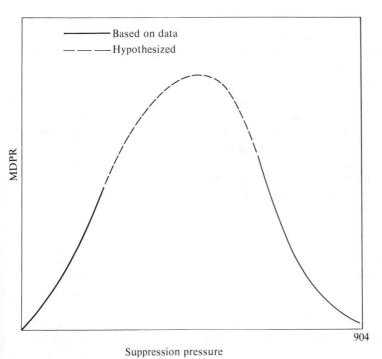

904

Suppression pressure

Figure 5-28 Cavitation-erosion rate vs. suppression pressure (or NPSH).

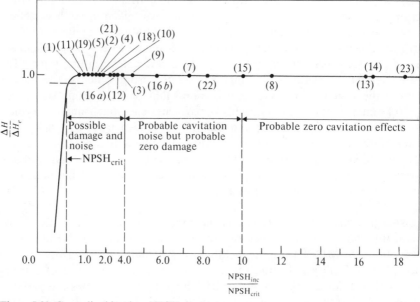

Figure 5-29 Generalized head vs. NPSH pump curve.

of an overall curve applying qualitatively to any cavitating device, showing that there exists an "optimum" damage-suppression pressure (or NPSH) for any device, i.e., a suppression pressure for which the damage rate will be a maximum. While the vibratory device results apply only to the low NPSH portion of the curve, most cavitating flow machines normally, but not always, operate in the high NPSH portion of the curve so that an increase in NPSH will usually reduce damage. However, as shown in Fig. 5-28, this is not always the case. For example, it has been reported that the maximum damage rate for centrifugal pumps often occurs above or near the fall-off of ΔH in the standard ΔH versus NPSH curve (Fig. 5-29). In fact some damage and cavitation may well occur at NPSH values at least four times the standard inception NPSH [154, 155] (Hydraulic Institute specifies 3 percent head fall-off as the "inception point"), so that in some cases an increase in NPSH will result in increased damage rate.

The schematic representation of Fig. 5-28 is easily justified on the following basis. It is clear that damage will be zero at very high NPSH, because under such conditions cavitation will not exist. Reducing NPSH produces a few bubbles once the actual inception point is reached, but these collapse relatively violently because of the relatively high surrounding pressure. Continued reduction of NPSH produces an increased number of bubbles but reduced collapse violence. Damage increases with reducing NPSH until the reduced collapse violence balances the increasing number and size of bubbles, producing a maximum damage rate at "damage optimum" NPSH. Further NPSH reduction below this "optimum" NPSH results in reduced damage until finally the damage rate approaches zero,

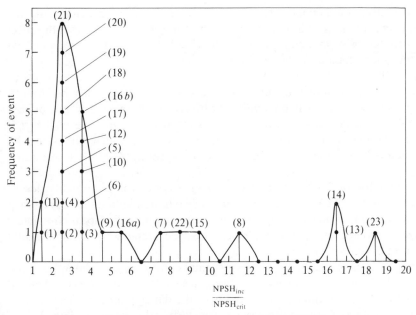

Figure 5-30 Frequency of event vs. ratio of inception to critical NPSH*.

as collapse violence becomes small, because of reduced surrounding suppression pressure.

Figures 5-29 and 5-30 show recent results [154, 155], indicating that in many centrifugal pumps possibly damaging bubble collapse exists at NPSH values at least four times that corresponding to the conventional head fall-off point. These data were obtained from ~20 pump tests reported from numerous laboratories where acoustic or visual observations (transparent casing) were made to allow the determination of actual inception without regard to the external measurement of head fall-off.

Time effects on damage rate Figure 6-35 represents schematically the conventional S-shaped cumulative damage and damage rate versus exposure time curve.†
The importance of exposure-time effects in erosion tests was perhaps first emphasized by Thiruvengadam [84], who presented similar idealized curves. In any case it is now well recognized that in both cavitation and droplet-impact erosion, the curves have the general trend shown, i.e., an initial "incubation period" where little or no damage occurs, followed by an increasing rate building up to a maximum rate period, after which the damage rate decreases, sometimes approaching an eventual approximately constant rate. Some tests, however, show more complex curves with two or more maximum rate periods. Further discussion is provided in Chap. 6.

† Pertinent either to liquid impact or cavitation.

Air-content effects on damage rate   Figure 3-20*b* summarizes schematically the presumed effect of gas (or air) content upon the damage rate. In general the effect is similar to that of suppression pressure (discussed above), in that moderate gas content produces maximum damage, whereas very high or very low gas content reduces the damage rate to near zero. However, only the central portion of this curve has yet been demonstrated experimentally, as explained further in Chap. 3. Very high gas contents will no doubt effectively "cushion" the vapor-bubble collapse, so that little damage will result. This result is partially due to the high internal gas content of the bubbles under such conditions, which will essentially prevent their collapse, and also to the increased compressibility of the surrounding liquid.

The reduction of damage rate at very low gas content is presumably due to the reduced number of gas "nuclei" and hence of cavitation bubbles. It is presumed that cavitation damage in many engineering situations would disappear completely at near-zero gas content, because the inherent tensile strength of the liquid would then prevent cavitation entirely.

Figure 3-20*b* is based upon total (rather than entrained) gas content, because in almost all cases only total gas has been measured. Nevertheless, entrained gas is probably more intimately involved with the damage process. The state of the art concerning entrained gas measurements is fully discussed in Chap. 3.

REFERENCES

1. Hammitt, F. G.: Erosive Wear Testing, in "Selection and Use of Wear Tests for Metals," *ASTM STP* 615, pp. 45–67, 1975.
2. Huang, Y. C.: "Numerical Studies of Unsteady, Two-Dimensional Liquid Impact Phenomena," Ph.D. thesis, Mech. Engr. Dept., University of Michigan; also available as University of Michigan ORA Report UMICH 03371-8-T, June, 1971; see also Y. C. Huang, F. G. Hammitt, and W. J. Yang: Hydrodynamic Phenomena During High-Speed Collision Between Liquid Droplet and Rigid Plane, *Trans. ASME, J. Fluids Engr.*, vol. 95, ser. D, no. 2, pp. 276–294, 1973.
3. Hwang, J. B.: "The Impact Between a Liquid Drop and an Elastic Half-Space," Ph.D. thesis, Mech. Engr. Dept., University of Michigan; also available as University of Michigan ORA Rept. UMICH 012449-5-T, March, 1975.
4. Hwang, J. B., and F. G. Hammitt: Transient Distribution of the Stress During the Impact Between a Liquid Drop and an Aluminum Body, *BHRA Third Int'l Symp. Jet Cutting Technology*, May 11–13, 1976, Chicago; also available as University of Michigan ORA Report UMICH 012449-8-T, June, 1975.
5. Hammitt, F. G., and F. J. Heymann: Liquid-Erosion Failures, in "Metals Handbook," vol. 10, 8th ed., pp. 160–167, American Society of Metals, Metals Park, Ohio, 1975.
6. Knapp, R. T., J. W. Daily, and F. G. Hammitt: "Cavitation," McGraw-Hill, New York, 1970.
7. Knapp, R. T., and A. Hollander: Laboratory Investigations of the Mechanism of Cavitation, *Trans. ASME*, vol. 70, pp. 419–435, 1948.
8. Hickling, R., and M. S. Plesset: Collapse and Rebound of a Spherical Bubble in Water, *The Physics of Fluids*, vol. 7, no. 1, pp. 7–14, 1964.
9. Ivany, R. D., and F. G. Hammitt: Cavitation Bubble Collapse in Viscous, Compressible Liquids-Numerical Analysis, *Trans. ASME, J. Basic Engr.*, vol. 87, ser. D, no. 5, pp. 977–985, 1965.

10. Hammitt, F. G.: Collapsing Bubble Damage to Solids, *Cavitation State of Knowledge, ASME*, pp. 87–102, 1969.
11. Robinson, M. J., and F. G. Hammitt: Detailed Damage Characteristics in a Cavitating Venturi, *Trans. ASME, J. Basic Engr.*, vol. 89, ser. D, pp. 161–173, 1967.
12. Kling, C. L., and F. G. Hammitt: A Photographic Study of Spark-Induced Cavitation Bubble Collapse, *Trans. ASME, J. Basic Engr.*, vol. 94, ser. D, no. 4, pp. 825–833, December, 1972.
13. Wagner, H. A., J. M. Decker, and J. C. Marsh: Corrosion-Erosion of Boiler Feed Pumps and Regulating Valves, *Trans. ASME*, vol. 69, pp. 389–403, May, 1947.
14. Decker, J. M., H. A. Wagner, and J. C. Marsh: Corrosion-Erosion of Boiler Feed Pumps and Regulating Valves at Marysville, 2nd Test Program, *Trans. ASME*, vol. 72, pp. 19–36, 1950.
15. Ackeret, J., and P. de Haller: Untersuchen uber Korrosion durch Wasserstoss, *Schweiz. Bauzeitung*, vol. 98, pp. 309–310, 1931.
16. Elliott, D. E., J. B. Marriott, and A. Smith: Comparison of Erosion Resistance of Standard Steam Turbine Blade and Shield Materials on Four Test Rigs, *ASTM STP 474*, pp. 127–161, October, 1970.
17. Wahl, N. E.: "Investigation of the Phenomena of Rain Erosion at Supersonic Speeds," Air Force Materials Lab. Rept. AFML-TR-65-330, WPAFB, Dayton, Ohio, October, 1965.
18. Schmitt Jr., G. F., and A. H. Krabill: "Velocity-Erosion Rate Relationships of Materials in Rain at Supersonic Speeds," Air Force Materials Lab. Rept. AFML-TR-70-44, WPAFB, Dayton, Ohio, October, 1970.
19. Schmitt Jr., G. F., W. G. Reinecke, and G. D. Waldman: Influence of Velocity, Impingement Angle, Heating, and Aerodynamic Shock Layers on Erosion of Materials at Velocities of 5500 f/s (1700 m/s), *ASTM STP 567*, pp. 219–238, December, 1974.
20. Steller, K.: Personal communication, Institute of Fluid Flow Machines, Polish Academy of Science, Gdansk, Poland, 1974.
21. Boetcher, H. N.: Failure of Metals due to Cavitation under Experimental Conditions, *Trans. ASME*, vol. 58, pp. 355–360, 1936.
22. Thiruvengadam, A.: "A Comparative Evaluation of Cavitation Damage Test Devices," Hydronautics, Tech. Rept. 232-2, November, 1963.
23. Shal'nev, K. K.: Experimental Study of the Intensity of Erosion due to Cavitation, in "Cavitation in Hydrodynamics," vol. 22, pp. 1–37, Nat'l Phys. Lab., Teddington, England, October, 1955.
24. Rasmussen, R. E. H.: Some Experiments on Cavitation Erosion in Water Mixed with Air, in "Cavitation in Hydrodynamics," vol. 20, pp. 1–25, Nat'l Phys. Lab., Teddington, England, October, 1955.
25. Wood, G. M., L. K. Knudson, and F. G. Hammitt: Cavitation Studies with Rotating Disk, *Trans. ASME, J. Basic Engr.*, vol. 89, ser. D, pp. 98–110, 1967.
26. Anonymous: "Standard Method of Vibratory Cavitation Erosion Test," ASTM, G 32-72, 1972, and G 32-76, 1976.
27. Numachi, F.: An Experimental Study of Accelerated Cavitation Induced by Ultrasonics, *Trans. ASME, J. Basic Engr.*, vol. 87, pp. 967–976, 1965.
28. Numachi, F., and M. Hongo: Ultrasonic Shock Waves Emitted by Cavitation at Perforation on Plate/Rept. 1, *Rept. Inst. High Speed Mech.*, vol. 30, p. 277, 1974.
29. Makarov, V. K., A. A. Kortnev, S. G. Suprun, and G. I. Okolelov: Cavitation Erosion Spectra Analysis of Pulse-Heights Produced by Cavitation Bubbles, *Proc. Sixth Non-Linear Acoustics Conf.*, July, 1975, (Odessa Poly. Inst.) Moscow.
30. Hammitt, F. G., J. B. Hwang, M. K. De, et al.: "Final Report for Argonne National Laboratory Project," University of Michigan ORA Rept. UMICH 013503-2-F, June, 1976.
31. Lord Rayleigh: On the Pressure Developed in a Liquid During the Collapse of a Spherical Cavity, *Phil. Mag.*, vol. 34, pp. 94–98, 1917.
32. Hammitt, F. G.: Collapsing Bubble Damage to Solids, *Cavitation State of Art Symp.*, June, 1969, ASME.
33. Hammitt, F. G., et al.: Initial Phases of Damage to Test Specimens in a Cavitating Venturi, *Trans. ASME, J. Basic Engr.*, vol. 87, ser. D, no. 2, pp. 453–464, 1965.

34. Hammitt, F. G.: Observations on Cavitation Damage in a Flowing System, *Trans. ASME, J. Basic Engr.*, vol. 85, ser. D, no. 3, pp. 347–359, 1963.
35. Robinson, M. J., and F. G. Hammitt: Detailed Damage Characteristics in a Cavitating Venturi, *Trans. ASME, J. Basic Engr.*, vol. 89, ser. D, no. 1, pp. 161–173, 1967.
36. Knapp, R. T.: Recent Investigations of the Mechanics of Cavitation and Cavitation Damage, *Trans. ASME*, pp. 1045–1054, October, 1955.
37. Kling, C. L.: "A High Speed Photographic Study of Cavitation Bubble Collapse," Ph.D. thesis, Nuclear Engr. Dept., University of Michigan, 1970; also available as University of Michigan ORA Rept. UMICH 03371-2-T, March, 1970.
38. Timm, E. E.: "An Experimental Photographic Study of Vapor Bubble Collapse and Liquid Jet Impingement," Ph.D. thesis, Chem. Engr. Dept., University of Michigan, 1974; also available as University of Michigan ORA Report UMICH 01357-39-T, June, 1974.
39. Naudé, C. F., and A. T. Ellis: On the Mechanism of Cavitation Damage by Nonhemispherical Cavities Collapsing in Contact with a Solid Boundary, *Trans. ASME, J. Basic Engr.*, vol. 83, ser. D, no. 4, pp. 648–656, 1961.
40. Benjamin, T. B., and A. T. Ellis: The Collapse of Cavitation Bubbles and the Pressures Thereby Produced Against Solid Boundaries, *Phil. Trans. Roy. Soc.*, ser. A, vol. 260, no. 1110, pp. 221–240, 1966.
41. Shutler, N. D., and R. B. Mesler: A Photographic Study of the Dynamics and Damage Capabilities of Bubbles Collapsing Near Solid Boundaries, *Trans. ASME, J. Basic Engr.*, vol. 87, ser. D, no. 2, pp. 511–517, 1965.
42. Ivany, R. D., F. G. Hammitt, and T. M. Mitchell: Cavitation Bubble Collapse Observations in a Venturi, *Trans. ASME, J. Basic Engr.*, vol. 88, ser. D, no. 3, pp. 649–657, 1966.
43. Eller, A., and H. G. Flynn: The Equilibrium and Stability of a Translating Cavity in a Fluid, *J. Fluid Mech.*, vol. 30, pt. 4, pp. 785–803, 1967.
44. Yeh, H. C., and W. J. Yang: Dynamics of Bubbles Moving in Liquids with Pressure Gradient, *J. Appl. Phys.*, vol. 19, no. 7, pp. 3156–3165, 1968.
45. Gibson, D. C.: "The Collapse of Vapour Cavities," Ph.D. thesis, Churchill College, Cambridge University, July, 1967.
46. Rattray, M.: "Perturbation Effects in Cavitation Bubble Dynamics," Ph.D. thesis, California Institute of Technology, Pasadena, Calif., 1952.
47. Plesset, M. S., and R. B. Chapman: Collapse of an Initially Spherical Vapor Cavity in the Neighborhood of a Solid Boundary, *J. Fluid Mech.*, vol. 47, no. 2, p. 283, May, 1971.
48. Mitchell, T. M., and F. G. Hammitt: Asymmetric Cavitation Bubble Collapse, *Trans. ASME, J. Fluids Engr.*, vol. 95, no. 1, pp. 29–37, March, 1973.
49. Korovkin, A. N., and Y. L. Levkovskiy: Closing of Cavitation Caverns Close to a Solid Wall (in Russian), *J. of Engr. Phys.*, vol. XII, no. 2, pp. 246–253, April, 1967.
50. Brunton, J. H.: The Physics of Impact and Deformation: Single Impact, *Phil. Trans. Roy. Soc.*, ser. A, vol. 260, no. 1110, pp. 79–85, 1966.
51. Barclay, F. J., T. J. Ledwidge, and G. C. Cornfield: Discussion of F. G. Hammitt, Damage Due to Cavitation and Sub-Cooled Boiling Bubble Collapse, *Proc. I. Mech. E.*, 1968–1969, vol. 183, pt. I.
52. Hickling, R.: Some Physical Effects of Cavity Collapse in Liquids, *Trans. ASME, J. Basic Engr.*, vol. 88, ser. D, no. 1, pp. 229–235, 1966; see also Effects of Thermal Conduction in Sonoluminescence, *J. Acous. Soc. Amer.*, vol. 35, no. 7, pp. 967–974, July, 1963.
53. Mitchell, T. M., and F. G. Hammitt: On the Effects of Heat Transfer Upon Collapsing Bubbles, *Nucl. Sci. and Engr.*, vol. 53, no. 3, pp. 263–276, 1974.
54. Ellis, A. T.: On Jets and Shock Waves from Cavitation, *Proc. Sixth Naval Symp.*, October, 1966, pp. 6-1 to 6-19, Washington, D.C.
55. Lauterborn, W., and H. Bolle: Experimental Investigations of Cavitation-Bubble Collapse in the Neighbourhood of a Solid Boundary, *J. Fluid Mech.*, vol. 72, no. 2, pp. 391–399, 1975.
56. Ellis, A. T.: Private communication, 1968.
57. Canavelis, R.: "Comparison of the Resistance of Different Materials with a Jet Impact Test Rig," Electricité de France Rept. HC/061-230-9, Chatou, France, November, 1967.

58. Delhaye, J. M., R. Semeria, and J. C. Flamand: Void Fraction, Vapor and Liquid Temperature. Local Measurements in Two-Phase Flow Using a Microthermocouple, *J. Heat Transfer*, pp. 365–370, August, 1973.

59. Sutton, G. W.: A Photo-Elastic Study of Strain Waves Caused by Cavitation, *J. Appl. Mech.*, vol. 24, no. 3, pp. 340–348, 1957.

60. Ellis, A. T.: "Parameters Affecting Cavitation and Some New Methods for Their Study," California Institute of Technology Hydrodynamics Lab. Rept. E-115.1, October, 1965.

61. Bober, W.: "An Analytical Investigation of the Pressure Field in a Cavitating Flow," ASME Paper 69-FE-34, 1969.

62. Johnson, V. E., and T. Hsieh: The Influence of Entrained Gas Nuclei Trajectories on Cavitation Inception, *Proc. Sixth Naval Hydrodynamics Symp.*, October, 1966, Washington, D.C.

63. Shal'nev, K. K.: Experimental Study of the Intensity of Erosion Due to Cavitation, in "Cavitation in Hydrodynamics," vol. 22, pp. 1-22, 37, Nat'l Phys. Lab., Teddington, England, October, 1955.

64. Hammitt, F. G.: Cavitation Damage Scale Effects—State of Art Summarization, *J. Hyd. Research (IAHR)*, vol. 13, pp. 1–18, 1975.

65. Young, S. G., and J. R. Johnston: "Effect of Cover Gas Pressures on Accelerated Cavitation Damage in Sodium," NASA TN, D-4235, 1967.

66. Hammitt, F. G., and D. O. Rogers: Effects of Pressure and Temperature Variation in Vibratory Cavitation Damage Test, *J. Mech. Engr. Sci.*, vol. 12, no. 6, pp. 432–439, 1970.

67. Hammitt, F. G., and N. R. Bhatt: Cavitation Damage at Elevated Temperature and Pressure, *1972 ASME Polyphase Flow Forum*, pp. 11–13.

68. Hobbs, J. M.: Experience with a 20-KC Cavitation Erosion Test, *ASTM STP* 408, pp. 159–179, 1967.

69. Stahl, H. A., and A. J. Stepanoff: Thermodynamic Aspects of Cavitation in Centrifugal Pumps, *Trans. ASME*, vol. 78, pp. 1691–1693, 1956.

70. Bonnin, J.: "Theoretical and Experimental Investigations of Incipient Cavitation in Different Liquids," ASME Paper 72-WA/FE-31, 1972.

71. Bonnin, J.: "Influence des Effects Thermiques sur l'Erosion de Cavitation (Essais Vibratoires)," Electricité de France, Direction des Etudes et Recherches, Chatou (Internal report), June, 1972.

72. Garcia, R., and F. G. Hammitt: Cavitation Damage and Correlations with Material and Fluid Properties, *Trans. ASME., J. Basic Engr.*, vol. 89, ser. D, no. 4, pp. 753–763, 1967.

73. Devine, R. E., and M. S. Plesset: "Temperature Effects in Cavitation Damage," CIT Rept. 85-27, Pasadena, Calif., April, 1964.

74. Hammitt, F. G., Y. C. Huang, et al.: A Statistically Verified Model for Correlating Volume Loss Due to Cavitation or Liquid Impingement, *ASTM STP* 474, pp. 288–322, 1969.

75. Heymann, F. J.: Discussion of Ref. 72, *ASTM STP* 474, pp. 312–315, 1969.

76. Heymann, F. J.: Toward Quantitative Prediction of Liquid Impact Erosion, *ASTM STP* 474, pp. 212–248, 1969.

77. Rao, B. C. Syamala, N. S. Lakshmana Rao, and K. Seetharamiah: Cavitation Erosion Studies with Venturi and Rotating Disk in Water, *Trans. ASME, J. Basic Engr.*, vol. 92, pp. 563–579, September, 1970.

78. Hammitt, F. G.: Damage to Solids Caused by Cavitation, *Phil. Trans. Roy. Soc.*, ser. A, vol. 260, no. 1110, pp. 245–255, 1966.

79. Kemppainen, D. J., and F. G. Hammitt: Effects of External Stress on Cavitation Damage, *Proc. Thirteenth Congr. IAHR*, August, 1969, Kyoto, Japan.

80. Shal'nev, K. K., R. D. Stepanov, and S. P. Kozyrev: Effect of the Stressed State of Metals on Resistance to Cavitation, *Soviet Physics-Doklady*, vol. 11, no. 9, pp. 822–824, March, 1967.

81. Garcia, R., and F. G. Hammitt: Cavitation Damage and Correlations with Materials and Fluid Properties, *Trans. ASME, J. Basic Engr.*, vol. 89, ser. D, no. 4, pp. 753–763, 1967; see also R. Garcia, "Comprehensive Cavitation Damage Data for Water and Various Liquid Metals Including Correlations with Material and Fluid Properties," Ph.D. thesis, Nuclear Engr. Dept., University of Michigan, 1966, Ann Arbor, Michigan.

82. Hoff, G., G. Langbein, and H. Rieger: Material Destruction Due to Liquid Impact, *ASTM STP* 408, pp. 42–69, 1966.

83. ASTM Committee G-2: "Standard Definitions of Terms Relating to Erosion by Cavitation and Impingement," 1973, 1978.

84. Thiruvengadam, A.: A Unified Theory of Cavitation Damage, *Trans. ASME, J. Basic Engr.*, vol. 85, ser. D, no. 3, pp. 365–376, 1963.

85. Hammitt, F. G., et al.: Initial Phase of Damage of Test Specimens in a Cavitating Venturi as Affected by Fluid and Material Properties and Degree of Cavitation, *Trans. ASME, J. Basic Engr.*, vol. 87, ser. D, pp. 453–464, 1965.

86. Heymann, F. J.: "Erosion by Cavitation Liquid Impingement and Solid Impingement," Westinghouse Electric Engineering Rept. E-1460, Mar. 15, 1968.

87. Hammitt, F. G., C. Chao, C. L. Kling, and D. O. Rogers: ASTM Round-Robin Test with Vibratory Cavitation and Liquid Impact Facilities of 6061-T 6511 Aluminum Alloy, 316 Stainless Steel, Commercially Pure Nickel, *ASTM Materials Research and Standards, MTRSA*, vol. 10, no. 10, pp. 16–36, 1970.

88. King, R. B.: Letter to F. G. Hammitt, June 13, 1968. RAE, Farnborough, England.

89. Petracchi, G.: Investigation of Cavitation Corrosion (in Italian), *Metallurgica Italiana*, vol. 41, pp. 1–6, 1944; English summary in *Engr's Digest*, vol. 10, pp. 314–316, 1949.

90. Plesset, M. S.: Cathodic Protection in Cavitation Damage, *Trans. ASME, J. Basic Engr.*, vol. 82, ser. D, pp. 808–820, 1960.

91. Nemecek, S.: K Otazce Urcovani Kavitacni Odolnosti, *Konference o Vodnich Turbinach*, 1958, Brno.

92. Nemecek, S.: K Otazce Mechaniky Kavitace, *I. Sbornik Praci Vysoke Skoly Strojni v Liberci*, pp. 128–146, 1959.

93. Steller, K.: Personal communications with F. G. Hammitt, Inst. Fluid Flow Mach., Polish Academy Science, Gdansk, Poland, 1976.

94. Hammitt, F. G.: Cavitation Damage Scale Effects—State of Art Summarization, *J. Hyd. Research (IAHR)*, vol. 13, no. 1, pp. 1–18, 1975.

95. Hammitt, F. G.: Effects of Gas Content upon Cavitation Inception, Performance, and Damage, *J. Hyd. Research (IAHR)*, vol. 10, no. 3, pp. 259–290, 1972.

96. Hobbs, J. M., and A. Laird: "Pressure Temperature and Gas Content Effects in the Vibratory Test," NEL Report 438, October, 1969; see also *1969 ASME Cavitation Forum*, pp. 3–4.

97. Young, S. G., and J. R. Johnston: "Effect of Cover Gas Pressures on Accelerated Damage in Sodium," NASA TN D-4235, November, 1967.

98. Young, S. G., and J. R. Johnston: "Accelerated Cavitation Damage of Steels and Superalloys in Liquid Metals," NASA TN D-3226, May, 1966.

99. Hammitt, F. G., and D. O. Rogers: Effects of Pressure and Temperature Variation in Vibratory Cavitation Damage Test, *J. Mech. Engr. Sci.*, vol. 12, no. 6, pp. 432–439, 1970.

100. Hammitt, F. G., and N. R. Bhatt: Cavitation Damage at Elevated Temperature and Pressure, *ASME Cavitation and Polyphase Flow Forum*, 1972, pp. 11–13.

101. Robinson, M. J.: "On the Detailed Flow Structure and the Corresponding Damage to Test Specimens in a Cavitating Venturi," Ph.D. thesis, Nuclear Engr. Dept., University of Michigan, 1965.

102. Hammitt, F. G.: Observations on Cavitation Damage in a Flowing System, *Trans. ASME, J. Basic Engr.*, vol. 85, ser. D, pp. 347–359, 1963.

103. Hammitt, F. G., et al.: Initial Phases of Damage to Test Specimens in a Cavitating Venturi, *Trans. ASME, J. Basic Engr.*, vol. 87, ser. D, pp. 453–464, 1965.

104. Rao, B. C. Syamala, and D. V. Chandrasekhara: "Internal Report, Size and Velocity Scale Effects on Damage in a Venturi," Civil Engr. Dept., Bangalore Institute of Technology, 1973.

105. Tullis, P., and R. Govindarajan: Cavitation and Size Scale Effects for Orifices, Proc. paper 9605, *J. Hyd. Div., ASCE*, vol. 99, no. HY3, pp. 417–430, March, 1973.

106. Canavelis, R.: Jet Impact and Cavitation Damage, *Trans. ASME, J. Basic Engr.*, vol. 90, pp. 355–367, September, 1968.

107. Hobbs, J. M.: "Factors Affecting Damage Caused by Liquid Impact," NEL Report 262, Glasgow, December, 1966.

108. Hobbs, J. M., and W. C. Brunton: "Comparative Erosion Tests on Ferrous Materials, Part I: Drop Impact Test," NEL Report 205, Glasgow, November, 1965.
109. Ripken, J. F.: "Comparative Studies of Drop Impingement Erosion and Cavitation Erosion," University of Minnesota Project Rept. 105, St. Anthony Falls Hydraulic Lab., April, 1969; see also J. F. Ripken: A Test Rig for Studying Impingement and Cavitation Damage, *ASTM STP* 408, pp. 3–21, 1967.
110. Lichtman, J. Z., and E. R. Weingram: The Use of a Rotating Disk Apparatus in Determining Cavitation Erosion Resistance in Materials, *ASME Symp. on Cavitation Research Facilities and Techniques*, 1964, pp. 185–196.
111. Shal'nev, K. K., I. I. Varga, and D. Sebestyen: Investigation of the Scale Effects of Cavitation Erosion, *Phil. Trans. Roy. Soc. (London)*, vol. 260, no. 1110, pp. 256–266, July, 1966.
112. Thiruvengadam, A.: "Theory of Erosion," Hydronautics Tech. Rept. 233-11, March, 1967.
113. Thiruvengadam, A.: Cavitation Erosion, *Appl. Mech. Rev.*, vol. 24, no. 3, pp. 245–253, March, 1971.
114. Thiruvengadam, A.: "Effects of Hydrodynamic Parameters on Cavitation Erosion Intensity," Hydronautics Tech. Rept. 233-14, November, 1970.
115. Thiruvengadam, A.: "Scaling Laws for Cavitation Erosion," Hydronautics Tech. Rept. 233-15, December, 1971.
116. Knapp, R. T.: Recent Investigations of Cavitation and Cavitation Damage, *Trans. ASME*, vol. 77, pp. 1045–1054, 1955.
117. Knapp, R. T.: Accelerated Field Tests of Cavitation Intensity, *Trans. ASME*, vol. 80, pp. 91–102, 1958.
118. Steller, K., Z. Reymann, and T. Krzysztofowicz: Evaluation of the Resistance of Materials to Cavitation Erosion, *Proc. Fifth Conf. on Fluid Machinery*, 1975, pp. 1081–1096, Akademiai Kiado, Budapest, Hungary.
119. Steller, K., T. Krzysztofowicz, and Z. Reymann: Effects of Cavitation on Materials in Field and Laboratory Conditions, in "Erosion, Wear, and Interfaces with Corrosion," *ASTM STP* 567, pp. 152–170, 1974.
120. Steller, K.: Wstepne Wyniki Badan nad Odpornoscia Tworzyw Sztucznych na Dzialanie Kawitacji, *Prace Instytutu Maszyn Przeplywowych*, vol. 61, 1973.
121. Thuss, W. (J. M. Voith GmbH): Letters to F. G. Hammitt, 1972–1973.
122. Giraud, H.: Etude Detaillée de la Comparaison entre Essais Industriels et Modèle dans Quelques Cas Bien Definis, *La Houille Blanche*, no. 3, pp. 299–312, 1966.
123. Hammitt, F. G., and N. R. Bhatt: Temperature and Pressure Effects in Vibratory Cavitation Damage Tests in Various Liquids, Including Molten Sodium, *Proc. Fifth Conf. on Fluid Machinery*, 1975, pp. 393–402, Akademiai Kiado, Budapest, Hungary; see also *ASME Cavitation and Polyphase Flow Forum*, 1975, pp. 22–25.
124. Hutton, S. P., and J. LoboGuerrero: The Damage Capacity of Some Cavitating Flows, *Proc. Fifth Conf. on Fluid Machinery*, 1975, pp. 427–438, Akademiai Kiado, Bundapest, Hungary.
125. Ball, J. W., J. P. Tullis, and T. E. Stripling: Predicting Cavitation in Sudden Enlargements, *J. of Hyd. Div.*, ASCE, vol. 101, no. HY7, pp. 857–870, July, 1975; see also J. W. Ball, J. P. Tullis, and T. E. Stripling: Incipient Cavitation Damage and Scale Effects for Sudden Enlargement Energy Dissipators, presented as *ASCE Hyd. Div. Speciality Conf.*, 1974, Tennessee.
126. Canavelis, R.: "Contribution a l'Etude de l'Erosion de Cavitation dans les Turbomachines Hydrauliques," Ph.D. thesis, University of Paris, Faculté de Science, 1966.
127. Meier, W., and H. Grein: Cavitation in Models and Prototypes of Storage Pumps and Pump Turbines, paper H3, *Trans. IAHR Symp.*, 1970, Stockholm.
128. Schiele, O., and G. Mollenkopf: Some Views on Different Cavitation Criteria of a Pump, *Proc. Cavitation, Fluid Mach. Group, Inst. Mech. Engrs.*, September, 1974, pp. 177–185, Heriot Watt University, Edinburgh.
129. Malyshev, V., and N. Pylaev: "Influence of Hydroturbine Size on Cavitation Pitting Intensity," *Proc. Cavitation, Fluid Mach. Group, Inst. Mech. Engrs.*, September, 1974, pp. 309–312, Heriot Watt University, Edinburgh.
130. Young, S. G., and J. R. Johnston: Accelerated Cavitation Damage of Steels and Superalloys

in Sodium and Mercury, in "Erosion by Cavitation or Impingement," *ASTM STP* 408, pp. 186–212, 1967.

131. Florschuetz, L. W., and B. T. Chao: On the Mechanics of Vapor Bubble Collapse—A Theoretical and Experimental Investigation, *Trans. ASME, J. Heat Transfer*, vol. 87, pp. 209–220, 1965.

132. Mitchell, T. M., and F. G. Hammitt: On the Effects of Heat Transfer Upon Collapsing Bubbles, *Nucl. Sci. and Engr.*, vol. 53, no. 3, pp. 263–276, March, 1974.

133. Bonnin, J.: "Incipient Cavitation in Liquids Other than Cold Water," Electricité de France Rept. 030, Chatou, France, January, 1971; also see *1971 ASME Cavitation Forum*, pp. 14–16.

134. Bonnin, J.: Debut de Cavitation dans des Liquides Differents, *EDF-Bull. de la Direction Recherches—Serie A Nucleaire Hydraulique, Thermique*, no. 4, pp. 53–72, 1970.

135. Bonnin, J.: Influence de la Temperature sur le Début de Cavitation dans l'Eau, *Societé France, Journées de l'Hydraulique, Paris*, vol. XII, pp. 1–7, 1972.

136. Nowotny, H.: "Destruction of Materials by Cavitation (in German)," VDI Verlag, Berlin, 1942.

137. Heymann, F. J.: "Preliminary Report on Liquid Impact Erosion, Round-Robin Test Program," Westinghouse Steam Turbine Division, Internal Rept., July, 1973.

138. Hammitt, F. G., and Y. C. Huang: Liquid Droplet Impingement Studies at the University of Michigan, Conf. Pub. no. 3, *Proc. Warwick Conf.*, Apr. 3–5, 1973, pp. 237–243, Inst. of Mech. Engrs., University of Warwick, Coventry.

139. Hwang, J. B., and F. G. Hammitt: High Speed Impact Between Curved Liquid Surface and Rigid Flat Surface, ASME paper no. 76-WA/FE-34, *Trans. ASME, J. Fluids Engr.*, vol. 99, ser. I, no. 2, pp. 396–404, June, 1977.

140. Hwang, J. B., F. G. Hammitt, and W. Kim: On Liquid-Solid Impact Phenomena, *1976 ASME Cavitation Forum*, pp. 24–27.

141. Kling, C. L.: "A High Speed Photographic Study of Cavitation Bubble Collapse," Ph.D. thesis, Nuclear Engr. Dept., University of Michigan, 1970; also as University of Michigan ORA Report UMICH 03371-2-T, March, 1970.

142. Timm, E. E.: "An Experimental Photographic Study of Vapor Bubble Collapse and Liquid Jet Impingement," Ph.D. thesis, Chem. Engr. Dept., University of Michigan, 1974; also as University of Michigan ORA Rept. UMICH 01357-39-T, 1974.

143. Mitchell, T. M.: "Numerical Studies of Asymmetric and Thermodynamic Effects on Cavitation Bubble Collapse," Ph.D. thesis, Nuclear Engr. Dept., University of Michigan, 1970; also as University of Michigan ORA Rept. UMICH 03371-5-T, December, 1970.

144. Mitchell, T. M., and F. G. Hammitt: Asymmetric Cavitation Bubble Collapse, *Trans. ASME, J. Fluids Engr.*, vol. 95, no. 1, pp. 29–37, March, 1973.

145. Pearsall, I. S., and P. J. McNulty: Comparisons of Cavitation Noise with Erosion, *1968 ASME Cavitation Forum*, pp. 6–7.

146. Varga, J. J., and Gy. Sebestyen: Determination of Hydrodynamic Cavitation Intensity by Noise Measurement, *Proc. Second Int'l JSME Symp. on Fluid Machinery and Fluidics*, September, 1972, pp. 285–292, Tokyo; see also J. J. Varga, Gy. Sebestyen, and A. Fay: Detection of Cavitation by Acoustic and Vibration-Measurement Methods, *La Houille Blanche*, no. 2, pp. 137–149, 1969.

147. Lush, P. A., and S. P. Hutton: The Relation Between Cavitation Intensity and Noise in a Venturi-Type Section, *Proc. Int'l Conf. on Pump and Turbines*, September, 1976, pp. 1–3, NEL, Glasgow.

148. Numachi, F.: Transitional Phenomena in Ultrasonic Shock Waves Emitted by Cavitation on Hydrofoils, *Trans. ASME, J. Basic Engr.*, vol. 81, p. 153, June, 1959.

149. Hammitt, F. G., and J. B. Hwang: Ultrasonic Cavitation Regime Pulse-Count Spectra as Related to Cavitation Erosion, *Proc. Seventh Int'l Non-Linear Acoustics Conf.*, August, 1976, Blacksburg, Va.

150. Hammitt, F. G., S. A. Barber, M. K. De, and A. N. El Hasrouni: Predictive Capability for Cavitation Damage from Bubble Collapse Pulse Count Spectra, *Proc. Conf. Scaling for Performance Prediction in Rotodynamic Machines*, Inst. Mech. Engrs., 6–8 Sept., 1977, University of Stirling, Scotland.

151. Hammitt, F. G., S. A. Barber, M. K. De, and A. N. El Hasrouni: Cavitation Damage Prediction from Bubble Collapse Pulse Count Spectra, *1977 ASME Cavitation and Polyphase Flow Forum*, pp. 25–28, 1977.
152. Hammitt, F. G., and M. K. De: Cavitation Damage Prediction, *Wear* (Elsevier) 52, 2, pp. 243–262, 1979; see also "Cavitation Erosion of Aluminum and Erosion Efficiency," *Wear*, 55, 2, pp. 221–234, 1979.
153. Walsh, W. J., and F. G. Hammitt: Cavitation and Erosion Damage Measurements with Radio-Isotopes, *Nuc. Sci. and Engr.*, vol. 14, no. 3, pp. 217–223, November, 1962.
154. Hammitt, F. G.: Detailed Cavitation Flow Regimes for Centrifugal Pumps and Head Versus NPSH Curves, *1975 ASME Cavitation and Multiphase Flow Forum*, pp. 12–15.
155. Wislicenus, G. F.: Discussion of Ref. 154 in *1975 ASME Cavitation and Multiphase Flow Forum*.
156. Hammitt, F. G. and G. Schmitt, Jr., Chapters 10 and 11 concerning cavitation and particle impact erosion in *ASME Wear Control Handbook* (in press, 1979).

SIX

LIQUID JET AND DROPLET IMPACT

6-1 BASIC THEORY AND ANALYSIS

6-1-1 General

In recent years, substantial efforts at the author's laboratory [1–7] and elsewhere [8–10] have been concentrated upon the computer modeling of droplet and jet impact. It is useful to present a brief review of the status of this field.

To a first approximation, the liquid-solid impact process, as it applies to material erosion, is essentially a nonsteady process, basically that classically termed "water hammer" and governed by [11]

$$\Delta P_{WH} = \rho_0 C_0 V \qquad (6\text{-}1)$$

according to the linear theory. The zero subscripts indicate ambient liquid properties, ΔP_{WH} is the resultant pressure rise, ρ the liquid density, C the pressure-wave celerity ($\sim$ sonic velocity) in the liquid, and V the perpendicular component of impact velocity.

This equation is almost exactly correct for an impact between two "semi-infinite" planes of liquid and solid. However, for drops or jets of finite size and/or curved leading surfaces, there will obviously be corrections related to droplet shape. In addition there are important corrections due to changing liquid properties, such as ρ and C, under the high pressures induced by the collision itself, as well as corrections due to the nonrigidity of the target material.

If the process were essentially "steady state," as would be the case for the impact of an elongated jet (many L/D ratios), the pressure along the impacted surface would very quickly fall from water-hammer pressure magnitude to that of stagnation pressure

$$\Delta P_{st} = \frac{\rho_0 V^2}{2} \tag{6-2}$$

which is in general much smaller if impact velocity is substantially less than acoustic speed of the liquid but is the highest pressure attainable in a steady process. When V approaches C_0, ΔP_{st} approaches $\Delta P_{WH}/2$, to a first approximation.†

6-1-2 Geometrical Effects

The perpendicular collision between flat semi-infinite liquid and solid planes is entirely analogous to the sudden deceleration of a liquid column in a "rigid" pipe, and hence the classical water-hammer equation describes the phenomenon correctly. However, the collision of finite flat or curved surfaces of liquid with a semi-infinite, rigid flat solid surface involves important differences. For a finite flat liquid surface, the effect of the angle of impact is much more difficult to analyze than the others, since the phenomenon is no longer "axially symmetric." Nevertheless, all of these situations have been investigated to some extent in the

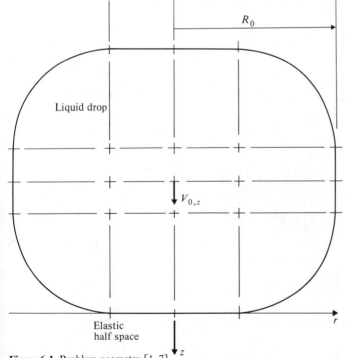

Figure 6-1 Problem geometry [1–7].

† See Eq. (6-23) for a more precise result.

U-M computer studies [1–7], as well as elsewhere. For simplicity, the present discussion will consider primarily our own work.

The situations investigated here for rigid flat surfaces include:

1. Spherical and combined flat and spherical shapes (Fig. 6-1).
2. Right circular cylinder impacting in a direction parallel to the axis.
3. Conical droplet impacting in a direction parallel to the axis. This geometry was chosen to allow the investigation of a constant contact angle of impact, as opposed to the situation with an impacting spherical drop, where the contact impact angle increases from zero to 90° as the collision proceeds.

The spherical and combined flat-spherical shapes (Fig. 6-1) were investigated, since these shapes match most closely many actual droplet shapes in such important impingement cases as aircraft rain erosion and turbine wet-steam erosion. Impact with the leading edge of a right circular cylindrical drop was included to model as closely as feasible impact with elongated jets such as the liquid microjet originating from cavitation bubble collapse, as well as to provide a simple "limiting case" where acute contact-angle effects which are present with spherical drops are absent. Finally, the conical drop was investigated as another relatively simple (axially symmetric) limiting case where an acute-angle effect is present but the angle is constant. This allows comparison with previous more-

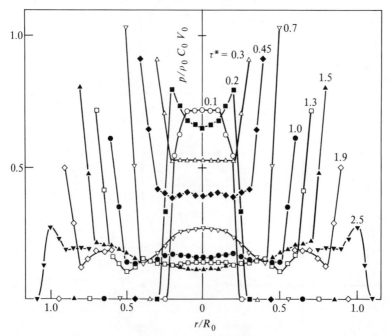

Figure 6-2 Pressure distribution on the liquid-solid interface ($z = 0$) at several instants after the collision of a spherical water drop and a rigid plane [4–7].

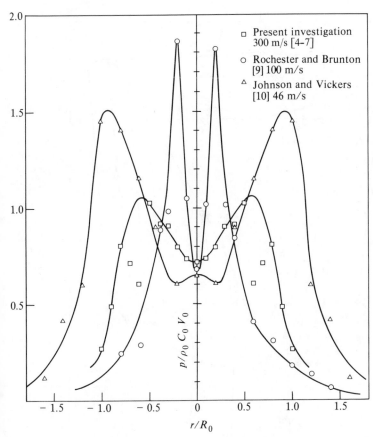

Figure 6-3 Distribution of maximum pressure on the liquid-solid interface ($z = 0$) [4-7].

simplified analyses which still take into account the contact angle of impact effect [8].

In general, the effect of the impact angle, e.g., as for a spherical drop, is to provide a nonuniform pressure distribution along the impact surface which reaches a maximum away from the axis and near the perimeter of contact, i.e., at increasing radius as the collision proceeds. While the water-hammer model [Eq. (6-1)] still provides a good approximation of the average pressure along the contact surface at any instant during the impact, the pressure at a point increases toward the outside of the contact region and is a minimum along the axis (Fig. 6-2). According to our model, it exceeds the classical water-hammer pressure only slightly. Some experimental information (Fig. 6-3) indicates that our calculation may be some-what conservative in this regard and that the actual peak pressure may be up to 1.5 times the water-hammer pressure [9, 10]. However, the difficulty of obtaining precise experimental information on this phenomenon is very considerable. Hence it must be admitted that the situation is not fully resolved at this time.

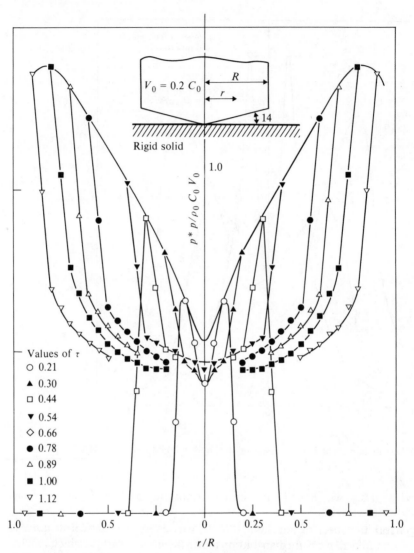

Figure 6-4 Pressure distribution on the interface ($z = 0$) at several instants following the collision of a conical water droplet and a rigid plane. $M_{liq} = 0.2$ [4–7].

The time of application of the high pressures is extremely short. Figure 6-2 indicates nondimensional times of less than unity. For water droplets of ~ 1 mm radius, the units of this nondimensional time are ~ 1 μs. The degree of damage to be expected from such very transient stressing of the material is unclear. Also, the conventional mechanical material properties cannot be expected to be valid for such high rates of loading, and the effective failure stresses may be much greater than expected.

Figure 6-4 depicts the surface pressures resulting from conical-droplet impact.

In this case also, the maximum contact pressure moves outward as the collision proceeds, finally reaching a value of ~ 1.4 times the water-hammer pressure. Again the duration of the entire process is ~ 1 μs for a droplet radius of ~ 1 mm.

6-1-3 Bulk and Elastic Modulus Effects

General Elastic modulus of the material impacted as well as bulk modulus of the liquid affect the results as well as the geometrical effects previously discussed. This point leads to the general conclusion that the results from the mathematical models here discussed depend only upon:

1. Mach number of impact (referred to liquid sonic velocity, that is, V/C_0).
2. Rigidity of impacted surface.
3. Geometry of collision, i.e., shape of the impacting droplet and impacted surface. The absolute droplet size is not required, except for the eventual computation of the absolute duration of effects. The resultant pressures are thus not a function of droplet size.

Bulk modulus of liquid Any meaningful analysis of the liquid-impact problem must obviously consider the liquid to possess a finite compressibility. An "incompressible-liquid" model (infinite bulk modulus) would lead to a calculated infinite contact pressure of zero duration, since the liquid sonic velocity would then be infinite. Thus the water-hammer model requires a pressure wave speed, C in Eq. (6-1) [1–3, 8]:

$$\frac{C}{C_0} = 1 + 2\,\frac{V}{C_0} - 0.1\left(\frac{V}{C_0}\right)^2 \tag{6-3}$$

For relatively weak water-hammer pressure, it is adequate to assume $C \cong C_0 =$ sonic speed of liquid. However, this is not a good approximation for the impact case unless the impact liquid Mach number is very small, as is the case for conventional water hammer. Equation (6-3) shows the pertinent correction to C_0 for water for impact velocities up to ~ 4500 m/s [1–3, 8]. For impact speeds of 300 m/s, i.e., liquid Mach number for water $\cong 0.2$, the shock-wave speed is ~ 1.4 times the acoustic speed. A correction to density ρ_0 must also be considered, but this is relatively unimportant except for very high impact Mach numbers.

Target material rigidity The discussions above have all assumed a fully rigid flat material surface, although it is well known that finite target elasticity (or plasticity) will reduce somewhat the contact pressures, and hence also the stresses induced in the target material. As originally shown by DeHaller [11], the first-order correction for material elasticity is

$$\Delta P = \frac{\rho_0 C_0 V}{1 + (\rho_0 C_0 / \rho_s C_s)} \tag{6-4}$$

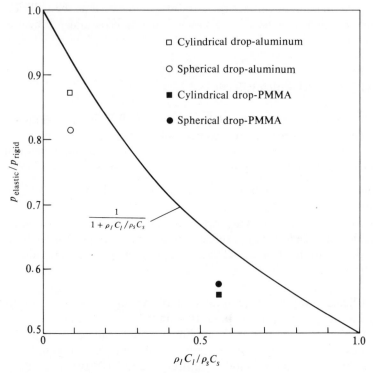

Figure 6-5 The reduction of impact pressure as a function of the ratio of acoustic impedance. $M_{liq} = 0.2$, liquid in water [5, 6, 11].

where ρ_s and C_s are the density and acoustic speed of the solid material, and their product is the "acoustic impedance." If the ratio of acoustic impedances between liquid and solid is appreciable, the reduction in impact pressure caused by the material elasticity effect can be large (Fig. 6-5).

The ratio of contact pressures between a collision of a liquid with an elastic solid and with a rigid solid [5, 6] for the various cases here computed in detail, using a full mathematical model of the impact and assuming elastic response of the material, are shown in Fig. 6-5. Spherical and cylindrical drops were considered with target materials of aluminum and Plexiglas for liquid Mach number 0.2. Figure 6-5 also shows the result from the classical DeHaller model [Eq. (6-4)] which appears to be somewhat conservative compared to the actual computer results, i.e., calculations using the DeHaller model will overestimate the contact pressures to some extent. It is noted that the reduction in contact pressure for aluminum (as compared to a rigid solid) for these cases is ~15 percent and for Plexiglas is ~45 percent.

The computer calculations for Fig. 6-5 also provide the impact stresses and strains in the material as a function of position and time. These are reported in detail elsewhere [4–7], and are also discussed in Sec. 6-2. However, this type o calculation can be made for any material- and liquid-impact situation. From such results, meaningful estimates of the probability and type of material failure to be expected could be made. However, it is not yet possible to estimate from such results the rates of material removal. Such an estimate would depend upon many factors, such as the type of material and material failure, importance of corrosion and other mechanisms, importance of very short-term impact loadings ($\sim \mu$s)† as opposed to conventional mechanical properties which are generally obtained under much slower rates of loading, and many other complicating factors.

6-2 MATHEMATICAL MODELS AND COMPUTER ANALYSES

6-2-1 General

The previous section has reviewed the general fluid-dynamic and material parameter interactions involved in the collision of a liquid jet with a solid-material surface, i.e., the "physics" of the liquid-solid impact process. This section will review these processes in further detail, including the details of the numerical analyses which have been performed. The following section will consider the experimental and photographic results which have been obtained and discuss how these relate to theoretical and numerical predictions.

Liquid-impact discussions in this book are limited by the scope of the book to the intermediate velocity range, i.e., the low to moderate subsonic range as referred to sonic velocity of the liquid. This is the general range of interest for such phenomena as cavitation bubble-collapse "microjets," steam-turbine droplet impact, and aircraft rain erosion, i.e., the range up to about 1000 m/s. As it happens, the impact problem is of maximum complexity in this general velocity range, since none of the major governing fluid or material mechanisms can be neglected there. For very low velocity it is, of course, possible to consider fluid and material elasticity using only linear approximations. For "hypervelocity" impact, i.e., the high supersonic range referred to liquid sonic speed, such as might be encountered with missile-reentry cone rain erosion, for example, or micrometeorite impact of spacecraft, many fluid and material properties can be neglected. For example, a liquid and a solid become essentially identical under these circumstances, since the relative kinetic energies of the collision greatly exceed material or liquid "binding energy," so that solid or liquid particles behave essentially as a highly compressible fluid. This general problem is attackable by computer techniques such as PIC (particle in cell; see, for example, Ref. 12), which is generally limited to cases where fluid, or material, compressibility is substantial. Though such problems are beyond the scope of this book, it is interesting to note that for hypersonic collision even chemical energies may become negligible compared to the kinetic energies. Thus a collision with a small stone or an equal mass of detonating TNT might be essentially equally damaging for such impact velocities.

† For ~ 1 mm droplet.

6-2-2 Controlling Parameters for Moderately Subsonic Collisions

For the low to moderate subsonic liquid range here considered, the pertinent form of the fluid equation of motion is the "nonsteady" form. It must at a minimum contain inertial and pressure terms. The inclusion of viscosity may or may not be necessary, depending upon the relative value of various other parameters, but it is unlikely that gravity or surface tension need be included. The nonsteady term is controlling during the initial portion of the collision. This is the portion important from the viewpoint of damage, where pressures of the general magnitude of water-hammer, rather than stagnation, pressure exist. For an elongated jet at least, the pressure later falls to the general level of the stagnation pressure, as would be expected. The problem of steady-state jet impingement has to a large extent been solved classically in general fluids texts. It is of much less complexity than the initial phase of jet impact, or of the droplet-impact problem in general.

The conservation of mass equation applicable to the liquid-impact problem in the moderate subsonic range must consider liquid compressibility. A "first-order" consideration of this factor is used in the conventional water-hammer analysis, originally proposed by St-Venant [13] and later by Cook [14]. This results in the pressure calculation of Eq. (6-1), and is often sufficient. However, as explained in Sec. 6-1, correction to this relation for moderate subsonic liquid Mach numbers is required for good results.

While the complete neglect of liquid compressibility would lead to trivial results, i.e., infinite pressure for zero time, meaningful results will be obtained in many cases if only target material compressibility is completely neglected, i.e., the "rigid-surface" approximation is used. However, depending upon the material-liquid combination and impact Mach number, target-material elasticity and/or plasticity must also be considered to obtain reasonably accurate results. These points will be explained in greater detail in the following sections.

6-2-3 Detailed Pertinent Differential Equations for Impact between Liquid and Rigid Solid

The detailed differential equations pertinent to the description of an axially symmetric collision between a water droplet of otherwise arbitrary shape and solid surface follow. The degree of rigidity and precise shape of the solid surface, as well as the precise shape of the liquid droplet, enter the problem only through the boundary conditions, and hence do not affect the basic differential equations. These, and numerical techniques pertinent to the axially symmetric liquid-solid impact problem in the velocity range of interest, are well described by two Ph.D. theses recently completed at the author's laboratory [1, 4], and the subsequent publications [2, 3, 5–7]. For convenience, the equations and some of the other pertinent material will be summarized here.

Under the pertinent assumptions, the equations governing the impact between a liquid droplet of spherical or cylindrical shape and a rigid solid surface are as follows.

The equation of continuity for the liquid phase is

$$\frac{\partial \rho}{\partial t} + \frac{\partial (\rho u)}{\partial z} + \frac{1}{r}\frac{\partial (r\rho v)}{\partial r} = 0 \tag{6-5}$$

The momentum equations for the liquid phase, neglecting viscous effects, are

$$\frac{\partial (\rho u)}{\partial t} + \frac{\partial (\rho u^2)}{\partial z} + \frac{1}{r}\frac{\partial (r\rho vu)}{\partial r} = -\frac{\partial p}{\partial z} \tag{6-6}$$

$$\frac{\partial (\rho v)}{\partial t} + \frac{\partial (\rho vu)}{\partial z} + \frac{1}{r}\frac{\partial (r\rho v^2)}{\partial r} = -\frac{\partial p}{\partial r} \tag{6-7}$$

Just as in the conventional "acoustic approximation" for nonsteady fluid problems, viscous effects are of second-order importance for this problem. This assumption was justified by numerical results where some viscous effects were considered. For example, the assumption of full-slip or zero-slip boundary conditions made little difference to the results. Also, in some cases an "artificial viscosity" was used to stabilize the numerical technique, and little effect otherwise upon the results was found.

A suitable form [1, 4, 15] of the equation of state for water (Tait's equation) is

$$\frac{p + B}{p_0 + B} = \left(\frac{\rho}{\rho_0}\right)^A \tag{6-8}$$

above, u and v are the axial and radial velocity components, respectively, for the cylindrical coordinates z and r. Time t is another independent variable, and ρ and p are the fluid density and pressure. The values of the two constants in the equation of state for water are chosen [15] as

$$A = 7.15 \quad \text{and} \quad B = 3.047 \text{ kbar} \tag{6-9}$$

Strictly speaking, the energy equation is also required. However, its neglect is justified by the fact that little temperature dependence of pertinent fluid properties occurs in this calculation, although it can be important in the particular case of stagnation temperature. For impact velocity in the range considered here, the liquid stagnation-temperature rise is not great ($\sim 45°C$ for 600 m/s water). However, for hypersonic impact, temperature effects would become more important.

6-2-4 Pertinent Numerical Techniques

Various numerical techniques are possible for the solution of the differential equation set represented by Eqs. (6-5) to (6-8). That named "ComCAM" (compressible cell-and-marker), developed at the author's laboratory by Huang [1–3, 16], was apparently the first and probably best known. The descriptions will thus consider particularly these studies. This program was somewhat modified and improved in the subsequent Ph.D. thesis of Hwang [4–7]. The modifications were particularly concerned with the handling of the interface condition between liquid and solid. Hwang also [4, 5] included the effect of such resilient target solids as aluminum and Plexiglas in his eventual very-comprehensive program. Further details of his study of elastic target effects will be given in a later section.

Regardless of the numerical technique used for the analysis, it is also highly desirable to convert the equations into a nondimensional form so that maximum applicability of the results will be attained. This was, of course, done by both Huang [1–3, 16] and Hwang [4–7]. The result was that pressure can be considered to be normalized by the conventional water-hammer pressure, velocity by sonic velocity, etc. It thus results that, for a fully rigid solid with a given liquid such as water, the only independent parameters for the problem are impact liquid Mach number and droplet shape. The results in terms of nondimensional time and position are then normalized pressure and velocity.

The method begins with an eulerian grid. Field variables such as density and velocity are directly associated with the cells of the grid. In addition, a series of lagrangian "marker particles" are assigned to the liquid. These are necessary to mark the free-surface movement. Numerical computation starts with particles located only along the surface. This is possible because fluid particles initially on the free boundary always remain on the free boundary [17]. "ComCAM" then combines eulerian and lagrangian techniques. In general, for eulerian calculations, a modified two-step Lax-Wendroff scheme [18] is used. It is not worthwhile to pursue the program details more fully here, since these can be found in full detail for those interested in the Ph.D. theses mentioned [1, 4] and related papers. Full program listings are also given in these theses.

6-2-5 Dynamic Equations for Solid Materials

General For a homogeneous isotropic elastic medium, the equations of motion are given by [19–21]

$$(\lambda + 2G)\nabla(\nabla \cdot \mathbf{u}) - G\nabla X \nabla X \mathbf{u} = \rho_s \frac{\partial^2 \mathbf{u}}{\partial t^2} \tag{6-10}$$

where $\mathbf{u} = \mathbf{u}(r, z, t)$ is the displacement vector (not velocity here), G are Lamé's constants, and ρ is the density. Written explicitly in terms of radial and axial components, $u_r\,(r, z, t)$ and $u_z\,(r, z, t)$, we have

$$(\lambda + 2G)\left(\frac{\partial^2 u_r}{\partial r^2} + \frac{1}{r}\frac{\partial u_r}{\partial r} - \frac{u_r}{r^2} + \frac{\partial^2 u_z}{\partial r\,\partial z}\right) + G\left(\frac{\partial^2 u_r}{\partial r^2} - \frac{\partial^2 u_z}{\partial r\,\partial z}\right) = \rho\frac{\partial^2 u_r}{\partial t^2} \tag{6-11}$$

$$(\lambda + 2G)\left(\frac{\partial^2 u_z}{\partial z^2} + \frac{1}{r}\frac{\partial v_r}{\partial z} + \frac{\partial^2 u_r}{\partial z\,\partial r}\right) + G\left(\frac{\partial^2 u_z}{\partial r^2} - \frac{\partial^2 u_r}{\partial z\,\partial r} - \frac{1}{r}\frac{\partial u_r}{\partial z} + \frac{1}{r}\frac{\partial u_z}{\partial r}\right) = \rho\frac{\partial^2 u_z}{\partial t^2}$$

$$\tag{6-12}$$

One of the Lamé's constants, G, is commonly known as shear modulus. The relations between Lamé's constants, Young's modulus E, and Poisson's ratio v are as listed below:

$$\frac{\lambda}{G} = \frac{2v}{1 - 2v}$$

$$G = \frac{E}{2(1 + v)}$$

$$\lambda = \frac{vE}{(1 + v)(1 - 2v)}$$

$$v = \frac{\lambda}{2(\lambda + G)}$$

$$E = \frac{G(3\lambda + 2G)}{\lambda + G}$$

Strain and stresses can be derived from displacements with the following relations:

Strain:

$$\varepsilon_r = \frac{\partial u_r}{\partial r} \qquad \varepsilon_z = \frac{\partial u_z}{\partial z} \qquad \varepsilon_\theta = \frac{u_r}{r}$$

$$\gamma_{rz} = \gamma_{zr} = \frac{1}{2}\left(\frac{\partial u_z}{\partial r} + \frac{\partial u_r}{\partial z}\right)$$

(6-13)

Stress:

$$\sigma_r = (\lambda + 2G)(\varepsilon_r + \varepsilon_\theta + \varepsilon_z) - 2G(\varepsilon_\theta + \varepsilon_z)$$

$$= (\lambda + 2G)\left(\frac{\partial u_r}{\partial r} + \frac{u_r}{r} + \frac{\partial u_z}{\partial z}\right) - 2G\left(\frac{u_r}{r} + \frac{\partial u_z}{\partial z}\right)$$

$$\sigma_z = (\lambda + 2G)(\varepsilon_r + \varepsilon_\theta + \varepsilon_z) - 2G(\varepsilon_r + \varepsilon_\theta)$$

$$= (\lambda + 2G)\left(\frac{\partial u_r}{\partial r} + \frac{u_r}{r} + \frac{\partial u_z}{\partial z}\right) - 2G\left(\frac{\partial u_r}{\partial r} + \frac{u_r}{r}\right)$$

(6-14)

$$\sigma_\theta = (\lambda + 2G)(\varepsilon_r + \varepsilon_\theta + \varepsilon_z) - 2G(\varepsilon_r + \varepsilon_z)$$

$$= (\lambda + 2G)\left(\frac{\partial u_r}{\partial r} + \frac{u_r}{r} + \frac{\partial u_z}{\partial z}\right) - 2G\left(\frac{\partial u_r}{\partial r} + \frac{\partial u_z}{\partial z}\right)$$

$$\tau_{rz} = \tau_{zr} = 2G\gamma_{rz} = 2G\gamma_{zr} = G\left(\frac{\partial u_z}{\partial r} + \frac{\partial u_r}{\partial z}\right)$$

The elastic half space is originally at rest. Thus, we have the following initial conditions:

$$u_r(r, z, 0) = u_z(r, z, 0) = 0$$

(6-15)

Boundary conditions on the axis of symmetry ($r = 0$) are

$$u_r(0, z, t) = 0 \tag{6-16a}$$

and

$$\frac{\partial u_z}{\partial r}(0, z, t) = 0 \tag{6-16b}$$

Calculational procedure For each time step of the numerical calculation, the response of the liquid phase is calculated first using the shape and velocity of the solid surface at the beginning of this time step as boundary conditions on the interface. The new surface shape and velocity are then calculated using Eqs. (6-11 and (6-12), assuming that the pressure distribution just calculated in the liquid phase is applied through this time step. The response of the liquid phase is then refined using the surface shape and velocity. This trial-and-error procedure is performed until the surface shape and velocity from two consecutive iterations are judged to be sufficiently close.

The necessary conditions for the liquid-solid interface, during the early stages of the impact, are the continuity of displacement and velocity. The liquid pressure is the only load on the solid surface. Liquid and solid must have the same normal displacement and velocity if they are to remain in contact. Mathematically, these boundary conditions are

$$\sigma_z(r, 0, t) \Leftarrow -p[r, u_z(r, 0, t), t] \tag{6-17}$$

$$\sigma_r(r, 0, t) \begin{cases} = 0 & \text{if full-slip} \\ \Leftarrow -\mu \dfrac{\partial v_r}{\partial z}[r, u_z(r, 0, t), t] & \text{if no slip} \end{cases} \tag{6-18}$$

$$v_z(r, Z_{\text{int}}, t) \Leftarrow \frac{\partial}{\partial t} u_z(r, 0, t) \tag{6-19}$$

$$\begin{cases} \dfrac{\partial}{\partial z} v_r(r, Z_{\text{int}}, t) = 0 & \text{if full slip} \\[2ex] v_r(r, Z_{\text{int}}, t) \Leftarrow \dfrac{\partial}{\partial t} u_r(r, 0, t) & \text{if no slip} \end{cases} \tag{6-20}$$

$$\frac{\partial}{\partial z} p(r, Z_{\text{int}}, t) = 0 \tag{6-21}$$

where $Z_{\text{int}} = u_z(r, 0, t)$ is the axial displacement of the solid surface. Arrows in the equations denote how, in the iteration discussed earlier, the quantities in one phase are fed into the other phase.

Nondimensional variables for liquid and solid phases follow. A dimensionless quantity (with superscript *) is defined as the quotient of a physical quantity and a corresponding characteristic value (with subscript c).

Nondimensionalization of liquid equations The following dimensionless quantities are defined for the liquid phase:

$$t^* = \frac{t}{t_{cL}} \qquad \rho^* = \frac{\rho_L}{\rho_c}$$

$$r^* = \frac{r}{r_c} \qquad z^* = \frac{z}{z_c}$$

$$R^* = \frac{R}{r_c} \qquad Z^* = \frac{Z}{z_c}$$

$$p^* = \frac{p}{p_c} \qquad B^* = \frac{B}{p_c}$$

$$v_r^* = \frac{v_r}{v_c} \qquad v_z^* = \frac{v_z}{v_c}$$

Nondimensionalization of solid-material equation The following dimensionless quantities are introduced for the solid material:

$$u_r^* = \frac{u_r}{u_c} \qquad u_z^* = \frac{u_z}{u_c} \qquad t_s^* = \frac{tC_1}{R_0} \tag{6-22}$$

where $C_1 = \sqrt{(\lambda + 2G)/\rho_s}$ is the longitudinal wave velocity for the solid.

6-2-6 Results of Numerical Analyses

While full details of these numerical analyses are given elsewhere [1–7], the more important results will be summarized here, as well as results from other related analyses.

Droplet geometry effects (rigid flat surface)

General As discussed in Sec. 6-1, detailed fluid-dynamic behavior during a liquid-droplet collision with a rigid flat plane depends heavily upon geometrical considerations such as droplet shape and angle of collision. It is somewhat less sensitive to the only other independent parameter involved (aside, of course, from various liquid properties), i.e., impact liquid Mach number. While the classical water-hammer pressure [Eq. (6-1)] provides a good preliminary estimate of the maximum interface pressure attained during the collision, substantial detailed variations, both positive and negative, from this estimate exist, as discussed briefly in Sec. 6-1. Since only maximum pressures and velocities are generally of interest from the viewpoint of erosion, which is the major practical interest in the collision process, it is only the very initial portion of the collision that must be considered. In fact, within a very few microseconds, the surface pressure falls for a droplet of 1 mm radius from the order of water-hammer pressure to that of stagnation

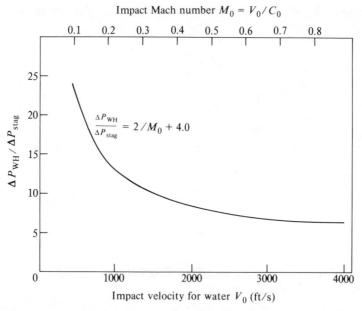

Figure 6-6 Ratio of water hammer to stagnation pressure rise vs. impact velocity [4, 7].

pressure, as will be shown in detail from the results of the numerical studies previously discussed [1–7].

For relatively very low velocities such as might be typical of pipeline water-hammer problems, the stagnation pressure is negligible from the viewpoint of damage, while the water-hammer pressure is, of course, important. However, the ratio between these pressures decreases for increased velocities. Figure 6-6 shows this ratio as a function of both impact liquid Mach number and velocity [7] for water. The following equation is pertinent:

$$\frac{\Delta P_{WH}}{\Delta P_{stag}} \cong \frac{2}{M_0} + 4.0 \qquad (6\text{-}23)$$

where $M_0 = V_0/C$ and V_0 = impact velocity. As shown, the ratio approaches infinity for very low impact velocity but 4.0 for very high velocity. In the usual range of interest for the droplet-impact problems here considered, the ratio is in the range 5 to 10.

Spherical droplet impact Impacting liquid droplets encountered in most pertinent situations can be reasonably well modeled by a spherical drop, which can also simulate the leading edge of a liquid jet. Figure 6-7 shows the shape history of an initially spherical water drop after perpendicular impact with a rigid flat plate at liquid Mach number 0.2 ($\sim$300 m/s), as calculated by Hwang [4–7]. The droplet deforming process is also very similar to that previously computed by Huang [1–3], except that the liquid surface adheres somewhat more closely to

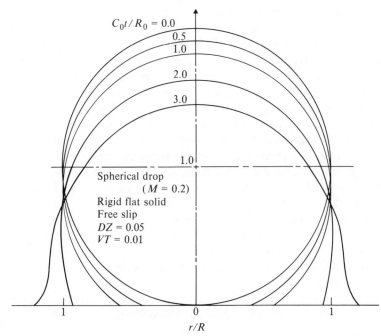

$C_0t/R_0 = 0.0$
0.5
1.0
2.0
3.0

1.0

Spherical drop
($M = 0.2$)
Rigid flat solid
Free slip
$DZ = 0.05$
$VT = 0.01$

1   0   1

r/R

Figure 6-7 Shape history of an initially spherical water drop following the collision with a rigid plane [4–7].

the plate. Differences between the Huang and the Hwang studies were less for cylindrical droplets, as discussed later.

Figure 6-2 shows the pressure distribution on the rigid plane ($z = 0$) at several instants after the initial contact. Maximum pressure occurs initially at the center of the contact area, but shifts to the edge after nondimensional time $C_0t/R_0 = 0.2$. Pressure decreases rapidly at the center after reaching a peak value of $0.7\rho_0C_0V_0$. The maximum pressure travels radially outward as the solid-liquid contact progresses in this direction. The peak pressure for the entire process is $\sim 1.0\rho_0C_0V_0$, and occurs along a circle of half the initial droplet diameter at $C_0t/R_0 = 0.7$. At the maximum time investigated ($C_0t/R_0 = 2.5$), the pressure is everywhere approaching the stagnation pressure, as would be expected. Note that to a first approximation $\Delta P_{stag}/\Delta P_{WH} \cong M_0/2$.

The occurrence of peak pressure at a location other than the center is consistent with recent experimental findings. The computed distribution of maximum pressure is shown in Fig. 6-3 along with measurements by Rochester and Brunton [22, 23] and by Johnson and Vickers [24]. These curves show qualitative rather than quantitative agreement, since the physical situations differed to some extent. Rochester and Brunton [22] projected a solid projectile against a 5-mm diameter water "disc," rather than against a sphere as considered in our calculation at the University of Michigan. Their impact liquid Mach number was only 0.07 rather than 0.2 for our calculation. Since the experiment used a two-dimensional rather than the three-dimensional geometry of our calculation, it is probable that the pressures measured would be greater, since the degrees of freedom for the escape of the liquid entrapped in the collision area are greater. Figure 6-3 shows that this was in fact the case. Heymann's earlier analysis [8], which also predicted higher pressures, was also for a two-dimensional situation. While his analysis does not provide the pressure distribution, it predicts an edge pressure of $2.8\rho_0 C_0 V_0$ for an impact liquid Mach number of 0.2, as used in our calculation. Thus its "coefficient" is much higher than for all other analyses or experiments. Heymann predicted this maximum pressure would occur at a radius of $0.1 R_0$.

Johnson and Vickers [24] measured the impact pressure for an approximately cylindrical water jet fired against a flat surface, so that their geometry differs considerably from our analysis. In their case, the pressure spreads over a relatively large area because the leading portion of their mushroom-shaped jet (shown schematically in Fig. 6-8) has a considerably larger radius than that of the jet body, R_0, upon which Fig. 6-3 is based, since no more accurate information on the jet leading-edge geometry exists. Their liquid jet Mach number (~ 0.03) was much less than that of either our calculation or the Rochester-Brunton measurement.

Engel [25] examined the damage marks on rubber coatings bonded to metallic bases produced by 2-mm diameter water drops at 450 to 800 m/s ($M \cong 0.3$ to 0.53).

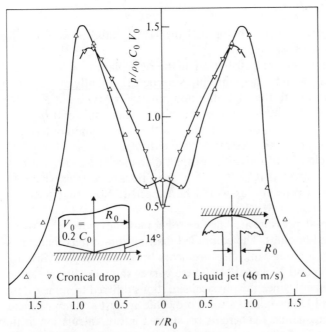

Figure 6-8 The distribution of maximum pressure on a rigid plane due to the impact of a conical drop ($V_0 = 0.2\,C_0$) and a liquid jet (46 m/s) [4, 24].

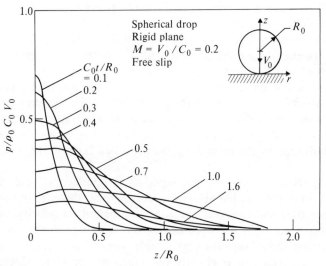

Figure 6-9 Pressure distribution on the axis of symmetry ($r = 0$) at several instants following the collision of a spherical drop and a rigid plane [4–7].

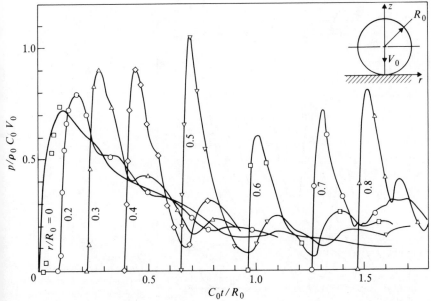

Figure 6-10 Transient pressure at selected locations on the interface following the collision of a spherical water droplet and a rigid plane [4-7].

She suggested that peak pressure developed when the contact-area radius reached $0.5R_0$. This result is reasonably consistent with our calculations [1–7].

Figure 6-9 shows our pressure distribution on the axis of symmetry at several instants after impact. In general the pressure magnitude decreases for greater distances from the impact surface during the early portion of the process, and is always much less than water-hammer pressure at the axis. Figure 6-10 depicts the pressure transient at various locations on the plane. Very rapid pressure-rise rates are indicated immediately following impact, but these very rapidly decay to a relatively steady value of $\sim 0.1\rho_0 C_0 V_0$. The "coefficient" is, as would be expected, about that indicated by the ratio $\Delta P_{stag}/\Delta P_{WH} \cong M_0/2$, since $M_0 = 0.2$ for these calculations.

Figure 6-11 shows radial velocity on the liquid-solid interface at several instants after collision. Maximum radial velocity, occurring near the edge of the contact area throughout the impact process, exceeds twice the impact velocity, as also confirmed photographically [26, 27]. The existence of a radial velocity substantially greater than the impact velocity is explicable, since this is actually a "spouting velocity" from a region of essentially water-hammer pressure which, as previously explained, substantially exceeds stagnation pressure.

Figure 6-12 shows lines of constant pressure as a function of water-hammer pressure, within an impacting spherical drop [1–3]. Results show computed regions of negative pressure where presumably cavitation might occur. Presumably this is due to the complex interplay of pressure and reflection waves. Photographic evidence of such cavitation has been obtained by Brunton and Camus [27] (Fig. 6-13) at locations quite similar to those predicted by our calculation.

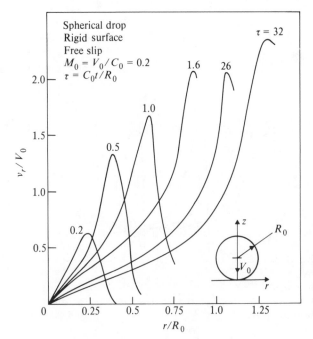

Figure 6-11 Radial velocity on the liquid-solid interface ($z = 0$) at several instants following the collision of a spherical water drop and a rigid plane [4–7].

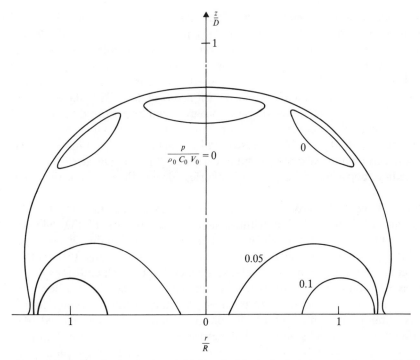

Figure 6-12 Isobar distribution in an initially cylindrical-spherical composite droplet with $R_1/R = 0.25$ and $L/D = 1$, at time $Ct/D = 1.5$ for liquid impact Mach number of 0.2 and for free-slip boundary condition [1–3].

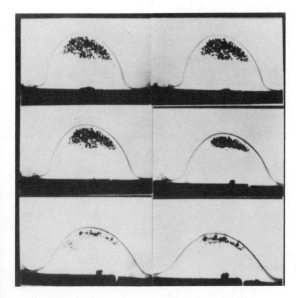

Figure 6-13 Photographs of cavitation in a water droplet following an impact on a solid plane (Brunton and Camus [27, 49].

Cylindrical and composite droplet impact   The perpendicular impact of cylindrical droplets, moving parallel to their axis, with flat plates was studied by both Huang [1–3] and Hwang [4–7], primarily because it represents the simplest case possible after the one-dimensional case (collision between liquid and solid semi-infinite planes). Also a complex droplet shape composed of a central cylindrical section surrounded by spherical segments can be envisioned (Fig. 6-1). By suitable choice of the controlling parameters, the computer program can be arranged to consider either purely spherical, purely cylindrical, or composite droplet shapes. Such composite droplets (Fig. 6-1) probably represent real droplet or jet impact more closely than either purely cylindrical or purely spherical shapes. Since the numerical results are intermediate between cylindrical and spherical drops, they are not included here.

Figure 6-14 depicts the cross section of an initially cylindrical drop ($L/D = 1.0$) at several instants after collision with a rigid plane. The top of the cylinder remains flat until $C_0 t/R_0 \cong 1.6$, which is ~ 1 μs for a 1-mm radius water drop. During this period the drop also begins to spread radially along the contact surface, giving rise to the high radial velocities generated by such collisions.

Figures 6-15 and 6-16 show the pressure distribution on the liquid-solid interface and on the axis of symmetry, respectively, at different instants after initial contact. As opposed to the spherical-droplet collisions already discussed, the contact pressure at the axis substantially exceeds the classical water-hammer pressure $\rho_0 C_0 V_0$, primarily because of the increase in C during the collision. This

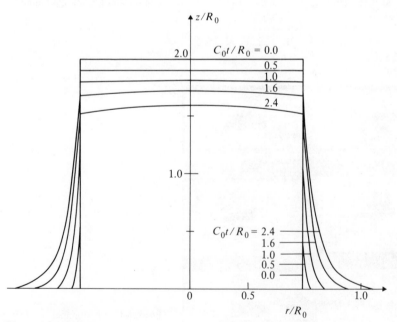

Figure 6-14 Shape history of an initially cylindrical water drop following the collision with a rigid plane [4].

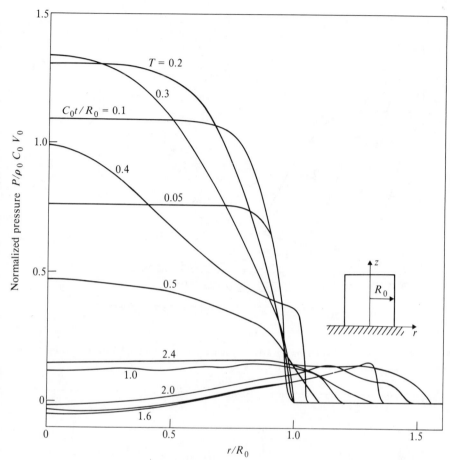

Figure 6-15 Pressure distribution on the liquid-solid interface $(z = 0)$ at several instants after the collision of a cylindrical droplet and a rigid plane [1–3].

effect would, of course, increase for higher impact liquid Mach numbers than that used here (0.2). However, for cylindrical impact there is no increase of contact pressure away from the origin as for the spherical case, except at the end of collision (Fig. 6-15), i.e., no "shaped-charge" effect. Rather the pressure attenuates radially, as would be normally supposed. Also, eventually the pressure becomes negative near the center, giving rise to the possibility here of cavitation, as already discussed for a spherical drop.

Figure 6-16 shows the propagation of the "shock wave," i.e., strong pressure gradient, into the interior of the droplet. The propagation speed is relatively constant up to nondimensional time, $\tau = C_0 t/R_0 = 0.5$. Thereafter it decreases somewhat. The wavefront attains significant thickness due to the existence of a "finite mesh size" and "artificial viscosity," both features of the numerical procedure used. "Artificial viscosity" is useful to stabilize the numerical calculation,

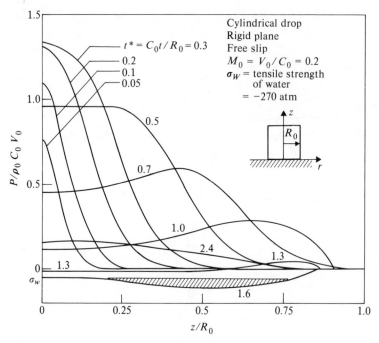

Figure 6-16 Pressure distribution on the axis of symmetry ($r = 0$) at several instants after the collision of a cylindrical water drop and a rigid plane [1–3].

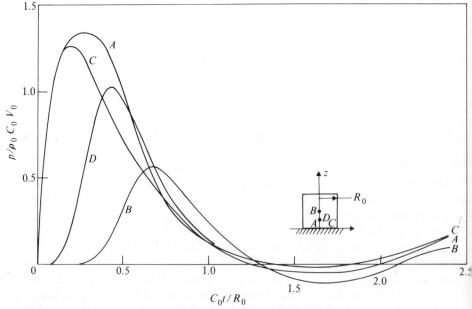

Figure 6-17 Transient pressure at selected locations inside an initially cylindrical droplet following the collision with a rigid plane. Liquid impact Mach number = 0.2 [1–3].

and also to give the capability of modeling to some extent the effect of real-liquid viscosity, especially for highly viscous liquids where this effect might become important. At $C_0t/R_0 = 1.6$ the pressure becomes negative over most of the axis, even falling below the presumed tensile strength of water (~ 270 bar; see Refs. 28 and 29) for an impact velocity of 300 m/s (with a liquid Mach number of 0.2). This could, of course, lead to local cavitation, as previously discussed, during droplet impact. Such cavitation could augment the erosion often associated with droplet impact. Calculations by Huang [1–3] and Glenn [30] also showed negative pressure on a rigid wall as well as inside a cylindrical drop. Also, as previously mentioned, there is photographic evidence [27] of cavitation within an impacting-liquid disc.

Figure 6-17 shows the pressure at selected locations as a function of time. The peak pressure propagates into the droplet with a speed of $\sim 2.5C_0$, and is sustained for a nondimensional time of ~ 0.8, which gives ~ 0.5 μs for $R_0 = 1$ mm. Con-

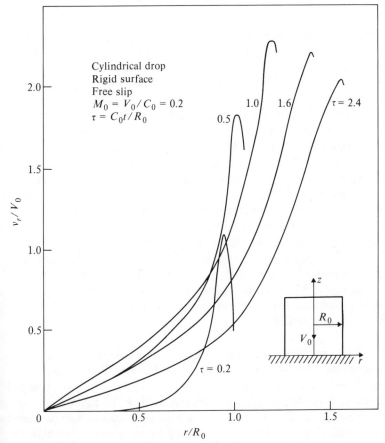

Figure 6-18 Radial velocity on the liquid-solid interface ($z = 0$) at several instants after the collision of a cylindrical water drop and a rigid plane [1–3].

Table 6-1 The maximum pressure for various impact situations, $M_{liq} = 0.2$ [4–7]

Solid material	$\rho_s C_s$, kg/m$^2 \cdot$s	$1 + \dfrac{\rho_0 C_0}{\rho_s C_s}$	Droplet shape	Maximum impact pressure $p_{max}/\rho_0 C_0 V_0$	$\dfrac{p_{elastic}}{p_{rigid}}$
Rigid	∞	1.0	Cylinder	1.34	1.0
			Sphere	1.06	1.0
Aluminum	1.74×10^6	1.086	Cylinder	1.17	0.873
			Sphere	0.861	0.812
PMMA (Plexiglas)	0.267×10^6	1.557	Cylinder	0.75	0.560
			Sphere	0.613	0.578

sidering the definition of nondimensional time, $\tau = C_0 t/R_0$, actual duration t is proportional to droplet radius and is inversely proportional to sonic velocity in the liquid.

Figure 6-18 shows the radial velocity along the liquid-solid interface at different instants. This is much greater near the edge of the contact area than elsewhere. As previously mentioned in connection with spherical droplets, the maximum radial velocity is several times the original impact speed. This also has been confirmed photographically [26, 27, 31, 32], as previously mentioned.

Table 6-1 summarizes the results of the computer calculations discussed in the foregoing, comparing the effects of droplet geometry and material rigidity, all for the liquid impact Mach number of 0.2, upon flat semi-infinite planes. It is there shown that the maximum pressure is greater for the cylinder than for the sphere, and is of course greater for a rigid than deformable solid, as discussed later. Also, according to the present results for the liquid Mach number 0.2, it never exceeds very substantially the classical water-hammer pressure, and in fact may be considerably less than ΔP_{WH} for the nonrigid materials investigated. For a higher liquid Mach number, the classical water-hammer pressure can be substantially exceeded, primarily because of the increase in liquid shock-wave speed under the compressed conditions.

Conical droplet While maximum pressure occurs at the axis for a cylindrical drop, it occurs near the edge of the contact surface for a spherical one. This outward movement of the maximum pressure point is associated with the expanding contact area caused by the curved liquid leading surface. To investigate this situation under somewhat simpler conditions, i.e., a constant contact angle between liquid and solid, a conical droplet shape was studied [4, 6], at the suggestion of Heymann [8]. Figure 6-19 depicts the shape history of this droplet and Fig. 6-4 the pressure distribution at several instants. For this case the peak pressure

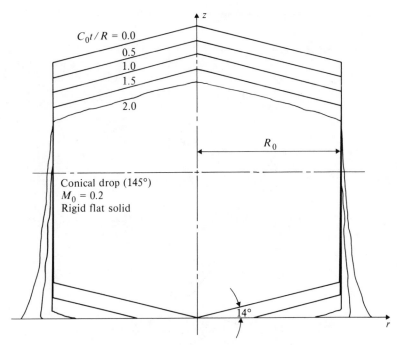

Figure 6-19 Shape history of an initially conical water drop following the collision with a rigid plane [4, 6].

increases linearly with radius up to $0.8R_0$, after which the pressure reduces, presumably due to the termination of the conical surface. The maximum pressure attained was $\sim 1.4\rho_0 C_0 V_0$, about the same magnitude reached with a cylindrical drop (Table 6-1). A direct comparison with Heymann's predicted "coefficient" [8] of 2.8 was not possible because the conical angle of $14°$, chosen to simplify the computer problem, was greater than his predicted critical angle. However, no theoretical or experimental information indicates the likelihood of a coefficient at $M = 0.2$ greater than ~ 1.5 (Table 6-1 and Fig. 6-4). This situation is very well summarized by the author's closure and Heymann's discussion of G. R. Johnson's recent paper [33].

Radial velocity again has a maximum near the outer edge of the contact surface, reaching maximum values similar to those attained for the other droplet shapes.

The distribution of maximum pressure calculated at the author's laboratory shows excellent agreement with the experimental data of Johnson and Vickers [24]. The different distributions near the axis (Fig. 6-3) can be attributed primarily to the different droplet geometries. While our calculation is for a conical droplet shape, the leading edge of their jet is smooth and flat at the center (Fig. 6-8).

Material rigidity effects As previously discussed, the inclusion of nonrigid material properties in the computation results in reduced maximum pressures

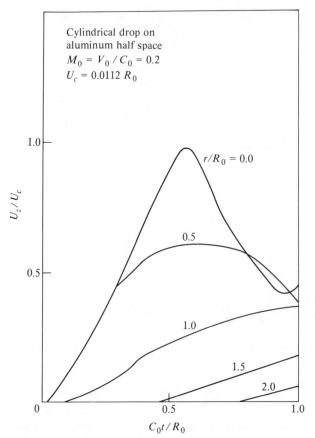

Figure 6-20 Transient displacement on the surface of a semi-infinite aluminum body following the impact by a cylindrical water drop [4–7].

during the collision (Table 6-1). The relative importance of this effect is best estimated by Eq. (6-4), where the "acoustic impedance ratio" between liquid and solid is considered. For fixed liquid properties, the effect obviously becomes of increased importance roughly as the product ρE of the material is reduced. Thus, as would be expected, the impact-pressure reduction for aluminum, and especially PMMA, is much greater than for steel. For this reason, we have computed the effect only for these materials (Table 6-1).

Figure 6-5, as already discussed to some extent, shows the reduction in maximum surface pressure due to surface elasticity as a function of the acoustic impedance ratio between liquid and solid. Our computed points for cylindrical and spherical drops upon aluminum and Plexiglas (PMMA) with water impact at the liquid Mach number of 0.2 are shown, as well as the curve predicted by the DeHaller model [Eq. (6-4)]. The DeHaller relation [11] provides a reasonable engineering approximation which is always conservative, i.e., it overestimates the surface stresses and pressures.

The transient surface displacements for aluminum are indicated in Fig. 6-20. The center displacement is a maximum at $C_0t/R_0 \cong 0.5$; it then rebounds rapidly. Figure 6-21 shows the surface shape at several instants after the initial contact, while Fig. 6-22 shows a typical distribution of principal stresses. Stresses are normalized by $\lambda + 2G$ whose value for aluminum is 1.12×10^5 MPa/m². Tensile stress (Fig. 6-23) on the surface first occurs outside the contact area, reaching a maximum value of 1080 bar (15,600 lb/in²) at $C_0t/R_0 \cong 0.5$. The position of maximum tensile stress then shifts outward gradually before starting to decay at $C_0t/R_0 \cong 0.70$. A somewhat larger tensile stress (1290 bar, or 18,700 lb/in²) was reached within the body near the axis at $C_0t/R_0 = 0.63$. The maximum compressive stress (Fig. 6-24) is much larger, reaching a value of ~ 6896 bar ($\cong 100,000$ lb/in²) on the surface near the axis. Figure 6-25 shows the distribution of shear stress at a given instant. The maximum attained is 3000 bar (43,500 lb/in²). It occurs near the axis and decreases with radial distance. Such stresses could certainly cause failure to many aluminum alloys in one impact (300 m/s), and to many other alloys through eventual fatigue mechanisms.

Other numerical studies Several pertinent droplet-impact numerical studies have been done in the author's laboratory at the University of Michigan; these have been emphasized in the foregoing, since they form the most unified comprehensive group of such studies available, to the author's knowledge. Earlier pre-computer analyses have only been mentioned when they appeared to be particularly pertinent to the discussion, since they do not particularly add to the

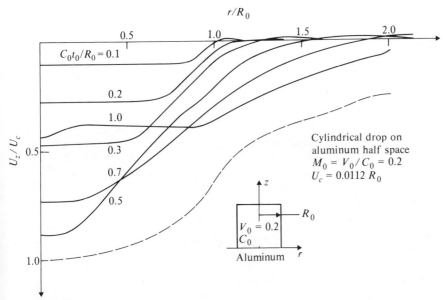

Figure 6-21 Surface deformation of a semi-infinite aluminum body at several instants after the collision with a cylindrical water drop [4–7].

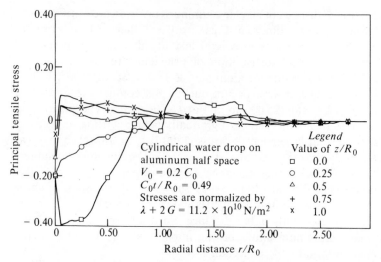

Figure 6-22 Distribution of principal tensile stress at $C_0 t / R_0 = 0.49$ after the collision of a cylindrical drop and a semi-infinite aluminum body [4–7].

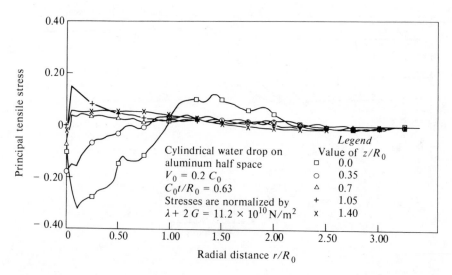

Figure 6-23 Distribution of principal tensile stress $C_0 t / R_0 = 0.63$ after the collision of a cylindrical drop and a semi-infinite aluminum body [4–7].

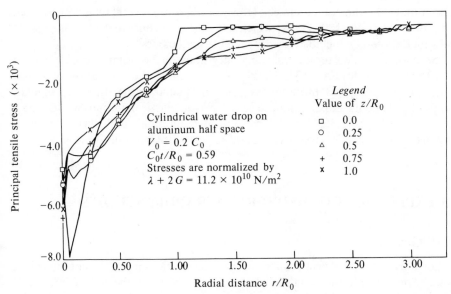

Figure 6-24 Distribution of principal compressive stress at $C_0 t/R_0 = 0.59$ after the collision of a cylindrical drop and a semi-infinite aluminum body [4–7].

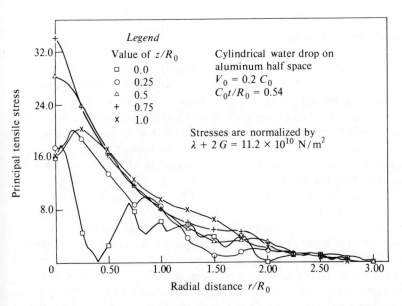

Figure 6-25 Distribution of maximum shear stress at $C_0 t/R_0 = 0.54$ after the collision of a cylindrical drop and a semi-infinite aluminum body [4–7].

present understanding of the subject. However, there are several relatively isolated computerized studies [30, 33–38] which are generally consistent with our results and which should be included for the sake of completeness.

Several very new studies regarding variations of the original problem have been reported by Johnson [33, 37, 38]. These include the generation of a new and improved computer program involving triangular, rather than rectangular, computing elements, impact-induced severe elastic-plastic deformations [33, 36]; previous work [4–7, 35, 36] had concentrated on small and elastic material responses. Reference 36 gives a very good review of this overall situation.

6-3 EXPERIMENTAL AND PHOTOGRAPHIC STUDIES

6-3-1 General

Section 6-2 has considered in detail the governing relations as well as computerized models applicable to the liquid-droplet impact problem for the subsonic range of liquid Mach numbers which apply to most problems of engineering interest. The effects of material elasticity have also been included, as well as some comparisons with experimental data. This section will consider all pertinent basic research experimental and photographic studies so far as is possible, in order to illustrate the research state of the art for this range of phenomena and the degree of agreement achieved between prediction and experiment.

6-3-2 Impact-Surface Pressures

General Experimental investigations of liquid-solid impact phenomena have involved many facets. From the viewpoint of hydrodynamics, efforts have been made to measure the impact pressure and to obtain high-speed photography of the impact process. As to its damaging effect on solid materials, there have been experiments to study the damage patterns and thus learn something of the surface pressures and velocities which may have produced the damage.

A high relative velocity between solid and liquid is obviously required for droplet-impact studies. This can be achieved either by propelling the liquid droplet or jet toward the material surface to be eroded, or by propelling the solid material toward a relatively stationary droplet or through a droplet environment, such as, for example, an artificial rain regime. Due to problems of droplet instability, sometimes described in terms of a critical Weber number, it is usually not possible to propel a spherical droplet of sufficient size and at sufficient velocity to model conditions of interest. However, approximately spherical droplets of $\gtrsim 1$ mm diameter are characteristic of such real-life applications as steam-turbine-blading droplet erosion, and aircraft or missile rain erosion. Also, droplets of known and regular geometry are required so that meaningful comparisons can be made with the results of numerical computations such as those described in the previous section.

Due to the instability of high-velocity liquid droplets, in general, experiments take the form either of impact between high-velocity solid targets and essentially stationary droplets or impact between projected liquid slugs of relatively undetermined or nonspherical geometry and a stationary target. A common form of such a jet, which has been used for many studies, has a relatively "mushroom"-shaped leading edge (Fig. 6-8), as previously discussed in connection with the Johnson and Vickers experiment [24]. Typical experiments have been described [9, 39–41] where the droplet has been propelled toward a stationary target, while others consider impact with liquid environments such as artificial rain [42, 43]. Cases are also considered [44–46] where high-velocity liquid jets are fired toward stationary targets.

Specific investigations Engel [25, 47, 48] has made extensive studies of waterdrop collision with solid surfaces, and was one of the first to study this particular problem from the viewpoint of aircraft rain erosion. Her experiments included starch-iodide tracer for radial liquid flow, high-speed photography and Schlieren pictures of the collision process, oscilloscope traces from high-response-rate pressure transducers, observation of time dependence of the droplet radius and flow velocity, and splash patterns as a result of water drops impacting surfaces with different degrees of resilience. Collision velocity was controlled by the height of liquid fall, so that this particular set of experiments was limited to quite low impact velocities (~ 10 m/s).

Jenkins and Booker [39, 40] studied the impact process of a water drop and a surface moving at high speed by firing a projectile at a water drop suspended on a fine web. High-speed shadowgraph and streak photography were used to determine the splash shape and radial velocity, respectively. Radial velocity for 1000 ft/s (328 m/s) impact was found to be ~ 1036 m/s.

Brunton and others at the Cavendish Laboratory at Cambridge University have made many important contributions over the years. For instance, Hancox and Brunton [45] studied the deformation of solids by liquid jet impact. Flatplate specimens mounted on the periphery of a high-speed rotating disc impacted a relatively low-velocity cylindrical water jet streaming parallel to the axis of rotation of the disc. The location where appreciable radial flow commenced was estimated from the distribution of shear pits on the surface. They found the angle of contact between the liquid surface and solid at the time when radial flow started to be $\sim 17°$ over a range of jet diameters and velocities. The impact load was measured using a barium-titanate pressure transducer which replaced the specimen for these runs.

Brunton and Camus [27, 49] examined the secondary flow processes within an impacting drop and found cavitation to exist on the liquid-solid interface and inside the drop. As previously mentioned, cavitation induced within impacting droplets has been predicted by computer models in the author's laboratory [1–4] and elsewhere [48]. Collapse of such cavitation bubbles [49] produced measurable shock waves and estimated liquid-collapse velocities greater than the initial impact velocity.

Rochester and Brunton [41] measured the distribution of maximum impact pressure and shear stress on the interface between a water drop and a solid surface. An oblong-shaped "bullet" with a flat front surface was fired against a 5-mm diameter water disc suspended between two Perspex plates. A very small piezoelectric ceramic was mounted on the front surface to measure local pressures and shear stresses. The results of this work have already been discussed herein for comparison with the computer model results. With impact velocity of 100 m/s, maximum pressure was $1.8\rho_0 C_0 V_0$ and occurred 0.5 mm from the axis. Pressure at the center was $\sim 0.7\rho_0 C_0 V_0$. The maximum in measured pressure coincided with the position of the edge at the instant outward jetting began. The impact angle at this point was $\sim 11°$ over a range of droplet diameters.

Johnson and Vickers [24] also measured the transient normal and shear stress distributions, in their case caused by perpendicular and inclined impact of water jets against nominally rigid surfaces. Their liquid jet was ~ 50 mm diameter with an impact velocity of 46 m/s. They found a maximum normal stress of $\sim 1.5\rho_0 C_0 V_0$ along a circle of the same radius as the impacting jet. This result differs from Rochester and Brunton [41], who found the maximum pressure relatively much closer to the axis. Johnson and Vickers [24] found the maximum shear stress to be $\sim 0.45\rho_0 C_0 V_0$ along a circle of 0.9 jet radius.

One can tentatively conclude from these results, and the computer model (which is in approximate agreement, as previously discussed), that:

1. The classical water-hammer pressure $\rho_0 C_0 V_0$ provides a relatively good engineering approximation for the impact pressure between a liquid curved surface and a flat solid plate.
2. The instantaneous pressure during the start of impact exceeds this value by a factor of ~ 1.5. While the exact value of this "coefficient" is somewhat in doubt at the moment, it apparently does not exceed 2, contrary to Heymann's earlier approximate analysis [8, 50].
3. The induced radial velocity is several (~ 2 to 3) times the impact velocity.
4. The duration of pressures in excess of the classical water-hammer pressure for droplets of the diameter range 1 to 3 mm (which are of most practical interest) is very short, probably only a fraction of a microsecond. Hence its damage-creating capability is in doubt.

6-3-3 Material Deformation, Stresses, and Damage from Liquid Impingement

The group at Cambridge University has also contributed very importantly to the study of surface damage due to liquid impingement. For example, Bowden and Brunton [51] and Bowden and Field [52] reported on the development and use of a momentum-exchange water-gun device, which has since become quite classic in the field. Modified versions have been used in many laboratories, including our own. Their water gun, as originally reported, produced a jet velocity of up to ~ 1200 m/s, with a jet of ~ 2 mm diameter. These investigations have combined the bombardment of surfaces with such jets with ultra-high-speed photography (in the order of 10^6 Hz). They thus confirmed the very short duration of the high-pressure portion of the impact (a few microseconds for droplets of a few millimeters diameter). The elongated liquid slug produced by these water guns (Figs. 6-26 and 6-27) is typically preceded either by a small "Monroe jet" [51, 52], at a velocity somewhat higher than that of the main jet, or by a mushroom-shaped leading edge (Fig. 6-8), which shows a schematic representation of this jet form. The actual form of jet obtained depends upon the position of the meniscus of the water slug in the nozzle portion of its container prior to firing [51, 52]. A concave-inward meniscus produces a Monroe jet, which is quite analogous to the explosive jet produced by an armor-piercing "shaped charge" ("Bazooka" of World War II), and represents a technique previously also used in mining.

As a result of the above and related work at Cambridge, numerous details of the form of surface failure for various materials were observed. In summary, the following features were observed:

1. Circumferential surface fracture
2. Subsurface flow and fracture

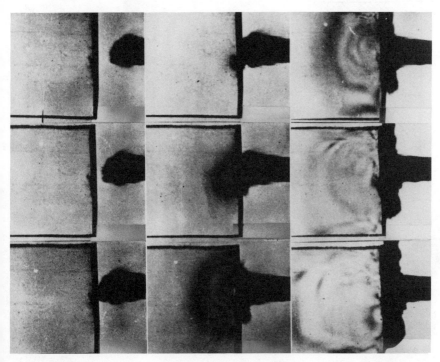

Figure 6-26 The impact of a 2 mm diameter water jet against a polymethylmethacrylate plate. Impact velocity 680 m/s. Framing interval 0.8 μs [51, 52].

3. Large-scale plastic deformation
4. Shear deformation around the periphery of the impact zone
5. Fracture due to the reflection and interference of stress waves

Hancox and Brunton [45] studied the damage to solids by multiple low-velocity impact (~40 to 180 m/s) of liquids using a wheel and jet apparatus, previously discussed. They followed the development of erosion damage with increasing numbers of impacts (as have various other investigators) for several materials of different characteristics, and studied the influence of impact velocity, grain size, surface roughness, and other parameters on erosion damage.

Other investigations have considered much higher velocity ranges typical of the aircraft rain-erosion problem. It is unnecessary to attempt to review very many of these here. However, as typical examples, Schmitt and his colleagues [42, 53, 54] performed a comprehensive investigation of a large number of plastic, ceramic, metallic, and composite materials for short-time exposure rain-erosion resistance at velocities of 1.5, 2.0, 2.5, 3.0, and 4.0 times the sound speed of air,

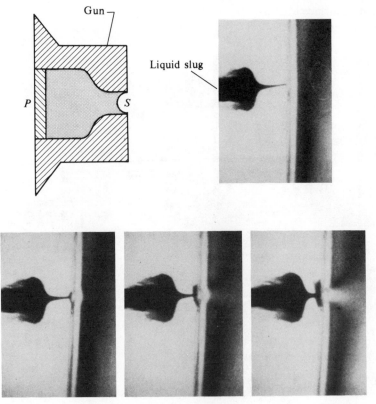

Figure 6-27 Formation of micro-Munroe jet when a shock wave falls on the concave surfaces S of a liquid. Velocity of main jet c. 650 m/s, of microjet c. 1900 m/s. Interval between frames, 0.8 μs [51–52].

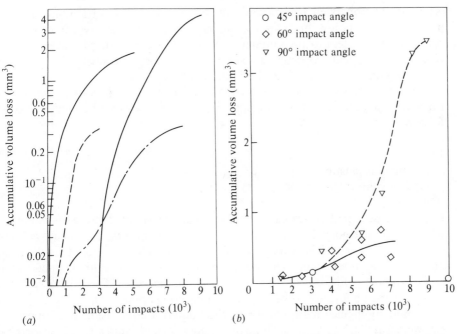

Figure 6-28 Typical results of liquid erosion resistance tests with the water-jet gun at the University of Michigan [56].

i.e., these values can be considered as conventional atmospheric air Mach numbers. Specimens for these tests were mounted on multistaged rocket sleds fired down a track through an artificial rain field at Holloman Air Force Base. The application of interest was, of course, aircraft and missile component rain erosion.

Studies of the high-velocity liquid-impact erosion resistance of various materials under different conditions have also been performed in various laboratories using either rotating-arm or disc devices, or liquid jet-guns, usually essentially modifications of the original Bowden-Brunton [51] momentum-exchange gun device. Typical results of such experiments are shown in Fig. 6-28. Aluminum (type 1100-F) and nickel (type 270) were tested in the author's laboratory using a repeating water-jet gun, of a design originated by Kenyon [55], but producing jets very similar to those of the Bowden-Brunton device. This device [56] can produce ~2 mm diameter jets of up to 600 m/s velocity. For the present tests (Fig. 6-28), impact velocities were 400 and 600 m/s. The typical "incubation period" is shown with little mass loss but noticeable surface deformation.

6-3-4 Impact between Liquid Drop and Elastic Half Space

It is generally believed at present that most ductile materials fail in shear, and brittle materials in tension. However, it also appears probable that ductile behavior is relatively suppressed here because of the very high rate of loading in both impact and cavitation. For this reason, as discussed under cavitation damage, brittle fractures seem to predominate, and the best correlation between erosion rate and mechanical properties is achieved with "ultimate resilience," i.e., "strain energy to failure," if failure is in the brittle mode.

When ductile solids are subjected to impact by a high-velocity object, or otherwise induced impulsive pressure, the deformation can be categorized as follows:

1. Elastic
2. Elastic plus plastic
3. Hydrodynamic, i.e., viscous

For elastic deformation (category 1) there is no apparent damage after impact, whereas for category 2 a permanent deformation occurs. In category 3 the combination of materials properties and impact conditions is such that the solid responds as a viscous fluid. This occurs primarily in "hypersonic impacts" such as, for example, collisions between meteoroids and spacecraft. This type of collision is not of concern for most other types of engineering applications.

For many engineering applications where liquid-solid collisions are important, e.g., wet-steam turbines, permanent damage does not start until after a considerable period of initial exposure, usually termed the "incubation period," as shown in Figs. 6-28 and 6-35. Permanent damage does not then start until stress hardening and other related processes have degraded the surface resistance of the material considerably. The solid material can be reasonably considered as an elastic medium during this incubation period.

The duration of the incubation period, of course, is related to the strength and duration of the impact pulse [57]. If this is too short, there is insufficient time for appreciable material flow due either to shear or normal stresses, so that many materials withstand elastically momentary stresses much greater than their static (conventional) yield stress would indicate. Under these conditions, ductile materials may fail in a brittle mode, as previously mentioned.

Fractures of brittle materials originate at inherent flaws and propagate with finite speed along the surface of principal tensile stress. If the impulse duration is too short, cracks may not be able to expand enough to become significant. However, regardless of the nature of deformation, detailed information on the stress distribution is needed for predicting the location and cause of failure. As a first approximation, the elastic deformation assumption can best be used.

Bowden and Field [52] suggested and verified that small cracks surrounding the main ring fracture of brittle materials were initiated by Rayleigh surface stress waves. While this simple model appeared to hold true for thick materials, they found also that the reinforcement of reflected longitudinal and transverse waves with Rayleigh surface waves caused further localized cracks.

Blower [58] assumed the impact of a spherical droplet to be equivalent to a constant uniform pressure $\rho_0 C_0 V_0$ over a circle whose radius increased in proportion to the square root of time. He then solved the elastic wave equations by a Laplace transform technique, and obtained in closed form the transient surface stress and deformation distribution for a homogeneous isotropic elastic half space, with Poisson's ratio equal to 0.3. He showed that a Rayleigh surface wave would appear at the moment when the radius of the loaded area is increasing with the Rayleigh wave speed. Behind the Rayleigh wavefront, he found an abrupt transition (singularity) which caused extreme radial and azimuthal tensions. Since their duration was extremely short, he did not think them responsible for the observed liquid-impact erosion, as Bowden and Field [52] had suggested.

Peterson [59], based on results of Huang [1–3] at the author's laboratory and Tyler [34] elsewhere, assumed as a first approximation that the impact pressure could be represented by a quasistatic, hemispherical distribution over a circular area. He then applied Love's solution [60], i.e., Hertz contact stress, to obtain the stress distribution in a semi-infinite homogeneous elastic solid. The maximum impact pressure, which would then occur at the axis, was determined by one-dimensional Rankine-Hugoniot relations. He found the principal compressive stress to be greatest on the surface at the axis. Shear stress was maximum at $0.45R_0$ below the surface along the axis, where R_0 is the radius of the loaded area. Obviously, however, this model does not include surface shear stress along the surface induced by the impact-induced high radial velocity previously discussed. Some investigators have attributed a portion of the damage to this feature. However, simple calculations made by the present author indicate that fluid-induced surface shear ($\tau = \mu\, \partial u/\partial y$) is not likely to be important in the damaging process. Of course, liquid impact of the radial velocity against surface asperities may generate a substantial portion of the water-hammer pressure, which might then obviously contribute significantly to the damage. According to the Peterson analysis [59], the surface under the impact area, i.e., the area of his hypothetical imposed load, will be subjected only to compressive stresses, and thus should remain relatively undamaged. On the other hand, he found [59] that surface tensile stresses immediately outside the loaded area would be sufficient to initiate circumferential cracks in certain materials. Photographs of impact-damaged surfaces, discussed elsewhere, verify this conclusion.

Blower's assumption [58] for the expansion rate of the loaded area agrees quite well with our numerical results [1–7] for the impact of a spherical drop and a semi-infinite solid body. However, his assumption of constant and uniform pressure does not agree well with our calculations or other measurements. The weight of evidence now confirms the existence of high edge pressures for the impact of a spherical drop. Experiments of Goodier, Jahsman, and Ripperger [61] on the impact of a steel ball with an elastic half space with impact velocity to 6000 m/s showed that the measured duration of contact and force-time relation agreed well with the prediction of the Hertz contact theory [60] used by Peterson [59]. Utility of the hertzian approach for liquid-solid impact, however, is much more limited because liquids are far from elastic over the pressure range involved and pertinent static force-deformation relations are lacking. Furthermore, the duration of the high-pressure portion of usual liquid-solid impact is only a few

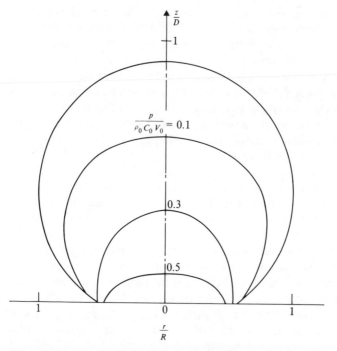

Figure 6-29 Isobar distribution in an initially spherical droplet at time $Ct/D = 0.5$, for impact Mach number of 0.2 on an elastic boundary, $P_s C_s / \rho_0 C_0 = 1.0$ [4, 5].

microseconds, whereas that for solid-solid impact, for which the Hertz contact theory was derived, are typically much longer, i.e., in the order of 100 μs.

Typical results from the U-M calculations [4, 5] in the form of isobar distributions at two different instants for the impact between a spherical water droplet and an elastic planar surface for collision speed of $0.2C_0$, i.e., liquid Mach number of 0.2, are shown in Figs. 6-29 and 6-30. Just as for impact between a spherical droplet and a rigid surface, maximum pressure is initially at the center and during the collision moves to the edge of the contact area. The edge of the contact area also forms a deeper depression than the center.

Timm [31, 32], working at the University of Michigan, used a million frames per second Beckman-Whitley rotating-mirror camera to study liquid-jet impingement upon various elastomeric coatings upon steel backings. Figures 6-31 and 6-32 show typical results for different elastomeric materials. He found significant differences in the velocity and direction of splash-back, depending upon the properties of the target material. Splash-back was minimal for the more rigid elastomerics and also for Plexiglas, but was quite pronounced (Fig. 6-31) for some of the rubber-coated materials. However, no correlation between the pattern of splash-back and erosion resistance of the materials could be found [31]. His pictures for the impact of a bullet upon Plexiglas indicated the propagation of elastic waves into the material.

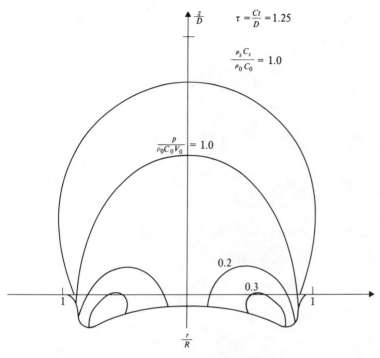

Figure 6-30 Isobar distribution in an initially spherical droplet at time $Ct/D = 1.25$, for impact Mach number of 0.2 on elastic boundary, $\rho_s C_s / \rho_0 C_0 = 1.0$ [1, 3].

Field et al. [62] took high-speed photographs of the impact between a water jet and Plexiglas, i.e., PMMA (polymethylmethacrylate) plastic. Their pictures show clearly the propagation of elastic waves (shock waves) into the plastic. Also shown is a detached shock wave in the plastic outside the impact area.

6-3-5 Miscellaneous Effects

Liquid relaxation and short time duration of liquid impact As already mentioned, the time duration of the important portion of liquid impact from the viewpoint of maximum pressures, and hence erosion effects, is extremely short—only a very few microseconds. Also as mentioned, this may give rise to effective material mechanical properties which differ substantially from those obtained in conventional material property tests. These can be regarded as semistatic by comparison. In general, the material strength properties under such highly dynamic loading conditions are greater than the conventional values, whereas ductility properties tend to be suppressed, with the result that failures often appear to be of the brittle-fracture type. Unfortunately these highly transient effects are not fully understood or documented, so that in this area of liquid-impact erosion research much yet remains to be done.

Run 510–BW–I–4
Jet velocity = 223 m/s
Target mtl = rubber B

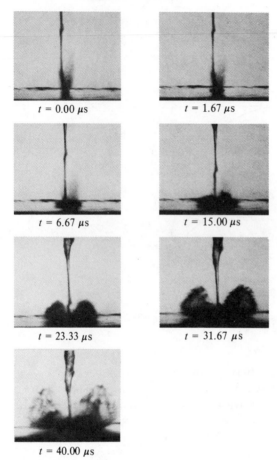

$t = 0.00 \ \mu s$

$t = 1.67 \ \mu s$

$t = 6.67 \ \mu s$

$t = 15.00 \ \mu s$

$t = 23.33 \ \mu s$

$t = 31.67 \ \mu s$

$t = 40.00 \ \mu s$

Figure 6-31 Water jet impact on soft rubber (large splash) [31, 32].

In addition to the material-oriented transient effects, the liquid behavior must also be taken into account. In both cavitation bubble collapse and liquid impact there is the possibility that fluid behavior, both liquid and vapor, may not follow equilibrium relations. For example, vapor evaporation or condensation, and subsequent heat transport, may not be sufficiently rapid to prevent a change in vapor pressure during the final portion of cavitation bubble collapse. This mechanism can restrain the growth or collapse of bubbles, reducing overall cavitation effects for either inception or damage. This particular situation, dubbed "thermodynamic effects" in the cavitation literature, is further discussed in Chap. 4 on cavitation bubble dynamics. However, there are also highly transient liquid effects which may affect liquid-solid impact. These are discussed in the following.

Run 4300 – BW– I – 2
Jet velocity = 223 m/s
Target mtl = rubber A

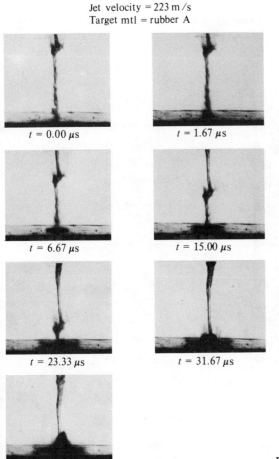

$t = 0.00\ \mu s$ $t = 1.67\ \mu s$

$t = 6.67\ \mu s$ $t = 15.00\ \mu s$

$t = 23.33\ \mu s$ $t = 31.67\ \mu s$

$t = 40.00\ \mu s$

Figure 6-32 Water jet impact on hard rubber (small splash) [31, 32].

Liquid relaxation effects   The duration of the final important portion of cavitation bubble collapse or that of the important initial portion of liquid-solid impact may be so short that, at least with certain liquids, nonequilibrium relaxation effects become important. This has been observed, for example, in impact tests by Kozirev and Shal'nev [63, 64] with certain resins, where the liquid appeared to behave as a brittle solid rather than as a conventional liquid during the impact process. Figure 6-33 is from a high-speed motion picture sequence showing this behavior, which they ascribe to the hypothesis that the relaxation time of the liquid is long compared to the duration of the impact. They believe that this type of phenomenon may occur in the final stages of cavitation bubble collapse and the resultant liquid microjet impact upon an adjacent material. Much research is still necessary

Bullet velocity = 24 m/s
Solution viscosity = 30,000 cP
Framing rate = 8000 frames/s
Exposure time = 60 μs

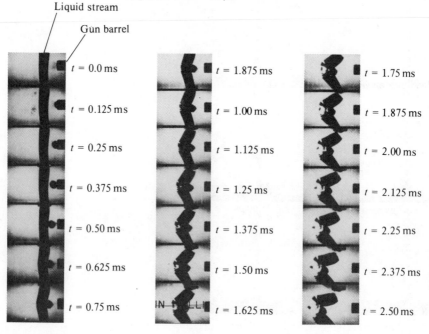

Liquid stream

Gun barrel

$t = 0.0$ ms	$t = 1.875$ ms	$t = 1.75$ ms
$t = 0.125$ ms	$t = 1.00$ ms	$t = 1.875$ ms
$t = 0.25$ ms	$t = 1.125$ ms	$t = 2.00$ ms
$t = 0.375$ ms	$t = 1.25$ ms	$t = 2.125$ ms
$t = 0.50$ ms	$t = 1.375$ ms	$t = 2.25$ ms
$t = 0.625$ ms	$t = 1.50$ ms	$t = 2.375$ ms
$t = 0.75$ ms	$t = 1.625$ ms	$t = 2.50$ ms

Figure 6-33 Brittle fracture of resin under impact (Shal'nev and Kozirev) [63, 64].

to evaluate the possible significance of this phenomenon, but it certainly appears possible that such nonequilibrium phenomena may be important in some instances.

Light-flash phenomena The observation of "sonoluminescence" in cavitation is well documented by numerous investigators and is discussed in further detail in Chap. 4 on cavitation bubble dynamics. The cavitation luminescence is generally ascribed to the very high temperatures resulting from the essentially adiabatic compression of the permanent gas and vapor trapped within collapsing cavitation bubbles. It is another example of nonequilibrium behavior due to the very short time duration of the phenomenon. However, somewhat similar liquid flashes have also been reported [65–67] for very high-velocity liquid-solid impact cases. The light is there believed to be due to essentially adiabatic compression of an air film trapped between the liquid jet and the solid. Again, further research is required to understand more definitely this phenomenon. However, it might represent a cushioning mechanism which would then reduce material damage.

6-4 OVERALL EROSION PERFORMANCE

6-4-1 General

The overall liquid-induced erosion performance of materials subjected particularly to cavitation has been discussed in detail in Sec. 5-3 of this book. Semiempirical relations were there presented relating measured erosion rates (volume loss rate per unit exposed area, i.e., MDPR = mean depth of penetration rate) and material mechanical and fluid properties. As discussed in the present section and also in Chap. 5, a consideration of the basic mechanisms of cavitation and liquid-impact damage show that these are in all probability very similar phenomena. In fact, historically, rotating-wheel jet impact tests have been used to provide an indication of cavitation, as well as droplet-impact, erosion resistance of materials. It is also well accepted that the general appearance of erosion resulting from these two phenomena is very similar. Hence, the data sets [68, 69] used for the generation of damage-predicting relations for cavitation damage particularly, but also presumably applicable to some extent to impact damage, were drawn from both cavitation and liquid-impact tests. The present section, emphasizing liquid impact particularly, discusses other pertinent features applicable only to impact effects, rather than cavitation. Examples of other such features are the effects of impact velocity, angle, and droplet or jet diameter. While such parameters apply also to cavitation damage, their effects cannot easily be studied except with impact tests, since in general their magnitudes are not known or controllable in cavitation tests.

A semirational approach to either the impact- or cavitation-erosion phenomenon can be made, as already discussed in Sec. 5-3. One such approach [68] results in Eq. (5-4), repeated here for convenience from Sec. 5-3. The nomenclature† is explained there:

$$\text{MDPR} = \frac{\eta}{\varepsilon} \frac{A_p}{A_e} \frac{\rho_{\text{eff}}}{2} V^3 \tag{6-24}$$

As explained in Sec. 5-3, only the first quotient in Eq. (6-24), that is, η/ε, and the effective density ρ_{eff} can be estimated at all for cavitation tests. The effective density ρ_{eff} is roughly that of the liquid, but it is the actual density of a mixture of liquid droplets and surrounding gas in a rain-erosion-type impact test. It can be readily calculated for such impact tests from the test parameters, but not for cavitation.

The efficiency of energy transfer between the impacting "microjet" (cavitation) or droplet and material surface η is a function of at least the "acoustic impedance ratio," while ε is a material property giving the necessary energy input to a material surface to cause removal of unit volume of the material. Hence for fixed "flow conditions" in either an impact or cavitation test, the ratio η/ε is assumed proportional to MDPR, i.e., rate of volume loss per unit area exposed.

---

† MDPR = mean depth of penetration rate
= volume loss rate/exposed area.

Thus, cavitation or impact tests could be used to evaluate "best-fit" relations between measured erosion rates and material and fluid properties. This is the course which has already been described in Sec. 5-3. The present section attempts to further clarify the relationship between measured erosion rates and the other parameters called to attention by semiempirical relations such as Eq. (6-24), and which are particularly pertinent to liquid impact rather than cavitation erosion.

6-4-2 Effects of Velocity and Angle

General considerations It has generally been accepted in the past that only the component of velocity normal to the surface [Eq. (6-24)] is important to the damage process. It is also generally believed, as the result of much previous test data, that damage rate is most often proportional to the fifth or sixth power of velocity rather than the third power, as shown by the semirational Eq. (6-24). However, such a higher-order velocity dependence may still be consistent with Eq. (6-24), since the efficiency of energy transfer between droplet and material surface η is quite probably also velocity dependent to some extent. There is also the possibility that the effect of the high radial velocity generated in the collision, discussed in Sec. 6-2, could affect the velocity dependence. This mechanism was not considered in the generation of the simplified model resulting in Eq. (6-24).

It has also been often considered that there is a "threshold velocity," presumably applied to the normal component, below which zero or negligible damage will result. Obviously the value of such a "threshold velocity" must depend upon duration of the test, so that such a simplified general model cannot be truly valid. If the "threshold velocity" model is used, then the velocity dependence is related to the energy increment above such a threshold. Such a concept would then assume a "threshold energy" necessary for damage to occur. This possibility also was not considered in the generation of Eq. (6-24). It is the purpose of the following discussion to examine the validity of these various concepts, based upon some relatively recent impact data, reported in detail in Ref. 68.

Specific test results For this purpose some of the data generated by recent rocket-sled tests at the Holloman Air Force Base have been examined to determine the suitability of certain of these semiempirical damage-predicting equations. The portion of the rocket-sled data selected for this analysis comprised ten groups of materials including ceramics, plastics, and metals. They were tested in the 6000-ft rain field at the Holloman Air Force Base at Mach numbers ranging from 1.5 to 5.0 and at various angles of impact ranging from 13.5 to 90°. The full details of this analysis have been reported previously [70]. However, some salient features will be repeated here for convenience.

An earlier report [71] based upon an experimental fit of rocket-sled data suggests that the erosion rate appears in an exponential form:

$$\text{MDPR} = C_1 e^{aV} \sin \theta + C_2 e^{bV} \tag{6-25}$$

where C_1, C_2, a, and b are constants depending on material properties, and θ = angle of impact (90° for perpendicular impact).

Baker, Jolliffe, and Pierson [72] proposed a relationship based on their impact data, which includes the concept of a threshold velocity below which damage is essentially zero:

$$\text{MDPR} = \begin{cases} \dfrac{K(V \sin \theta - V_0)}{\sin \theta} & \text{for } V \sin \theta > V_0 \\ 0 & \text{for } V \sin \theta \leq V_0 \end{cases} \tag{6-26}$$

More recently Hoff, Langbein, and Reiger [73] have suggested a modification of Eq. (6-26) whereby the denominator is squared:

$$\text{MDPR} = \begin{cases} \dfrac{K(V \sin \theta - V_0)^n}{\sin^2 \theta} & \text{for } V \sin \theta > V_0 \\ 0 & \text{for } V \sin \theta \geq V_0 \end{cases} \tag{6-27}$$

Equation (6-25) is simply a curve-fitting expression, not based on any physical model. Equation (6-26), on the other hand, assumes basically that MDPR is proportional to the difference between the normal component of the impact velocity and some "threshold velocity," all raised to some power n. A similar assumption has often been made in the cavitation literature [74–76], where damage was usually [74, 75] found to be proportional to the sixth power of the flow velocity, based upon tests on a soft aluminum ogive in a water tunnel in this particular case. In Eq. (6-26) $\sin \theta$ has been added to the denominator to take some account of the damage due to shear from the high radial velocity after impact, which increases for oblique collisions, as discussed to some extent in Sec. 6-2. Actually, since in the rocket-sled-type test the specimen impacts a reduced number of raindrops if the impact is not normal, it might be argued that an additional $\sin \theta$ is required in the numerator, cancelling that in the denominator. This latter variation was not tested in the present numerical analysis.

Equation (6-27) is identical to Eq. (6-26) except that $\sin^2 \theta$ appears in the denominator. This term can be derived logically from a model assuming energy flux on the target to be the predominant mechanism [73], if it is also assumed that the efficiency of energy transfer between the impacting drop and target is a function of $V \sin \theta$ only [Eq. (6-24)]. However, it seems unlikely that this is strictly the case, so that Eqs. (6-26) and (6-27) remain semiempirical in nature, and to be tested only in terms of a data fit.

Computer correlation results   A comprehensive analysis [70] of the rocket-sled data was made using Eq. (6-27). For each material a least-mean-square-fit regression analysis was made to determine the best value of the threshold velocity V_0, the amplitude constant K, and the velocity exponent n [70]. Table 6-2 shows the effects of the velocity threshold V_0, K, and the velocity exponent n for one of the materials.

This table shows particularly the effect of the choice of the threshold velocity

Table 6-2 Effect of V_0 on values of K and n for epoxy laminate [68, 70] (Eq. 6-27)

V_0, ft/s	n	$K \, (\times 10^5)$
0	6.44	25.7
200	6.36	27.9
400	6.24	29.7
600	6.08	32.3
800	5.87	34.8
1000	5.59	36.7
1200	5.22	39.6
1400	4.73	43.4
1600	4.09	41.3
1800	3.28	40.1
2000	2.28	28.4

V_0 on the best values for the velocity exponent n and the amplitude constant K. The effect on K of varying the assumed V_0 between 0 and 2000 ft/s (~ 650 m/s) is small, but the velocity exponent varies from 6.44 to 2.28, depending on the essentially arbitrary choice of "threshold velocity." No particular value of V_0 was in fact indicated by these tests. A plot of MDPR versus normal impact velocity V shows a small or zero MDPR for small velocities, and then a rapidly increasing MDPR for larger velocities. Such a curve can be fitted almost equally well by various combinations of V_0 (including zero) and the velocity exponent n, as the present calculations show. Since, strictly, it is unlikely that there will be *zero* damage for repeated impacts at *any* velocity, it may be permissible to avoid the concept of threshold velocity entirely. If it is used, it must obviously be a function of the number of impacts per second and the test duration as well as velocity. It may be necessary to define an arbitrarily small but finite limit for MDPR, which will then define the threshold velocity. Figure 6-34 shows two

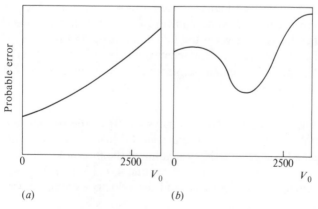

(a)   (b)

Figure 6-34 Typical curves for probable error in erosion as a function of threshold velocity [68, 70].

typical sketches for the probable relations between probable error and choice of threshold velocity for the present data. For those materials exhibiting behavior of the type shown in Fig. 6-34a, the optimum choice for threshold velocity is zero. For other materials, as in Fig. 6-34b, a definite optimum V_0 appears.

For some of the present materials, "best values" of V_0, K, and a were computed from both Eqs. (6-26) and (6-27). Table 6-3 shows the results for an inorganic laminate and a thermal plastic. A more complete description is given in Ref. 70. While Eq. (6-27) calls for an exponent 2 for the $\sin \theta$ term, the effect of exponents ranging from 1.0 to 2.5 was examined [$n = 1$ corresponds to Eq. (6-26)]. It is noted that, for these materials, the choice of a affects the best choice of threshold velocity (and of course n, which is not listed). However, it affects the minimum probable error only slightly. From these data it appears that a choice of $a = 1$, desirable for the sake of simplicity, would not significantly reduce the "goodness" of the correlation. The effect on probable error of assuming zero threshold velocity (also desirable for simplicity) is shown in the last column. It is noted that the additional error so introduced is not large.

Generalized erosion model The limited success achieved in correlating the rocket-sled test data using Eqs. (6-26) or (6-27) leads to the conclusion that a more basic mathematical model is required. However, the lack of good correlation is partly due to the type of data used. It is not entirely permissible to compare damage attained after a fixed exposure period for materials of widely differing resistances, as was necessary for these rocket-sled tests, since only a mean MDPR can then be computed for materials in very different portions of their MDPR versus time

Table 6-3 Results of evaluation of equation MDPR $\sin^a \theta = [K(V - V_0) \sin \theta]^n$ for various values of a [68, 70], (Eqs. 6-26 and 6-27)

Inorganic laminates

a	Threshold velocity, ft/s	Minimum probable error, μ/s	Probable error for $V_0 = 0$, μ/s
1.0	1100	82	146
1.5	1000	88	143
2.0	900	95	141
2.5	800	101	140

Thermal plastics

1.0	350	7.3	7.7
1.5	200	7.3	7.4
2.0	100	7.2	7.3
2.5	0	7.2	7.2

(or number of impacts) curves. It is thus necessary to use data wherein the total MDPR versus exposure curve is available, so that only comparable portions of this curve would be compared. The generation of a semirational erosion model has been described in detail in Sec. 5-3 which is pertinent to both cavitation and droplet impact. This results in the first equations of both Secs. 5-3 and 6-3.

6-4-3 Material Property Correlations

General As already discussed, material mechanical property correlations for predicting erosion resistance of given materials to liquid-droplet impact or cavitation erosion are essentially identical, at least in the present state of the art. These have already been discussed in detail in Sec. 5-3 under the subject of cavitation erosion resistance. The extent to which damage resistance of materials can be realistically related to the conventional material mechanical properties, which represents unfortunately a relatively low capability today, is there discussed. It was shown that "ultimate resilience" and hardness are apparently the properties of greatest use in this regard at present, but that standard deviation about predicting relations is of the order × 3. Whether or not such an imprecise prediction is, or is not, of engineering utility is of course a matter of personal judgment.

This relatively unsatisfactory situation exists today, even for material-fluid combinations where the role of corrosion cannot be more than negligible. For "corrodible" materials, the degree of uncertainty is of course much greater. However, a detailed examination of corrosion effects is beyond the scope of this book, which is related primarily to highly intensive fluid mechanical effects. Nevertheless, it is well known that the combined effects of cavitation, or impingement, along with corrosion are usually much greater than their summation if acting singly. The cause of this accelerated effect is easy to envisage. Corrosion roughens and weakens the surface, accelerating impingement or cavitation erosion, and the mechanical effects of cavitation or impingement quickly remove the protective coating, which normally would inhibit further corrosion.

Another interesting interrelation which exists between impact and cavitation is that liquid impact can often induce cavitation within the impacting droplet, as explained in previous sections of this chapter, while it is presently presumed that cavitation damage is to a substantial extent the result of liquid microjet impact upon the damaged surface. This microjet is induced by the collapse of cavitation bubbles near the surface, as fully explained in the discussions of cavitation bubble dynamics in this book.

Characteristic erosion-time curves Much earlier erosion work was based only on single runs (and necessarily even some recent work such as that using a rocket sled), previously described. These provide only a given volume loss after a fixed time interval [71]. However, it has more recently become generally agreed that the erosion-time curves induced in either cavitation or impact tests are not linear with time. This was probably first emphasized by the work of Thiruvengadam [77], and

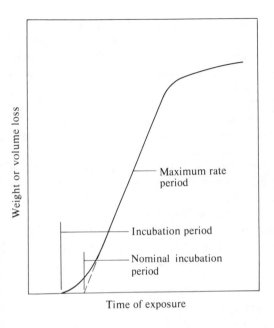

Weight or volume loss

Maximum rate period

Incubation period

Nominal incubation period

Time of exposure

Figure 6-35 Typical cavitation or liquid-impact "S-shaped" erosion curve [91].

is now well documented and described in the definitions, etc., promulgated by the ASTM Committee G-2 [78]. However, there is still much disagreement in this matter, and it appears that cavitation or impingement curves generated by different types of test devices and different material-fluid combinations differ importantly in their characteristic erosion-time curves. However, there appears to be no systematic difference between cavitation- and impingement-erosion tests in this regard. In general, "S-shaped" curves are generated (Fig. 6-35), characterized by an initial period of negligible or low damage rate, "incubation period." This is followed by a period of increasing rate, then a period of approximately constant maximum erosion rate, and finally a period of decreasing, or sometimes oscillating, rate. Which of these periods are of greatest importance, of course, depends upon the application.

Some pertinent theoretical work has also been done in recent years to explain the shape of these curves, particularly by Heymann [69], Engel [79, 80], and others. It is possible to justify any of the experimentally observed curve shapes by a mathematical model, assuming material failure to be primarily a high-strain fatigue mechanism [69].

Recent impact-erosion studies Several recent relatively comprehensive studies have been made to attempt to understand in more detail material reactions to droplet impact, and thus attain the ability to predict a priori material resistance to such attack. These studies have included detailed basic computer modeling of the fluid and material behavior, such as the work at the author's laboratory of Hwang [4–7], already discussed, and work elsewhere [33, 35–38] where the influence of a nonrigid (elastic) material was considered and stresses and strains therein examined. In addition, various quasibasic studies of material reaction, where erosion rates were correlated with theoretical material models, have been made. In some cases various nonmetallic materials have been studied because of their importance in such applications as rain erosion of high-speed aircraft and missile components. Relatively recent work by Engel [79, 80] has developed a quasi-empirical failure model for various nonmetallics for this purpose. More recently Springer and his colleagues at the University of Michigan [81-84] have developed such models for "homogeneous," coated, and fiber-reinforced materials. A detailed description of these and other such hypothetical failure models is beyond the scope of this book, which is concerned more with fluid behavior than material reaction. The latter is, of course, a major topic in itself. Various recent studies of this general type are found in ASTM hard-cover books (see Ref. 81, for example). Other related, relatively recent studies are reported in the series of *Meersburg Conference Proceedings* (see Ref. 82, for example). It is probable that continued conferences will appear in the future in these two series.

6-5 LIQUID-IMPACT TEST DEVICES

6-5-1 General: Applications Including Useful Jet Cutting

Liquid-impact test devices have to some extent been described in this chapter in Sec. 6-3, concerning experimental and photographic studies, and also in Sec. 5-1, on cavitation test devices, as well as in Chap. 2. Historically, conventional liquid-impact tests have also been considered to be cavitation tests. In fact, local cavitation may well occur in many apparent impact situations, and damaging cavitation bubble collapse apparently often involves liquid microjet impact. Hence the two phenomena are at this point intimately related. It seems useful here to describe briefly the various types of nominal liquid-impact test devices. These can best be divided according to whether the high relative impact velocity is generated primarily by motion of the specimen or of the liquid slug, of whatever geometry it may be.

For the testing of general erosion resistance of materials, as well as for purposes of basic research, devices in which the liquid slug is relatively stationary are most numerous and also most important. However, these devices are today by no means standardized,† since they are not sufficiently numerous and are usually designed to deal with relatively specialized needs in different specific applications. This lack of standardization is also at present the case in devices designed to produce jet cutting or controlled erosion, such as to assist in coal mining, tunneling, oil-well drilling, wood cutting, and numerous other applications. Also, standard tests are lacking for these jet-cutting devices, some of which rely partially on cavitation and/or mechanical cutters.

† A round-robin test program was recently conducted by the ASTM Committee G-2 [85], from which it was concluded that standardization at this time was not feasible.

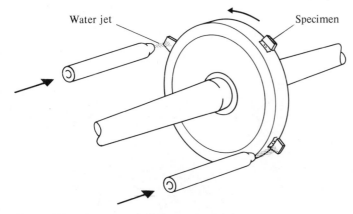

Water jet   Specimen

Figure 6-36 Jet impact damage device (schematic).

6-5-2 Moving-Target Devices

Rotating-disc tests for hydraulic-turbine application   The earliest and simplest form of this type of liquid-erosion testing device is the rotating disc (or wheel), to the periphery of which test specimens are attached. These have been employed since the 1930s [11, 86, 87]. The rotating specimens generally cut a relatively low-velocity cylindrical water jet, so that the relative impact velocity is primarily the result of the specimen velocity, generally ~ 100 m/s. It is apparent that this geometry (Fig. 6-36) would often generate local cavitation around the test specimens, so that the resultant erosion most probably included a substantial component of cavitation. In fact, the impact velocity was such that damage from impact itself of some of the resistant materials which were tested, such as various stainless steels, would not otherwise have been expected. The application to which this early testing was addressed was generally hydraulic turbines and pumps.

Rotating discs for steam-turbine application   Relatively similar test devices operating at higher velocities ($\lesssim 500$ m/s) have been developed and utilized by various steam-turbine companies [87], particularly in England. The application of interest for these studies was the wet-steam water-droplet erosion problem encountered in the low-pressure end of large steam turbines. This general problem is further discussed in Chap. 7.

Aircraft and missile rain erosion   During the post-World War II period, the problem of rain erosion of aircraft and missile components has become of importance. This has resulted in the development of even higher-velocity test devices, wherein material specimens are driven through an artificial rain field. In most cases this has been accomplished by "propellor arms," i.e., essentially zero-lift rotating propellor arms, to which the specimens to be tested are attached. These

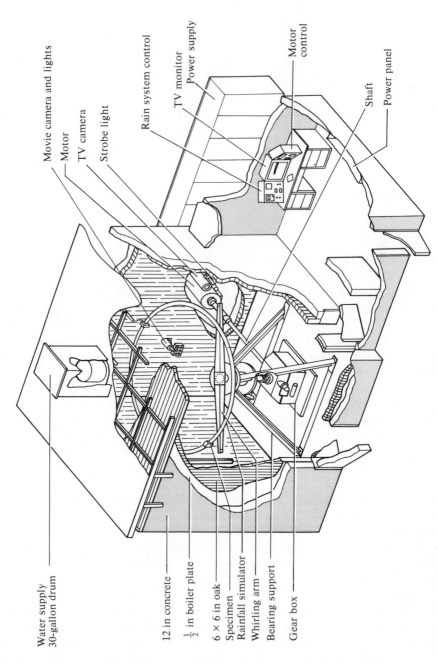

Water supply
30-gallon drum

Movie camera and lights
Motor
TV camera
Strobe light

Rain system control
TV monitor
Power supply

Motor
control

Shaft
Power panel

12 in concrete
$\frac{1}{2}$ in boiler plate
6 × 6 in oak
Specimen
Rainfall simulator
Whirling arm
Bearing support
Gear box

Figure 6-37 AFML rotating-arm apparatus, Dayton, Ohio [42].

"arms" (Fig. 6-37) are generally of relatively large radius, up to 14 ft in the Bell device [43], in order to minimize the centrifugal loading on the specimens. The highest-velocity machine to have been operated to the present is that at Bell Aerospace [43] with a maximum design speed of Mach 3, that is, ~ 1000 m/s. Numerous lower-speed devices exist in various laboratories in the United States, England, Germany (Dornier), and elsewhere. These machines are generally distinguished from the previously described steam-turbine test devices by the fact that the required test times are very much shorter, since the aircraft and missile components to be tested, such as radomes, for example, are of much less resistant materials than are the steam-turbine blades. A "radome," for example, must be primarily transparent to radar, and hence metallic materials are excluded. A recent book and reports by Springer [81–84] summarized much recent information on the erosion resistance of these types of materials, and presented predicting models.

Ultra-high-velocity tests for rain erosion have been conducted using a "rocket sled" propelled through an artificial rain field [42, 53, 54] at the Holloman Air Force Base. For this device velocities up to about air Mach 5 [88] have been reached. This test, of course, does not permit intermittent examination of the test specimens, as do the rotating-arm or disc tests previously discussed. It does, however, allow the simultaneous testing of numerous materials at different angles of impact in a single run, thus assuring that all experience the same rain field. This type of test also eliminates the extraneous centrifugal loading involved in the rotating-arm and disc tests.

Basic liquid-impact research test devices For purposes of basic research into liquid-impact phenomena, Fyall [26] and others at the Royal Aircraft Establishment at Farnborough developed a device wherein the target is fired as a "bullet" against a suspended stationary spherical liquid droplet. Control of the droplet shape and size is much more precise for this device than for any of those involving artificial rain fields previously described. Falling drops achieve some reasonable degree of spheroidicity, but not to the extent achieved by the Fyall device. Also, since the droplet is completely stationary, the potential for high-speed photography is maximized. Problems of droplet stability preclude the acceleration of a liquid droplet of desired size (~ 1 mm radius) against a stationary target. However, the Fyall device is not well adapted to the study of material erosion under repeated impact, since it is essentially a single-shot device. It also introduces the problem of target recovery without damage, but a decelerating system for this purpose was successfully developed.

A somewhat similar device wherein a target "bullet" was fired against a liquid disc supported between parallel plates was developed and used by Brunton and others [22, 27, 41, 49, 51, 89] at the University of Cambridge. This work has been previously discussed in this chapter. It does not, of course, model impact with a spherical drop, but rather introduces a droplet disc geometry, i.e., impact between a flat plate and the side of a liquid cylinder. However, it does allow precise control of droplet shape and size, as well as the inclusion of a very small pressure transducer in the "bullet." Again, recovery without damage of the bullet is required. This device [22, 41], as well as that of Fyall [26], does provide precise photographic and other information for direct comparison with such computer models as those already discussed in Sec. 6-3.

6-5-3 Stationary-Target Devices

As already mentioned, for reasons of droplet stability (i.e., a critical Weber number exists), it does not appear possible to propel a spherical droplet of pertinent size and velocity against a stationary target. For this reason, devices of this type are less adapted to liquid-impact erosion tests, since a "realistic" droplet shape cannot easily be achieved. However, in general, facilities of this type tend themselves to be much less complex and expensive. Also, devices for purposeful jet cutting are almost necessarily of the stationary-target type, so that standardized test devices for such machines must also propel the jet or other liquid slugs against essentially stationary targets.

The difficulty of uncontrolled and perhaps undesirable droplet shape was overcome to some extent by the momentum-exchange liquid-gun device, pioneered and developed by Bowden and Brunton [49, 51, 89] and discussed previously in Sec. 6-3. This device can produce a high-velocity jet with an approximately spherical leading edge. Hence it can model quite well impact with spherical droplets, since it is only the initial portion of the impact which is of primary importance from the viewpoint of erosion. It can also produce a microjet (Monroe jet) of even higher velocity than the main jet (depending upon the initial position of the liquid meniscus). Maximum velocities attained with such devices are reported to be up to ~ 1000 m/s [51]. As originally developed, this is a single-shot device, not adapted to multiple-impact erosion studies but rather to studies of the basic impact phenomenon.

The provision of repeated impacts upon a target with a device providing quite similar droplet shape, but somewhat reduced velocity capability (~ 550 m/s), was achieved by the "water-gun" device developed by Kenyon [55] and later used also in the author's laboratory at the University of Michigan for erosion testing

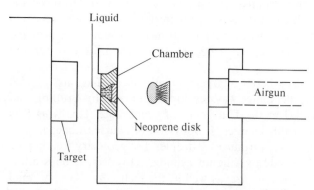

Figure 6-38 Diagram of method for producing a high-velocity liquid jet [49, 51, 89].

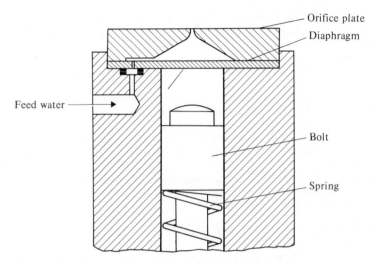

Figure 6-39 Schematic of Kenyon-Michigan jet water gun [46, 55, 56].

of various materials as well as high-speed photographic studies of droplet impact [46, 56]. Typical results are shown in Figs. 6-28, 6-31, and 6-32. The Kenyon and Michigan devices separate the water cavity from which the jet is to be ejected from the driving mechanism by a thin steel diaphragm, rather than a rubber washer as in the Bowden-Brunton device. The necessary shock for liquid propulsion is provided in the Kenyon-Michigan device by a steel bolt driven by a heavy spring, rather than a pellet from an air or gas gun or conventional rifle, as in the Bowden-Brunton device. Figures 6-38 and 6-39 show schematic representations of these two systems, respectively. The Kenyon-Michigan device allows automatic operation to provide essentially a water "machine gun," capable of producing about 30 shots per minute. These devices are thus obviously easily adapted to material resistance screening tests, but unfortunately do not provide a very simple and realistic droplet shape from the viewpoint of the aircraft or turbine erosion problem, as do the rotating-arm, disc, or rocket-sled devices, previously discussed. However, both Bowden-Brunton and Kenyon-Michigan devices do provide economic and simple bench-type facilities, useful for some types of impact-erosion testing and study. Their cost is generally an order of magnitude less than most of the rotating devices. Many of these devices are discussed in a relatively new *Metals Handbook* chapter [90], and a new ASME Wear Control Handbook [91].

REFERENCES

1. Huang, Y. C.: "Numerical Studies of Unsteady, Two-Dimensional Liquid Impact Phenomena," Ph.D. thesis, Mech. Engr. Dept., University of Michigan, Ann Arbor, Mich., 1971; also available as University of Michigan ORA Rept. UMICH 03371-8-T, June, 1971.

2. Huang, Y. C., F. G. Hammitt, and W. J. Yang: Hydrodynamic Phenomena During High-Speed Collision Between Liquid Droplet and Rigid Plane, *Trans. ASME, J. Fluids Engr.*, vol. 95, ser. D, no. 2, pp. 276–294, 1973.

3. Hammitt, F. G., and Y. C. Huang: "Liquid Droplet Impingement Studies at University of Michigan," Conf. Publication no. 3, pp. 237–243, Inst. of Mech. Engrs., University of Warwick, Coventry, April 3–5, 1973.

4. Hwang, J. B.: "The Impact Between a Liquid Drop and an Elastic Half-Space," Ph.D. thesis, Mech. Engr. Dept., University of Michigan, 1975; also available as University of Michigan ORA Rept. UMICH 012449-5-T, March, 1975, Ann Arbor, Michigan.

5. Hwang, J. B., and F. G. Hammitt: Transient Distribution of the Stress During the Impact Between a Liquid Drop and an Aluminum Body, *Proc. Third Int'l Symp. on Jet Cutting Technology*, May 11–13, 1976, Chicago.

6. Hwang, J. B., and F. G. Hammitt: "High Speed Impact Between Curved Liquid Surface and Rigid Flat Surfaces," University of Michigan ORA Rept. UMICH 012449-10-T, October, 1975, Ann Arbor, Michigan; also in *Trans. ASME, J. Fluids Engr.*, vol. 99, ser. I, no. 2, pp. 396–404, June, 1977.

7. Hwang, J. B., F. G. Hammitt, and W. Kim: On Liquid-Solid Impact Phenomena, *1976 ASME Cavitation Forum*, pp. 24–27.

8. Heymann, F. J.: On the Shock Wave Velocity and Impact Pressure in High-Speed Liquid-Solid Impact, *Trans. ASME, J. Basic Engr.*, vol. 90, ser. D, pp. 400–402, 1968.

9. Rochester, M. C., and J. H. Brunton: Surface Pressure Distribution During Drop Impingement, *Proc. Fourth Int'l Conf. on Rain Erosion and Assoc. Phenomena*, May, 1974, pp. 371–393, Meersburg, Germany.

10. Johnson, W., and G. W. Vickers: Transient Stress Distribution Caused by Water-Jet Impact, *J. Mech. Engr. Sci.*, vol. 15, no. 4, pp. 302–310, 1973.

11. DeHaller, P.: Untersuchungen uber die Durch Kavitation Hervorgerufenen Korrosionen, *Schwiez Bauzig*, vol. 101, p. 243, 1933.

12. Harlow, F. H.: The Particle-in-Cell (PIC) Computing Method for Fluid Dynamics, *Methods in Computational Physics*, vol. 3, pp. 319–343, 1964.

13. St-Venant, A. J. C. de Barré: Theorie de l'Elasticité des Corps Solides, *J. Mathématiques (Louiville)*, p. 387, 1867.

14. Cook, S. C.: Erosion by Water Hammer, *Proc. Roy. Soc. (London)*, ser. A, vol. 119, pp. 481–488, 1928.

15. Tait, P. G.: Report on Some of the Physical Properties of Fresh Water and Sea Water, Rept. on Scientific Results of Voyage of HMS *Challenger, Phys. Chem.*, vol. 2, pp. 1–71, 1888.

16. Huang, Y. C., and F. G. Hammitt: The Compressible Cell and Marker Numerical Method (ComCAM) for Compressible Fluids Problems, *Proc. Symp. on Applications of Computers to Fluid Dynamic Analysis and Design, Computers and Fluids*, January, 1973, New York; also available as University of Michigan ORA Rept. UMICH 03371-20-T, 1973, Ann Arbor, Michigan.

17. Lamb, Sir Horace: "Hydrodynamics," Dover, 1965.

18. Lax, P., and B. Wendroff: System of Conservation Laws, *Comm. on Pure and Appl. Math.*, vol. XIII, pp. 217–237, 1960.

19. Edwin, W., W. Jardetzky, and F. Press: "Elastic Waves in Layered Media," McGraw-Hill, New York, 1957.

20. Wasley, R. J.: "Stress Wave Propagation in Solids: An Introduction," Marcel Dekker, New York, 1973.

21. Timoshenko, S. P., and J. N. Goodier: "Theory of Elasticity," third ed., McGraw-Hill, New York, 1970.

22. Rochester, M. C., and J. H. Brunton: Surface Pressure Distribution During Drop Impingement, *Proc. Fourth Int'l Conf. on Rain Erosion and Assoc. Phenomena*, May, 1974, pp. 371–393, Meersburg, Germany.

23. Rochester, M. C., and J. H. Brunton: The Influence of Physical Properties of the Liquid on

the Erosion of Solids, in "Erosion, Wear and Interface with Corrosion," *ASTM STP* 567, pp. 128–151, 1974.

24. Johnson, W., and G. W. Vickers: Transient Stress Distribution Caused by Water-Jet Impact, *J. Mech. Eng. Sci.*, vol. 15, no. 4, pp. 302–310, 1973.

25. Engel, O. G.: Damage Produced by High-Speed Liquid-Drop Impact, *J. Appl. Phys.*, vol. 44, no. 2, pp. 692–704, February, 1973.

26. Fyall, A. A.: Single Impact Studies with Liquid and Solids, *Proc. Second Conf. on Rain Erosion*, August, 1967, Meersburg, Germany (edited by A. Fyall of the Royal Aircraft Est., Farnborough, England).

27. Brunton, J. H., and J. J. Camus: The Flow of Liquid Drop During Impact, *Proc. Third Int'l Conf. on Rain Erosion and Assoc. Phenomena*, Aug. 11–13, 1970, pp. 327–352, Elvetham Hall, England.

28. Briggs, L. J.: Maximum Superheating of Water as a Measurement of Negative Pressure, *J. Appl. Phys.*, vol. 26, no. 8, pp. 1001–1003, August, 1955.

29. Apfel, R. E.: The Tensile Strength of Liquids, *Scientific American*, vol. 227, no. 6, pp. 58–71, December, 1972.

30. Glenn, L. A.: On the Dynamics of Hypervelocity Liquid Jet Impact on a Flat Rigid Surface, *J. Appl. Math. and Phys.*, vol. 25, pp. 383–398, 1974.

31. Timm, E. E.: "An Experimental Photographic Study of Vapor Bubble Collapse and Liquid Jet Impingement," Ph.D. thesis, Chem. Engr. Dept., University of Michigan, 1974; also available as University of Michigan ORA Rept. UMICH-01357-49-T, 1974.

32. Hammitt, F. G., E. E. Timm, J-B. Hwang, and Y. C. Huang: Liquid Impact Behavior of Various Non-Metallic Materials, in "Erosion, Wear and Interface with Corrosion," *ASTM STP* 567, pp. 197–218, 1974.

33. Johnson, G. R.: Liquid–Solid Impact Calculations with Triangular Element, *Trans. ASME, J. Fluids Engr.*, vol. 99, no. 3, pp. 758–760, September, 1977.

34. Tyler, L. D.: The Flow Dynamics of a Liquid Drop During High-Speed Impact, *Bull. of Am. Phys. Soc.*, vol. 16, no. 11, p. 1319, 1971.

35. Blower, R. M.: On the Response of an Elastic Solid to Droplet Impact, *J. Inst. Math. and Its Applications*, vol. 5, no. 2, pp. 167–193, 1969. See also, Adler, W. F.: Liquid Drop Collisions on Deformable Media, *J. Material Sci.*, vol. 12, no. 6, pp. 1253–1271, June 1977; see also "Analytical Modeling of Liquid and Solid Particle Erosion," [36].

36. Air Force Materials Lab. Rept. AFML-TR-73-174, Wright-Patterson Air Force Base, Ohio, September, 1973.

37. Johnson, G. R.: Analysis of Elastic-Plastic Impact Involving Severe Distortions, *Trans. ASME, J. Appl. Mech.*, vol. 43, ser. E, no. 3, pp. 439–444, September, 1976.

38. Johnson, G. R.: A New Computational Technique for Impulsive Loads, *Proc. Third Int'l Symp. on Ballistics*, March, 1977, Karlsruhe, Germany.

39. Jenkins, D. C., and J. D. Booker: The Impingement of Water Drops on a Surface Moving at High Speed, in E. G. Richardson (ed.), "Aerodynamic Capture of Particles," p. 97, Pergamon Press, New York, 1960.

40. Jenkins, D. C.: An Experimental Method for Studying the High-Speed Impact of a Liquid Drop on a Liquid Surface, *Ingenieur Archiv*, vol. XXXVI, pp. 280–284, 1967.

41. Rochester, M. C., and J. H. Brunton: The Influence of Physical Properties of the Liquid on the Erosion of Solids, in "Erosion, Wear and Interface with Corrosion," *ASTM STP* 567, pp. 128–151, 1974.

42. Schmitt Jr., G. F., and A. H. Krabill: "Velocity-Erosion Rate Relationships of Materials in Rain at Subsonic Speeds," Air Force Material Lab. Rept. AFML-TR-70-44, Wright-Patterson Air Force Base, Ohio, October, 1970.

43. Wahl, N. E.: Design and Operation of Mach 3 Rotating Arm Erosion Testing Apparatus, *Proc. Third Int'l Conf. on Rain Erosion and Assoc. Phenomena*, Aug. 11–13, 1970, pp. 13–42, Royal Aircraft Est., Farnborough, England.

44. Hammitt, F. G., J. B. Hwang, et al.: Experimental and Theoretical Research on Liquid Droplet Impact, *Proc. Fourth Int'l Conf. on Rain Erosion and Associated Phenomena*, May 8–10, 1974, pp. 319–345, Meersburg, Germany.

45. Hancox, N. L., and J. H. Brunton: The Erosion of Solids by the Repeated Impact of Liquid Drops, *Phil. Trans. Roy. Soc. (London)*, ser. A, vol. 260, no. 1110, pp. 121–143, 1966.

46. Hammitt, F. G., J. B. Hwang, et al.: Cavitation and Droplet Impingement Damage of Aircraft Rain Erosion Materials, *Proc. Third Int'l Conf. on Rain Erosion and Associated Phenomena*, Aug. 11–13, 1970, pp. 907–935, Elvetham Hall, England.

47. Engel, O. G.: Water Drop Collision with Solid Surface, *J. Res. Nat. Bur. Stand.*, vol. 54, no. 5, p. 281, May, 1955.

48. Engel, O. G.: Pits in Metals Caused by Collision with Liquid Drops and Soft Metal Spheres, *J. Res. Nat. Bur. Stand.*, vol. 62, no. 6, pp. 229–246, 1959.

49. Brunton, J. H., and J. J. Camus: The Flow of Liquid Drop During Impact, *Proc. Int'l Conf. on Rain Erosion and Associated Phenomena*, Aug. 11–13, 1970, p. 327, Elvetham Hall, England.

50. Heymann, F. J.: On the High-Speed Impact Between a Liquid Drop and a Solid Surface, *J. Applied Physics*, vol. 40, no. 13, pp. 5113–5122, 1969.

51. Bowden, F. P., and J. H. Brunton: Deformation of Solids by Liquid Impact at Supersonic Speeds, *Proc. Roy. Soc. (London)*, ser. A, vol. 263, pp. 433–450, 1961; also see *Proc. Roy. Soc. (London)*, ser. A, vol. 260, no. 1110, pp. 79–85, 1966.

52. Bowden, F. P., and J. E. Field: The Brittle Fracture of Solids by Liquid Impacts, by Solid Impacts, and by Shock, *Proc. Roy. Soc. (London)*, ser. A, vol. 282, p. 331, 1964.

53. Schmitt, Jr., G. F.: Erosion Rate-Velocity Dependence for Materials at Supersonic Speeds, *ASTM STP* 474, pp. 323–352, 1969.

54. Schmitt, Jr., G. F., W. G. Reinecke, and G. D. Waldman: Influence of Velocity, Impingement Angle, Heating; and Aerodynamic Shock Layers on Erosion of Materials at Velocities of 5500 ft/s (1700 m/s), *ASTM STP* 567, pp. 219–238, 1973.

55. Kenyon, H. F., and P. H. Dawson: "Erosion by Water Jet Impacts," Associated Electrical Industries, Rept. TP/R 5626, Manchester, England, 1969.

56. Hammitt, F. G., J. B. Hwang, and Linh N. Do: Interrupted Jet Water Gun Impact Erosion Studies on Metallic Alloys, *1974 ASME Cavitation and Polyphase Flow Forum*, pp. 24–26.

57. Kolsky, H.: "Stress Waves in Solids," Dover, 1963.

58. Blower, R. M.: On the Response of an Elastic Solid to Droplet Impact, *J. Inst. Math. and Its Applications*, vol. 5, no. 2, pp. 167–193, 1969.

59. Peterson, F. B.: Some Consideration of Material Response Due to Liquid-Solid Impact, *Trans. ASME, J. Fluids Engr.*, vol. 95, ser. A, no. 2, pp. 263–270, 1973.

60. Love, A. E. H.: The Stress in a Semi-Infinite Solid by Pressure on Part of the Boundary, *Trans. Roy. Soc. (London)*, ser. A, vol. 228, pp. 377–420, 1929.

61. Goodier, W. E., W. E. Jahsman, and E. A. Ripperger: An Experimental Surface Wave Method for Recording Force Time Curves in Elastic Impacts, *Trans. ASME, J. Appl. Mech.*, vol. 26, no. 1, pp. 3–7, 1959.

62. Field, J. E., J. J. Camus, et al.: Impact Damage Produced by Large Water Drop, *Proc. Fourth Int'l Conf. on Rain Erosion and Allied Phenomena*, May 8–10, 1974, Meersburg, Germany (edited by A. Fyall of the Royal Aircraft Est., Farnborough, England).

63. Kozirev, S. P., and K. K. Shal'nev: Explanation of Solid Damage Due to Free Drop Impact, Jet Impact, and Cavitation by Relaxational Hypothesis, *Proc. Fourth Conf. on Fluid Mechanics*, 1972, Budapest, Hungary.

64. Kozirev, S. P., and K. K. Shal'nev: Interpretation of Experimental Data in Witness of Relaxational Hypothesis of Cavitation Erosion, *Proc. Fifth Conf. on Fluid Mechanics*, 1975, Budapest, Hungary.

65. De Corso, S., and R. Kothmann: Characteristics of the Light Flash Produced upon Impact of a Liquid with a Surface, *ASTM Materials Research and Standards*, vol. 5, no. 10, pp. 525–528, 1965.

66. Gaydon, A. G.: Light Emission from Shock Waves and Temperature Measurements, *Proc. Ninth Int'l Shock Tube Symp.*, 1973, pp. 11–22, Stanford University, California.

67. Hoff, G. R.: Personal communication to F. G. Hammitt, Dornier-System GmbH, Friedrichshafen, Germany, 1971.

68. Hammitt, F. G., et al.: Closure to a Statistically Verified Model for Correlating Volume Loss Due to Cavitation or Liquid Impingement, *ASTM STP* 474, pp. 319–322, 1970.
69. Heymann, F. J.: Toward Quantitative Prediction of Liquid Impact Erosion, *ASTM STP* 474, pp. 212–244, 1970.
70. Mitchell, T. M., and F. G. Hammitt: "Preliminary Analyses Applied to a Portion of Holloman AFB Rocket Sled Data on Rain Erosion Materials," University of Michigan ORA Rept. UMICH 01077-4-T, October, 1968, Ann Arbor, Michigan.
71. Tatnall, G., K. Foulke, and G. Schmitt Jr.: "Joint Air Force–Navy Supersonic Rain Erosion Evaluation of Dielectric and Other Materials," Rept. NADC-AE-6708, 1967.
72. Baker, W. C., K. H. Jolliffe, and D. Pierson: The Resistance of Materials to Impact Erosion Damage, A Discussion on Deformation of Solids by the Impact of Liquids, *Phil. Trans. Roy. Soc.*, ser. A, vol. 260, no. 1110, pp. 193–203, July, 1966.
73. Hoff, G., G. Langbein, and H. Reiger: "Investigation of the Angle-Time Dependence of Rain Erosion," Dornier-System GmbH, Progress Rept. 62269-7-002050, Friedrichshafen, Germany, March, 1968.
74. Knapp, R. T.: Recent Investigations of Cavitation and Cavitation Damage, *Trans. ASME*, vol. 77, pp. 1045–1054, 1955.
75. Knapp, R. T., J. W. Daily, and F. G. Hammitt: "Cavitation," McGraw-Hill, New York, 1970.
76. Hammitt, F. G., et al.: Initial Phase of Damage of Test Specimens in a Cavitating Venturi as Affected by Fluid and Material Properties and Degree of Cavitation, *Trans. ASME, J. Basic Engr.*, vol. 87, ser. D, pp. 453–464, 1965.
77. Thiruvengadam, A.: A Unified Theory of Cavitation Damage, *Trans. ASME, J. Basic Engr.*, vol. 85, ser. D, no. 3, pp. 365–376, 1963.
78. ASTM: "Standard Definitions of Terms Relating to Erosion by Cavitation and Impingement," Designation G 40-73, 1973, 1979.
79. Engel, O. G.: "Basic Research on Liquid-Drop-Impact Erosion," NASA Rept. GESP-253, July, 1969.
80. Engel, O. G., and A. J. Pietutowski, "Investigation of Composite-Coating Systems for Rain-Erosion Protection," Naval Air Systems Command Contract N00019-71-C-0108, University of Dayton, Dayton, Ohio, 1972.
81. Springer, G. S., and C. B. Baxi: A Model for Rain Erosion of Homogeneous Materials, *ASTM STP* 567, pp. 106–127, June, 1973.
82. Springer, G. S., C.-I. Yang, and P. S. Larsen: Analysis of Rain Erosion of Coated Materials, *Proc. Fourth Int'l Conf. on Rain Erosion and Associated Phenomena*, May 8–10, 1974, Meersburg, Germany (edited by A. A. Fyall and R. B. King of the Royal Aircraft Est., Farnborough, England).
83. Springer, G. S., and C-I. Yang: "Analysis of Rain Erosion of Coated and Uncoated Fiber Reinforced Composite Materials," Rept. AFML-TR-74-180, Contract F33615-72-C-1563, August, 1974.
84. Springer, G. S.: "Erosion by Liquid Impact," Scripta Publishing Company, distributed by Halsted-Wiley, New York, 1976.
85. Heymann, F. J.: Personal communication concerning the Liquid Impact Round Robin, for the ASTM Committee G-2, 1975–1977.
86. Nowotny, H.: "Destruction of Materials by Cavitation" (in German), VDI-Verlag, Berlin, 1942; trans. available F. G. Hammitt: University of Michigan ORA Rept. UMICH 03424-I-15, June, 1962, Ann Arbor, Michigan.
87. Christie, D. G., and G. W. Hayward: Observations of Events Leading to the Formation of Water Drops Which Cause Turbine Erosion, *Phil. Trans. Roy. Soc. (London)*, ser. A, vol. 260, no. 1100, pp. 183–192, 1966.
88. Letson, K. N., and P. A. Ormsby: "Rain Erosion Testing of Slip Cast Fused Silica at Mach 5," ASME paper 76-ENAS-6, 1976.
89. Bowden, F. P.: The Formation of Microjets in Liquids under the Influence of Impact or Shock, *Proc. Roy. Soc. (London)*, ser. A, vol. 260, no. 1110, pp. 94–95, 1966.
90. Hammitt, F. G., and F. J. Heymann, Liquid-Erosion Failures, in "Metals Handbook," vol. 10, ed. 8, pp. 160–167, American Society of Metals, Metals Park, Ohio, 1975.
91. Hammitt, F. G., and G. F. Schmitt, Jr., Chapters 10 and 11 concerning cavitation and particle impact erosion in ASME Wear Control Handbook (in press, 1979).

SEVEN

FLOW OF WET-STEAM AND RELATED PHENOMENA

7-1 INTRODUCTION

Wet-steam flow as it occurs in large steam turbines is a most important flow application today. It has assumed added prominence in recent years because of the relatively low temperature steam supply conditions, usually lacking in superheat and reheat, for water-cooled nuclear reactors, and also because of the ever-increasing rating of power plants and turbine rotors. The net result is an aggravated turbine moisture condition in the low-pressure stages, leading to both severe erosion problems, since the droplets impact the turbine blades with very high relative velocity (500 to 600 m/s), and to an overall loss of efficiency due to the behavior of the nonhomogeneous two-phase mixture. While these problems have been of growing importance in fossil-fueled plants over the years, they are generally less severe than for nuclear plants. This is particularly true as power outputs are increased, increasing turbine rotor diameter and blade length at the low-pressure end, and hence also the linear blade speeds due to fixed rotating speed requirements.

The wet-steam turbine problem outlined above has been included in the scope of this book because of its close relationship to the cavitation problem which is the main subject of the book. This relationship can be briefly summarized as follows. First, there is the erosion provoked by each phenomenon, which is very similar in both appearance and formative mechanisms. As discussed in Chaps. 5 and 6, cavitation erosion is thought to evolve significantly "microjet" liquid impact, and liquid-impact erosion in many cases appears to include local cavitation provoked by the jet or droplet impact. Second, there is the inverse similarity of the flow regimes, i.e., cavitation is the flow of a liquid continuum with an entrained discontinuous vapor or gas phase, and wet-steam flow involves a continuous vapor phase with an entrained discontinuous liquid phase.

While the wet-steam (or other vapor) turbine application is certainly the oldest and also one of the most important, there are other important applications for this type of flow regime. The turbine wet-steam flow phenomenon involves the following important aspects. First, there is the formation of a thin liquid film upon the stationary blades involving deposit there of microdroplets (~ 0.01 to 0.1 μm) formed by nucleation in the main steam flow upon the blading surfaces after it crosses the "Wilson line" [1, 2], i.e., somewhat after attaining the equilibrium saturated vapor line. While such a film may be partially removed by centrifugal effects from a rotating blade, it accumulates upon stationary blades affected only by the interface shear between the relatively stationary liquid and the high-velocity steam. Second, then, is the disruption of this thin film and its shedding in relatively large droplets (~ 1 mm), due to the shear from the high-velocity steam, into the wake downstream of the stationary blades. Third, is the impact between these relatively large, low-velocity liquid droplets, before they can be substantially accelerated by "drag" with the high-velocity steam, and the next downstream rotating-blade row. The relative collision velocity is thus essentially that of the rotating blade, and important erosion, often limiting to blade life, results. Thus the wet-steam turbine problem involves both the study of thin liquid films under a high-velocity steam (or gas) flow and also the droplet-erosion problem [3–8]. This latter is largely influenced by the kinematics of the turbine design, i.e., the pertinent "vector diagrams." Aside from the turbine problem, there are obviously other important technological problems associated with both thin liquid-film flows and droplet impact per se.

Thin liquid films under a high-velocity vapor flow are important in numerous cases of "film dry-out" associated with high void fraction boiling. These involve both steady-state and stability problems, as does the steam-turbine case. However, they differ from the turbine application in that significant heat transfer is involved, i.e., a diabatic rather than adiabatic flow condition is involved. A situation involving this flow phenomenon is encountered in the "loss of coolant accident" and consequent activation of the emergency cooling systems in liquid-cooled nuclear reactors.

A very different type of application involving diabatic flow of a thin liquid film under a high-velocity gas flow is that encountered with transpiration cooling of reentrant missile nose cones [9–11]. Here a liquid film is injected near the nose of the cone along the surface. Its subsequent evaporation provides the necessary coolant effect. Again the stability of the film and the avoidance of "dry-spots" is of primary importance.

The reentrant nose cone and other components, such as radomes, of high-speed missiles or aircraft also involve an important droplet-impact erosion problem, i.e., "rain erosion." Another important incidence of such rain erosion is observed with helicopter blades, and also gas-turbine compressor blades used to drive jet aircraft and helicopters. In the latter case, dust erosion is similarly involved. These erosion problems differ importantly from the wet-steam turbine application in that the required lifetimes are very much less, although the allowable materials are much less erosion resistant and the impact velocities in some cases, e.g., reentry missile, are much higher.

7-2 THIN-FILM FLOWS—GENERAL BACKGROUND

As briefly discussed in the foregoing, liquid-film flows are encountered in a variety of applications. These can best be divided for consideration according to several distinguishing characteristics such as:

1. Adiabatic or diabatic, i.e., is heat transfer an important feature?
2. Steady or nonsteady behavior of major interest? As in all engineering, understanding of the steady-state nature of a phenomenon is always of major interest. However, in many thin-film flow applications, the stability of the film is of major importance. The possible formation of dry-spots, etc., e.g., in boiling, or the type of wave formation at the film surface which may lead to film rupture and may affect the liquid particle distribution in the wake downstream, may be features of major importance.
3. The downstream particle spectrum distribution and the critical Weber number describing maximum entrained liquid-droplet size may be controlling parameters in the various droplet-impact erosion applications. Studies of this type are of primary importance in the wet-vapor turbine problems.
4. Is liquid film driven primarily by interface shear or by gravity, or is it necessary to consider both mechanisms? In many, but not all applications, gravity may be of negligible importance, in which case the orientation of the film, i.e., horizontal or vertical, need not be considered.

7-3 SPECIFIC RESEARCH RESULTS AND PRESENT STATE OF THE ART

7-3-1 General

It is obviously impossible to cover in detail in this chapter the entire field of "wet-steam flow and related phenomena," which is its nominal subject. The previous sections of this chapter have attempted to indicate those portions of that overall subject which are within the scope of this book. It is then concerned primarily with those portions of the overall subject of multiphase flow involving liquid erosion-producing phenomena such as cavitation and droplet impact. For example, we will not include such interesting subjects as the compressible one-dimensionable flow of wet vapors, leading to condensation shocks, conventional compressible-flow shocks, and so on. We will concern ourselves primarily only with those aspects necessary for the full study of the turbine-blade erosion problem, and this will then include the behavior of shear and/or gravity-driven thin liquid films, both adiabatic and diabatic. The associated film-stability problems, and the generation and behavior of secondary particles in the blade wake, will thus be included.

7-3-2 Primary Liquid Particles in Wet Steam

The nucleation of "primary" liquid particles in expanding wet-steam flow, such as through the low-pressure stages of a steam turbine, occurs just downstream of the "Wilson line." For conventional fossil-fuel plants using both superheat and reheat, this phenomenon is in general limited to the last few stages of the turbine. However, for present-day water-cooled nuclear plants, lacking in both superheat and reheat, there is the possibility of primary droplet nucleation throughout the turbine, even including the high-pressure end. However, research concerning high-pressure wet-steam flow is only very recent and as yet quite limited [12–15]. Most previous work is thus concerned with low-pressure wet-steam flow [16–19].

The in-depth study of primary nucleation in wet-steam flow is beyond the scope of this book and this chapter. However, the mechanisms of the liquid-film formation on the turbine blading will be considered.

The measurement of primary particle size is very difficult because of their extremely small size as well as their high velocity, since they are entrained in the steam. From what information exists [12–19], their mean diameter appears to lie within the range ~ 0.1 to 1.0 μm. This is too small to allow the use of visible light scattering for measuring directly their population and size distribution, because of the unfavorable relationship between particle diameter and the wavelength of visible light. As discussed in Chap. 3, the light-scattering technique is highly useful for measuring the size and population distribution of entrained gas nuclei concerned with cavitation nucleation, and also for measurement of "secondary" particle distributions in wet-steam flow, i.e., particles shed in the wake behind stationary blades. Thus no established and feasible procedure appears to exist at this time for the measurement of primary liquid particle distributions in wet-steam flows.

Nevertheless, from whatever experimental and theoretical information exists [12–19], the mean diameter of the entrained liquid particles in the wet steam appears to be extremely small—well less than 1 μm. A question of major significance, then, yet to be resolved, concerns the mechanisms by which the very small primary particles attain the blading surface to generate the thin liquid film, giving eventual rise to the larger "secondary" particles which can cause blade erosion. A complete understanding of these mechanisms could allow eventual design improvements, which would then reduce or alleviate the erosion problem. Several possible mechanisms have been investigated [16–21]:

1. Diffusion, including turbulent effects and possible brownian motion.
2. Centrifugal effects, since streamline curvature of the main steam flow is involved.

Present information appears to indicate that ordinary diffusion through the turbulent and laminar portions of the boundary layer are of major importance [18, 20, 21] although brownian motion itself probably does not provide a major contribution. It is also likely that in some cases at least centrifugal effects can provide an important contribution. The study of these detailed transport mechanisms may also provide information on the portion of total moisture carried by the main steam flow which can be expected to deposit eventually on the blades. In general, it appears that the portion so transported is small, perhaps in the order of 5 percent. Nevertheless, this is sufficient to cause the important turbine-blading erosion problem.

7-3-3 Liquid-Film Behavior

The study of the behavior of the thin liquid film on the surface of a stationary blade in the wet-steam turbine case includes both steady-state and stability problems involving the nonsteady behavior of the film. These will be considered separately in the following.

Steady-state

General and theoretical Much work has been done in recent years to measure and compute the steady-state liquid-film behavior pertinent to the steam-turbine problem, as well as to other applications involving thin liquid films. For most cases of importance, the films are driven by shear generated by the high-velocity steam flow. The relative velocity between steam and liquid is essentially that of the steam, since the liquid velocity is relatively very small. Since the liquid film is also thin, of the order of 10 to 300 μm, its thickness Reynolds number is ~ 10 to 10^2. Thus the film flow is definitely well within the laminar range, at least in the steam-turbine problem, while the steam flow is generally well within the turbulent range. The liquid flow is then essentially a "Couette flow," i.e., constant shear, linear velocity profile. Thus, if the interface shear could be accurately estimated, the liquid-film velocity profile could be precisely computed. The estimation of this shear is thus one of the principal problems to be resolved [22–31]. From the simplest viewpoint, a "friction factor" f is required. However, inadequate information exists concerning the effective friction factor between a gaseous and liquid phase with high relative velocity, even if surface waves were not generated, which is generally not a realistic assumption [22, 23, 26–31]. Thus the effective "relative roughness" is not known.

Film-thickness measurement techniques Microprobes for film-thickness measurement, mounted flush with the wall surface to be studied, can conceivably use various sensing techniques at least in principle, e.g., electrical conductance, capacitance, or inductance of the liquid film could be used for thickness-sensing purposes. Most, but not all, investigators [16, 28–40], have in fact used the electrical conductivity gage. Some, however, have used gages sensitive to capacitance [37,

38]. In either case, it appears that the gage output is linear with film thickness for relatively small thicknesses. However, the gage tends to "saturate" if the thickness is too great for the particular gage design considered. For either type of gage the response is virtually instantaneous, so that nonsteady effects such as surface waves can be measured.

Another possible type of measurement would be that provided by transverse photography, where the film thickness could be observed. However, it seems doubtful that the precision of such a measurement would be good compared to that obtained by the electrical surface gages already discussed. Of course, surface photography for the purpose of studying the wave patterns is a useful technique which will be discussed later.

Still another technique which might be pertinent for the measurement of film thickness, particularly in cases where relatively thick films can be used, is the use of high-intensity x-rays normal to the surface [39]. Similar use of "soft" gamma rays is another possibility, but "hard" gammas are probably not practical because of the reduced absorption coefficient for the higher energy radiation, and also the increased personnel hazards which might be involved.

Returning then to the conductance and/or capacitance gages, these are probably the most practical possibility for most cases, particularly where films are quite thin. However, there are still some relative advantages and disadvantages to be considered. From experiences in the author's laboratory [34–36] it appears that a salt addition of the order of 0.5 percent may be necessary to provide film water of adequate conductivity to achieve good gage operation. Salt, or any additive, has the disadvantage of altering to some extent (probably slight) the physical properties of the fluid to be studied, and in the case of a brine solution of creating a corrosion problem, which with pure (condensed) water is probably negligible. Such an additive is not required for the capacitance gage.

Another reputed advantage [37, 38] for the capacitance gage is less temperature sensitivity. However, some care for temperature effects upon gage output is no doubt necessary [16, 28–40] for both types. For either type of gage, ac excitation is required to avoid spurious effects due to static charge. The most suitable frequency range for the exciting circuit, i.e., where the response is least temperature dependent, is of the order 1 to 10 kHz. In any case there may always be some output dependence upon exciting frequency.

Finally, relative simplicity of gage manufacture appears to favor the conductivity gage. However, since little systematic comparisons have yet been made between the various types of gages, it is not possible at this time to state which may be definitely preferable in all cases.

Experimental and theoretical results Experimental and theoretical results concerning the steady-state behavior of shear and gravity-driven thin liquid films have been obtained in recent years from various laboratories throughout the world. These include laboratories in England (primarily the group of Prof. J. D. Ryley at Liverpool University, but also other scattered results from elsewhere in the United Kingdom), Poland [Institute of Fluid Flow Machinery (IFFM), Polish Academy of Science, Gdansk], Czechoslovakia (Skoda Laboratories at

Pilsen), the United States [University of Michigan (U-M), University of Delaware (Prof. J. Moszynski), Virginia Polytechnic Institute and State University (VPI)], and others. Most active at the moment appear to be the groups of J. D. Ryley, IFFM, and U-M. All of these groups are also studying the transient wave behavior of the film as well as its steady-state behavior, as discussed later. With the exception of the work at VPI, which was related to transpiration cooling of reentry nose cones, as previously discussed, and involved water-air behavior rather than water-steam, all of these studies have involved primarily adiabatic conditions. However, mass transfer has in some cases also been included. It is impossible to consider here all the detailed results obtained. However, some idea

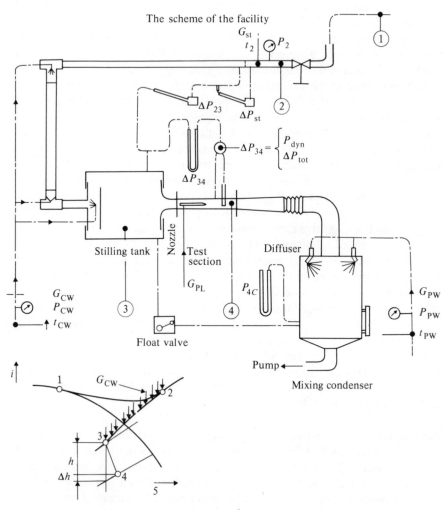

Figure 7-1 Schematic of steam tunnel facility [34–36].

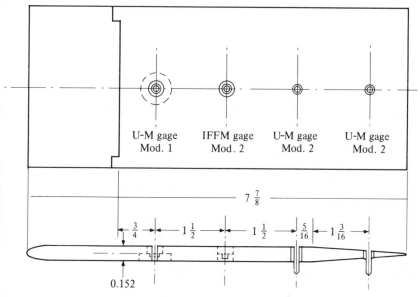

Figure 7-2 Schematic of blade with gases (dimensions in inches) [34–36].

of these can be given, and the work at Michigan [34–36] will be reviewed in more detail, since it is relatively typical.

The experiments at U-M have been conducted in a low-pressure steam tunnel, shown schematically in Fig. 7-1. Minimum test-section pressure is ~0.2 bar (abs) and maximum velocity ~Mach 1. The transparent test section is rectangular (~8 cm by ~8 cm). A flat tapered blade (Fig. 7-2) is inserted parallel to the flow direction, equally distant from the upper and lower walls. A transverse slot (Fig. 7-2) is used to inject a liquid film for some of the experiments. In others, a film condensed from the wet steam is used. Condensation is induced by an internal liquid coolant in those cases where the film is not injected. This type of facility is relatively typical of the other laboratories involved in this type of research.

The blades inserted in the test section have been fitted with electrical conductivity gages installed flush with the surface, as previously described, for the measurement of liquid-film thickness. Figure 7-3 shows schematically the gage design used. These are quite similar to those used previously at IFFM [33] and at the University of Liverpool [16].

Figure 7-4 shows typical results of measured film thickness versus steam velocity for various injected film flow rates. The trends of reducing film thickness for increased steam velocity, or for reduced injected film flow rate, are as would be expected on theoretical grounds [25–27, 34–36, 41]. The theoretical model used assumes the existence of Couette flow for the film, which is driven by interfacial shear between high-velocity steam and relatively low-velocity liquid. Presumably this type of calculation has been made at all the laboratories which

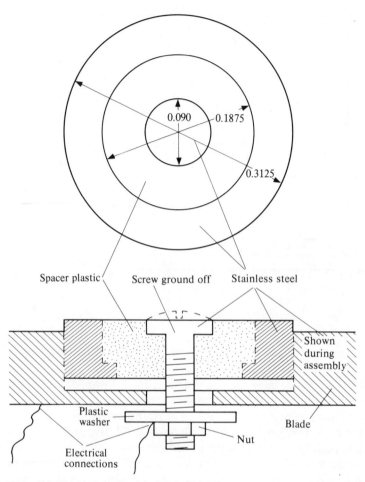

Figure 7-3 Electrical conductivity gages [34–36].

have been engaged in this type of study. The degree of agreement achieved between theory and experiment depends upon the estimate of interfacial shear. It could be argued, since there exists no really good method for estimating this shear, that the film-thickness measurement experiment here described could be taken as a valid method for measuring the interfacial shear, i.e., for computing the effective "friction factor" between steam and liquid.

Evaluation of the required relative roughness necessary to achieve agreement between measured and calculated film thicknesses indicates a required relative roughness in some cases in excess of the average film thickness. Figure 7-5 shows typical data comparing measured and theoretical film thicknesses for an assumed hydraulically smooth liquid-film surface for a fixed film-injection flow rate per unit blade width [41]. This disagrees somewhat with earlier, more approximate data [36] achieved at Michigan, although the trends are the same.

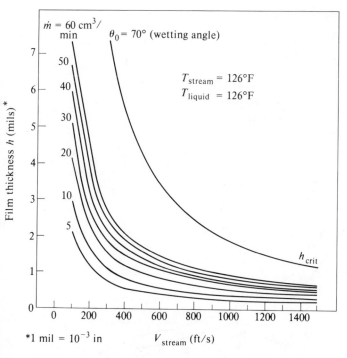

Figure 7-4 Summary of University of Michigan film thickness vs. steam velocity [41, 45].

Only relatively few research studies have as yet been performed to evaluate the effective friction factor between liquid films and a high-velocity gas or vapor stream. This factor obviously depends upon film wavelet characteristics, just as it depends upon relative roughness in conventional pipe flow. Presently available results [30, 31, 36, 37, 40, 41] indicate that the effective friction for such wavy film flow is somewhat greater than for comparable pipe flow. A factor of $3\sqrt{2}$ has been suggested [30, 31] for application to the wavelet height to achieve equivalence with relative roughness in pipe flow. As previously indicated, the latest U-M results [41, 45] indicate even greater effective roughnesses may apply.

In the earlier U-M experiments [34–36], the f-factor was estimated from the standard Moody friction chart, assuming an "hydraulically smooth" surface. This would presumably provide the smallest possible value for f, since surface wavelets, observed photographically and discussed later, provide a somewhat roughened surface. However, as shown in Fig. 7-5, the theoretical film thicknesses so computed are more than those measured by a factor of ~ 1.5. Hence, to achieve agreement, an f-value greater than that corresponding to an hydraulically smooth interface would need to be assumed. Since the Moody chart is based on flow along solid rather than liquid surfaces, admittedly such published f-values are uncertain. Newer studies elsewhere have been addressed to the effects of wavy surfaces on the f-factor [29–31]. These have also resulted in the development of a special

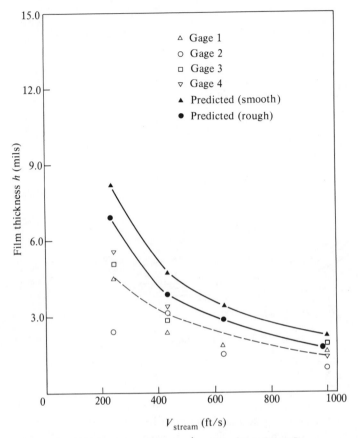

Figure 7-5 Film thickness vs. V steam, $\dot{Q} = 40$ cm^3/min [41, 45].

technique for directly measuring interfacial shear from the gas-phase velocit
profile [42–44]. A direct measure of shear from pressure drop is usually no
possible because it is too small for accurate measure. However, there may b
other reasons for the disagreement in the U-M tests, as discussed next.

Aside from the uncertainty in the interfacial shear, i.e., the friction facto
the most probable source of error in the U-M, tests [34–36, 41, 45] appears t
be the possibility of nonuniform transverse distribution of the film flow. Th
transverse distribution could not be measured in the relatively preliminary experi
ments here described [34–36, 41, 45], since the film-thickness gages were all locate
in the plane of the test-section axis. Nonuniform transverse distribution of th
film flow could result from significant nonuniformity in the injection slit widt
which is possible since the opening was only a few mils.† It could also be due t
wall effects associated with the noninfinite width of the test section, i.e., the flo
regime is not entirely two dimensional. Other possible causes of the disagreemer

† 1 mil = 10^{-3} in.

between theory and measurement could be the inevitable errors in various of the measured parameters such as film flow rate, steam velocity, temperature and pressure, and, probably most importantly, inaccuracy in the film-thickness measurement, due to possible errors in calibration.

Nonsteady behavior and film breakdown

General background Phenomena associated with surface wavelet formation compose the most important nonsteady feature of film flow regimes, since such wavelets, if of sufficient amplitude, will result in film breakdown and the formation of "dry spots." These can be disastrous in certain boiling applications. Wavelet patterns may also strongly affect the size and population spectra of wake droplets, thus importantly influencing the blading-erosion problem. The wavelet pattern can include both axial and transverse components, so that with sufficient wave amplitude the film can divide into axial rivulets of liquid with intervening dry areas. At other times, the dry spots can be simply closed dry patches, rather than elongated filaments. These points will be discussed in more detail in the next section.

The surface waves involved in thin-film flows bear some analogy to wind-driven shallow-water sea waves, but the analogy is not perfect, since interfacial tension for thin films (i.e., "capillary waves") is usually the predominant restoring mechanism rather than gravity, as with conventional ocean waves. Of course, capillary waves also exist in conjunction with gravity waves in ocean applications. Such capillary waves are very similar in appearance to gravity waves, but the scale is of course much smaller. Another possible mechanism in the present case is "shear waves," i.e. waves where the driving and restoring flow is interfacial shear.

Various attempts to utilize mathematical models to delineate the critical parameters controlling the formation of such wavelet patterns have been made, as described briefly in the next sections. However, it has not yet been possible to predict their behavior with adequate and usable engineering precision.

Measured wavelet patterns Observations of wavelet patterns have been made in various laboratories. However, no comprehensive understanding of the pattern forms has as yet been resolved. Recent photographs obtained in the author's laboratory [26, 41, 45] will be discussed as presenting typical results from the viewpoint of the steam-turbine blading application. These photos were taken for various conditions of steam velocity and film flow rate upon the blade shown in Fig. 7-2 in the steam tunnel of Fig. 7-1. The steam condition in all cases is ~ 0.2 bar (abs), saturated.

Figure 7-6 shows axial "dry patches" and filaments for a condition of relatively low steam velocity and film flow rate. These are the conditions conducive to the generation of such filaments and dry spots.

Figure 7-7 shows a type of wavelet formation which we have called "symmetric," i.e., symmetric about a line parallel to the direction of steam flow. In this case the steam velocity and film flow rate (in terms of the volume flow rate

1 cm   --→ Flow direction

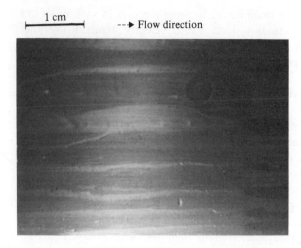

Figure 7-6 Dry patches $V_s = 79$ m/s, $\dot{q} = 5/8$ cm³/cm-min [41, 45].

per transverse distance) are less than for the case of filament-film breakdown in Fig. 7-6. The symmetrical wave pattern of Fig. 7-7 disappears over the tapered trailing edge (extreme right of photo).

Figure 7-8 shows the type of "nonsymmetric" wave generation which occurs for a much-increased film flow rate but relatively low steam velocity. Figure 7-9 shows how the nonsymmetrical wave pattern develops as steam velocity and film flow rate are further increased. This photo also shows particle shedding in the wake. Figure 7-10 shows film breakup and shedding of particles into the wake for the maximum steam velocity and film flow rate which was used. It is obvious from examination of Figs. 7-6 to 7-10 that no very simple framework exists to describe the relationship between wave pattern and the other experimental parameters. However, an attempt to systematize these results [26, 41] is shown in Figs. 7-11 to 7-13. Somewhat related results were reported previously by Tailby and Portalski [43].

1 cm   --→ Flow direction

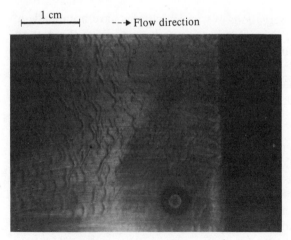

Figure 7-7 Symmetric waves $V_s = 130$ m/s, $\dot{q} = 15/16$ cm³/cm-min [41–45].

1 cm   – – ➔ Flow direction

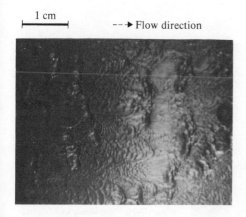

Figure 7-8 Nonsymmetric waves $V_s = 54$ m/s, $\dot{q} = 7.5$ cm³/cm-min [41–45].

Figure 7-11 shows dimensionless wavelength versus film-thickness Reynolds number. The data were drawn into two fairly close groupings corresponding to the "symmetric" and "nonsymmetric" wave patterns, previously described, by normalizing the wavelength by film thickness, measured by the electrical conductivity gages, previously described. In all cases, dimensional and dimensionless, wavelength decreases strongly with increased film Reynolds number, and also with increased film flow rate. For the present experiments, the observed wavelength lies in the range ~ 1 to 10 mm. The frequency of these waves can be estimated from the oscilloscope traces of the thickness-gage output. This was found to lie within the range ~ 10 to 1000 Hz. From such data the surface wave speed of propagation can be estimated from the relation,

$$\text{Speed} = \text{frequency} \times \text{wavelength}$$

From the data above, it then appears that the wave speed from these tests lies within the range 0.01 to 1 m/s.

Preliminary calculations [41] indicate that the observed celerity of the larger waves agree approximately with the classical shallow-water gravity-wave model,

1 cm   – – ➔ Flow direction

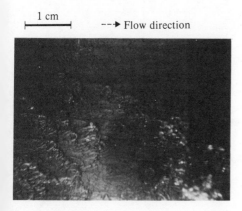

Figure 7-9 Nonsymmetric waves $V_s = 130$ m/s, $\dot{q} = 10$ cm³/cm-min [41–45].

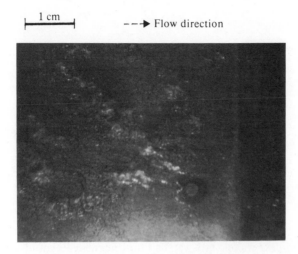

1 cm

---→ Flow direction

Figure 7-10 Film breakup and shedding $V_s = 302$ m/s, $\dot{q} = 12.5$ cm^3/cm-min [41, 45].

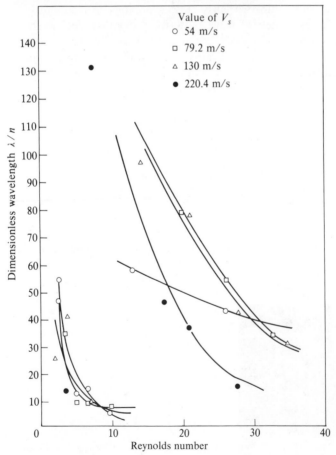

Value of V_s
○ 54 m/s
□ 79.2 m/s
△ 130 m/s
● 220.4 m/s

Figure 7-11 Dimensionless wavelength vs. Reynolds number [45].

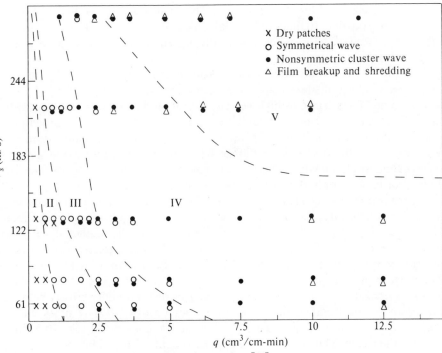

Figure 7-12 Transition map of liquid-film and steam flow [45].

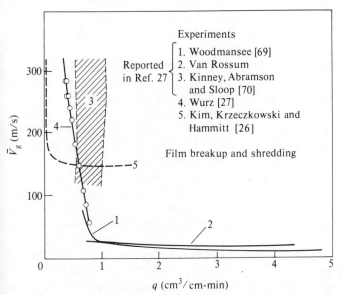

Figure 7-13 Transition line for film breakup (regions IV and V of Fig. 7-12) [45].

including the effects of interfacial tension. The small wavelets, however, show a celerity approximately 10 times that explainable by these models. It may be that a shear-wave mechanism is involved, but only a preliminary analysis has yet been completed [45].

Figure 7-12 is a transition "map" for this flow regime [26, 41, 45]. Several somewhat similar transition maps have been published previously [46, 47], but concern liquid films driven by high-velocity air, rather than steam as in the present case. The steam-water case is certainly more complex because of the added possibilities of heat and mass transfer, even with only very small temperature differentials, as in the present case. Other basic differences between the air and steam-tunnel studies involve differences in gas (or vapor) phase velocity, density, and viscosity. Generally the air velocities are much smaller (~ 100 versus ~ 400 m/s), because of the much greater ease of obtaining large velocities with a relatively modest steam facility and the range of interest of the pertinent applications. However, this is not always the case, as illustrated by the studies devoted to missile reentry [9–11, 48]. The air density is usually much greater, since the air pressure is approximately atmospheric, and that of the steam is ~ 0.1 to 0.2 bar. The most important parameter for comparison is probably the product of the absolute viscosity and the gas-phase velocity. This does not differ grossly between the air and steam facilities, since the absolute air viscosity is ~ 1.7 times that of the steam [49]. Thus the product μV_{gas} for the usual air facility is ~ 0.5 times that for the steam. Thus the tests are at least roughly comparable on this basis.

Another difference, however, is involved in terms of Reynolds number, which is greater in most of the air applications because of the greater density, i.e., $Re_{air} \sim 2.5 \, Re_{steam}$, for fixed test-section size, leading to a somewhat reduced gas-phase boundary-layer thickness for air and an increased turbulence. This increase of Reynolds number is, of course, especially important for the high Mach number tests [9–11, 48], and Mach number effects themselves are important in these tests.

Our transition "map" (Fig. 7-12) divides the regime of steam velocity versus "specific" liquid flow rate, i.e., volume flow per unit transverse length, into five regions. These range from dry patches and filaments for the region nearest to the origin (region I), through primarily "symmetric waves" (region II), a region of both "nonsymmetrical and symmetrical waves" (region III), a region of predominantly "nonsymmetrical waves" (region IV), and finally a region including film breakup and shredding (wave crests torn away into droplets) as well as nonsymmetrical waves (region V). Thus the transition between regions IV and V represents the critical condition for one form of film instability leading to macro-droplet entrainment as the wave crests are entrained into the high-velocity steam. Another form of film instability is that leading to the formation of axial filaments or dry spots of one form or another. The critical condition for this transformation is shown by the line separating regions I and II. This transition will be discussed further in the next section.

Figure 7-13 compares our transition line between regions IV and V with the results of various previous investigators, which are summarized in Ref. 26. Obviously the agreement between these various investigations is only quite rough at present, so that much additional research on this subject is obviously still required.

Film-stability analyses Various film-stability analyses exist in the literature [50–59]. Unfortunately none of these has as yet succeeded in closely matching experimental observations. These have been based on various and different mathematical models, and different models of film breakup have been investigated. References 53 and 54, for example, are concerned with laminar film breakup into axial filaments, such as shown in Fig. 7-6. For this type of film breakdown leading to dry spots, interfacial tension obviously plays a predominant role. The computing model in this case [53, 54] in fact is based upon a conservation of kinetic and interfacial energies before and after film breakdown. Since the interfacial tensions are most important in predicting film-breakdown parameters, it becomes important to determine accurately the wetting angles between condensate water and solid materials of interest, such as steam-turbine blading materials. Much detailed research [45, 60, 61] has in fact been done to attempt to resolve this problem, even in fluids such as sodium [61]. Incidentally, it is possible to achieve agreement between computing models [53, 54] and experimental results by proper choice of wetting angle. However, the required wetting angles for this purpose are generally not within the plausible range.

The mathematical model for film rupture and the formation of dry spots, discussed in the foregoing [53, 54], assumed a particular mode of film failure, i.e., into axial filaments. This form of failure was in fact observed in U-M recent tests [26, 41, 45] (Fig. 7-6), but it is certainly not the only possible mode of film failure leading to dry spots (or "dry-out"). This failure model was adiabatic, as it was applied to the turbine-blading case, but obviously other models involving local heat transfer would be suggested for dry-out problems in boiling [57–59]. However, a relatively comprehensive review concerning the heat-transfer applications is not within the scope of this book. The point of the discussion here is only to indicate the large variety of possible film-stability models. The models discussed previously [53, 54] assume the predominance of transverse waves whose troughs are aligned parallel to the direction of flow, so that breakdown into axial filaments results. However, the photos discussed previously (Figs. 7-6 to 7-10) indicate that, in most cases, wave patterns include other components also. Thus, if predominantly transverse waves lead to film rupture, an entirely different calculating model would be suggested. Also, the inclusion of gravity with shear effects may be necessary [46, 53]. The present discussion indicates the general complexity of the subject and that much analysis and experiment is yet to be done before the subject can be considered resolved.

Furthermore, other entirely different types of film breakdown may be involved in some cases, such as the tearing-off of wave crests (Fig. 7-10) and entrainment of liquid particles. Although this particular type of film breakdown does not lead to "dry-out" problems, it may be important in other applications: it is the type of film failure correlated by Fig. 7-13, involving region V of the "map" previously discussed (Fig. 7-12). This latter type of film breakdown does importantly affect the droplet-particle spectra in the wake, discussed in the next section, as well as mass or heat transfer between liquid film and high-velocity steam or gas.

High shear and Mach number effects   Wavelet patterns discussed in the foregoing generally result from steam flow conditions typical of the low-pressure stages of large steam turbines, that is, 0.1 to 0.3 bar (abs) and velocities up to ~ Mach 1. Of course, slightly supersonic velocities may exist in some cases, but generally the steam flow condition of interest is subsonic. Thus the wave patterns discussed are typical of interfacial shears and steam velocities of this typical regime. However, some applications of thin-film flows, such as that of transpiration cooling of re-entry nose cones, previously discussed, may involve considerably higher shears and highly supersonic velocities, giving quite different wavelet behavior [9–11, 48].

The tests for this application [9–11, 48] were conducted in the "hypersonic" air tunnel at New York University (NYU) on behalf of the research group at VPI. The shear was considerably greater than in the typical low-pressure steam tunnel, as previously discussed, since the velocity was very high.

A comparison of the wavelet pattern results from the NYU hypersonic wind-tunnel nose-cone tests [9–11, 48], wherein shear for the nose-cone application was modeled with those from the U-M steam-tunnel tests [26, 41], shows that a stable liquid film could not be maintained at all for subsonic velocities at atmospheric test-section pressure for the nose-cone tests at NYU [9–11, 48], i.e., the film was instantly and completely torn away by the high-velocity air. This complete film disruption did not occur at all for the U-M low-pressure steam tests (Figs. 7-6 to 7-10), although in the case of lowest shear and film injection rate, local axial filament-type dry spots were observed (Fig. 7-6).

It thus appears that there are too many differences, geometrical (nose cone versus flat plate) and other, between the NYU and U-M tests to resolve the discrepancies observed.

Another very interesting feature resulting from the NYU tests was the vast difference obtained between subsonic and supersonic airstreams. While, as mentioned above, no stable film could be maintained at all for the subsonic tests, a stable film was maintained for supersonic velocity. This very significant difference in performance, from the viewpoint of the transpiration cooling application, was attributed [9–11, 48] to the entirely different behavior of supersonic flow, as opposed to subsonic, over film roughnesses, i.e., wavelets. Apparently no comparable tests have yet been made for supersonic low-pressure steam flow.

Finally, it should be noted that some research is in progress [12–15] on high-pressure wet-steam flow concerned with the high-pressure end of large turbines for cases where superheat is absent, as in some types of nuclear plants. However, apparently no wavelet-pattern studies pertinent to the present discussion are as yet available.

Droplet spectra in wake
1. *General.* The droplet size, velocity, and population distribution in the wake region downstream of a stationary blade is of major importance to the erosion

problem for the next rotating row. If a wake droplet size, velocity, and population spectrum is assumed, then it is a relatively simple problem of turbine-blade kinematics, i.e., application of vector diagrams, plus an evaluation of droplet acceleration in the high-velocity steam, to determine the regions of possible erosion and the droplet-impact regime in terms of size, number, and relative velocity. A few such studies have already been made [3–8], as previously discussed. Given the results of the essentially kinematics problems described above, it is still not possible to predict with any certainty the rate of erosion in the area attacked. Reference can be made, however, to liquid-impact tests on the materials of interest, such as may be reported in the literature, and use made of the relatively approximate curve-fitting relations which have been published. This purely erosion problem has already been discussed in detail in Chap. 6 and hence will not be further considered here.

The solution outlined above for the blading-erosion problem requires, first, either assumed or measured droplet size, velocity, and population spectra for the wake region behind the static-blade row in question. It is the knowledge pertaining to these spectra, and the related research, which is the subject of this section.

The droplets in the wake region are predominantly "secondary," relatively large (~ 0.1 to 1 mm) droplets, formed from the thin films already discussed upon the upstream stator blades. The thin films, as already mentioned, are formed from "primary" droplets originating in the main steam flow, several orders of magnitude smaller than the secondary droplets, and hence relatively harmless from the viewpoint of erosion. The "secondary" droplets, on the other hand, are primarily responsible for erosion of the next rotating row [3–8]. These are formed from the shear-driven thin liquid film by various mechanisms. One such method, which has already been mentioned, is the entrainment of wavelet crests for cases involving relatively large wavelets. This is shown in Fig. 7-10. Another mechanism which may be important in some cases involves film breakdown into axial rivulets (Fig. 7-6), which then shed into the downstream wake as sufficient liquid accumulates at the trailing edge of the blade. Droplet size no doubt depends upon rivulet depth and width, as well as stream velocity and perhaps other parameters to a smaller extent. Another mechanism is shown in Fig. 7-9, where a portion of the film including large irregular waves sheds more or less together, while other portions of the trailing edge do not generate wake droplets at that particular moment. Here the size of the overall shed "slug" appears to depend primarily upon wave amplitude, which itself no doubt depends upon various other parameters as yet unresolved. In this particular case the individual particles composing the slug are ~ 1 mm. A detailed study of other photos from Refs. 26, 41–45 and other similar sources will reveal still other possible detailed mechanisms which may sometimes apply.

2. *Wake droplet spectra.* The maximum droplet size has been estimated on theoretical grounds [62] for the last stage of a typical turbine to be ~ 0.3 mm. Other investigators using measurements from experimental turbines [7, 63]

observed droplets greater than 1 mm, as did the U-M tunnel investigation [64].

A photographic study of wake droplet spectra [64] from the wet-steam tunnel operating under the conditions already described has attempted to generalize maximum droplet size in terms of a droplet Weber number:

$$We_{max} = \frac{\rho_g V_\infty^2 \, d_{max}}{\sigma} \tag{7-1}$$

where V_∞ refers to the average steam velocity outside the wake, σ is the water surface tension, and ρ_g is the gas or vapor density. Droplet diameter d_{max} was measured 22 cm downstream of the trailing edge, where the droplet size is fairly constant. Related results by Weigle and Severin [65] and Krzeczkowski and Hammitt [66], using an air tunnel, and Valha [14, 19, 67, 68], using both a steam and air tunnel, are compared.

Based on our analysis of several wake droplet photographs, a droplet size-distribution function has been established [64]:

$$F(d) = \frac{N(d)}{N \, \Delta d} \tag{7-2}$$

where d = droplet size
         Δd = droplet-size interval = 200 μm for present experiment
         $N(d)$ = droplet number of size enclosed in region $(d - \Delta d/2, d + \Delta d/2)$
         N = average total number of droplets visible in test area

From the size-distribution function [Eq. (7-2)], a mass-distribution function was also calculated [64]. Experiments were conducted for Mach numbers† of 0.35, 0.55, and 0.75, although limits of the photographic system used (1 μs flash) prevented much detailed droplet-size information at the highest Mach number from being obtained. Hence further work with a better flash, perhaps with a duration of only a few nanoseconds, would be desirable, as well as some system to measure the droplet velocities in the wake region.

Figures 7-14 to 7-16 are typical photos of the wake region and trailing edge of the blade section (Fig. 7-2) from our steam-tunnel tests [64], showing the effect on droplet size and number of steam Mach number and injected film flow rate. The quality of these photos (1 μs flash) is such that droplet sizes can be easily measured. The pictures from a region further downstream [64] are not as clear, since the droplet velocity is greater due to drag from the high-velocity steam.

Figure 7-17 indicates typical raw data of droplet diameter versus distance downstream from the trailing edge. A small data band indicates the envelope of d_{max}. It decreases strongly with downstream distance up to ~22 cm. Figure 7-18 summarizes the droplet diameter distance data for Mach numbers of 0.35, 0.55, and 0.75, and for a large range of injected film flow rates. The decreasing trend with distance is the same for all conditions studied, and continues to distances of at least 22 cm.

---

† Based on steam sonic velocity.

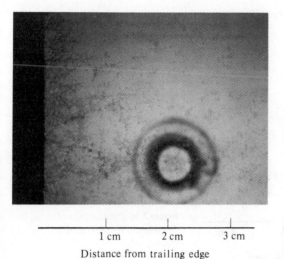

1 cm	2 cm	3 cm

Distance from trailing edge

Figure 7-14 Photographs of liquid-film disintegration into droplets at the trailing edge, $M = 0.35$, $\dot{q} = 5$ cm^3/cm-min [45, 64].

Figure 7-19 compares the present data with that of previous investigators [64, 65, 67, 68] in terms of critical Weber number versus Mach number. The agreement with Valha [67, 68] who used steam instead of air, which was used by Weigle and Severin, is quite good. However, the data of Weigle and Severin, from atmospheric air [65], do not agree well with the steam results. This may not be surprising when it is considered that the properties of the carrier fluid (steam or air) do not enter into the calculation of the assumed critical Weber number formulation. It may be surprising (Fig. 7-19) that critical Weber number increases somewhat, at least in the subsonic range, for increasing Mach number, approaching a value of ~ 50.

1 cm	2 cm	3 cm

Distance from trailing edge

Figure 7-15 Photographs of liquid-film disintegration into droplets at the trailing edge, $M = 0.55$, $\dot{q} = 25$ cm^3/cm-min [45, 64].

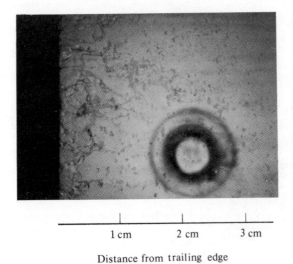

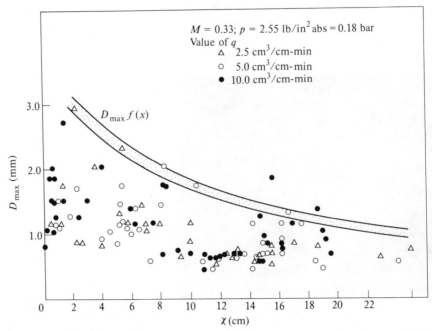

Figure 7-16 Photographs of liquid-film disintegration into droplets at the trailing edge, $M = 0.35$, $\dot{q} = 5$ cm^3/cm-min [45, 64].

Figure 7-17 Maximum droplet size as a function of the distance from the trailing edge, $M = 0.35$ [45, 64].

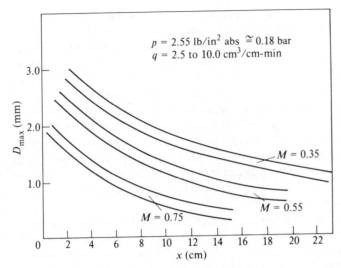

Figure 7-18 Maximum droplet size as a function of the distance from the trailing edge at three Mach numbers, M [45, 64].

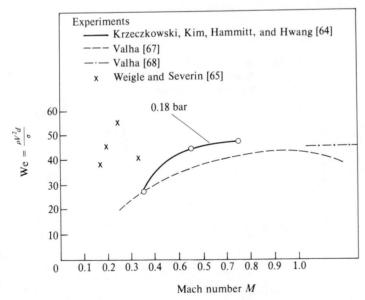

Figure 7-19 The Weber number of maximum droplets far downstream from the trailing edge [45, 64].

Figure 7-20 shows the typical experimental form for the droplet diameter distribution function $F(d)$ [Eq. (7-2)] at two different distances downstream of the trailing edge. Figure 7-21 shows the corresponding mass-distribution function for the same Mach number (0.55). It is shown that the most probable diameter (Fig. 7-20) is ~ 0.20 mm, whereas the most probable diameter for the mass distribution (Fig. 7-21) is ~ 0.6 mm. This difference between Figs. 7-20 and 7-21 is, of course, due to the d^3 dependence of droplet mass. The mass distribution is probably more important from the viewpoint of erosion.

3. *Droplet shapes in wake region.* A detailed photographic study of droplet shapes in an airstream was made recently at IFFM, Gdansk [65]. Figures 7-22 to 7-25 show photographically droplet distortion under the effects of a bypassing airstream. The nonsymmetrical shapes attained are quite similar to those found in cavitation bubble collapse discussed in Chap. 4.

4. *Shock wave droplets and rain erosion.* The problem of droplet distribution in the wake downstream of stator blades in steam turbines and its relationship to erosion of the next rotating row [5–8, 62–70] is somewhat analogous to the problem of droplet distribution after passing through an air shock front and the consequent rain erosion of supersonic aircraft or missile components, previously mentioned. Most cases involving important "rain erosion" involve passage of the component to be eroded through the air at highly supersonic or hypersonic velocities. Thus a bow shock wave, through which the impacting particles must pass before impacting the component to be eroded, exists. Such passage through a shock wave may degrade the liquid-particle spectrum significantly, thus reducing the rain-erosion problem which otherwise would exist. Some designs have been considered, in fact, where a leading "sting" might be used to reduce the erosion rate [71]. The whole problem is somewhat analogous

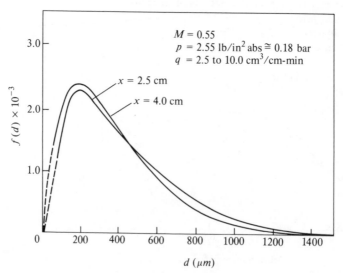

Figure 7-20 Droplets size-distribution function at $M = 0.55$, $x = 2.5$ cm and $x = 4.0$ cm. $x =$ distance downstream of blade trailing edge [45, 64, 66].

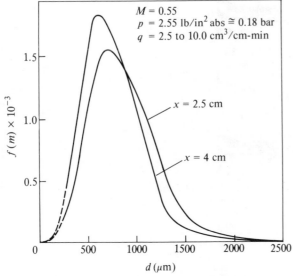

Figure 7-21 Droplets mass-distribution function at $M = 0.55$, $x = 2.5$ cm, and $x = 4.0$ cm. $x =$ distance downstream of blade trailing edge [45, 64, 66].

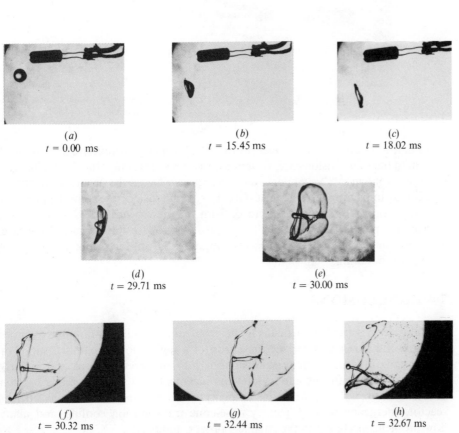

Figure 7-22 Droplet distortion under the effects of a bypassing airstream [65].

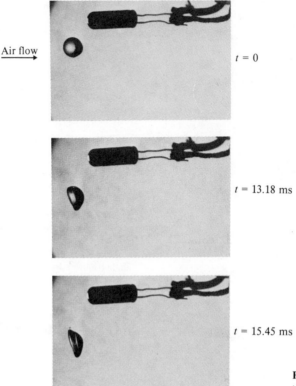

Air flow →

$t = 0$

$t = 13.18$ ms

$t = 15.45$ ms

Figure 7-23 Liquid-droplet formation in airstream I [65].

to that found in steam turbines, in that the "primary" droplet regime, natural rain in the rain-erosion case, is significantly altered through the fluid dynamics of the problem to a "secondary" droplet regime, which is the droplet distribution actually impacting the eroded surface. For "rain erosion" the secondary droplet distribution is somewhat degraded, from the viewpoint of erosion potential, from the primary distribution. The opposite is true for the steam-turbine case, where the secondary distribution is "upgraded" from the primary in terms of erosion potential.

7-4 CONCLUSIONS

This chapter has treated particularly the problem of thin liquid-film flow regimes along a solid surface, under the impulsion of a high-velocity, parallel, vapor or gas stream. This problem was introduced into the present book particularly because of the steam-turbine droplet impingement and rotating-blade erosion problem. However, it is pertinent to many other applications, such as nuclear reactor emergency cooling, reentry nose-cone transpiration cooling, and many flow problems pertinent in the chemical process industries.

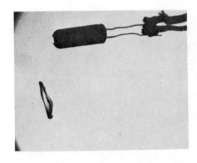

Air flow $t = 18.02$ ms

$t = 27.71$ ms

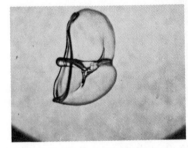
$t = 30.00$ ms

Figure 7-24 Liquid-droplet formation in airstream II [65].

In most cases the problem involves both steady-state and nonsteady film behavior, such as wavelet behavior and dry-spot formation, resulting in critical film-breakdown situations, which are important for various reasons and in different ways, in the different applications. In the primary application (from the viewpoint of this book) of the steam or vapor turbine-blading erosion problem, the film is formed on stationary blades from original very small "primary droplets" in the main steam flow after it crosses the "Wilson line" [1, 2]. The liquid film is then shed primarily from the trailing edge of stationary blades under effects of the high-velocity parallel main steam flow into the wake region in the form of relatively large "primary drops." These relatively stationary droplets then impact the next downstream rotating-blading row, causing an important erosion problem, since the impact velocity is essentially that of the rotating blades. Thus the problem involves: film formation; film behavior, both steady and nonsteady; film shedding and breakdown; droplet size and trajectory in the wake; and liquid-droplet impact erosion. The last phase of the problem was treated primarily

Air flow

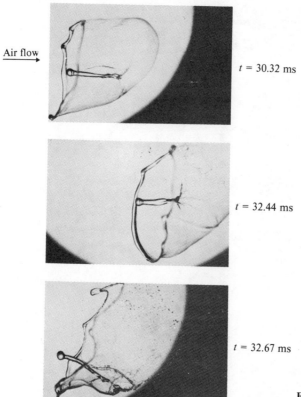

$t = 30.32$ ms

$t = 32.44$ ms

$t = 32.67$ ms

Figure 7-25 Liquid-droplet formation in airstream III [65].

in Chap. 6; the other problems, with the exception of that of primary film formation, which is not treated in this book, have been discussed in the present chapter.

The solution of this entire range of problems involves both highly sophisticated experimental and photographic studies, and highly complex mathematical models. As described in this chapter, no phase of the overall problem can be considered as reasonably completely solved at this time, and there is much on-going research pertinent to all these phases at present. However, reasonable agreement has been achieved for prediction of the steady-state film behavior. Much less success has been achieved concerning film-breakdown criteria, or for the droplet size and trajectory spectra in the wake.

REFERENCES

1. Wilson, C. T. R.: *Phil. Trans.*, ser. A, vol. 189, pp. 265–307, 1897.
2. Keenan, J. H.: "Thermodynamics," John Wiley, New York, 1941, p. 429.

3. Gardner, G. C.: Events Leading to Erosion in Steam Turbines, *Proc. Inst. Mech. Engrs.*, vol. 178, pt. 1, no. 23, pp. 593–623, 1963–1964.
4. Christie, D. G., and G. W. Hayward: Observation of Events Leading to Formation of Water Droplets Which Cause Turbine Blade Erosion, *Phil. Trans. Roy. Soc.*, ser. A, vol. 260, no. 1110, pp. 183–192, 1966.
5. Krzyzanowski, J., B. Weigle, and H. Severin: Semi-Empirical Criterion of Erosion Threat in Modern Steam Turbines, *Trans. ASME, J. Engr. for Power*, vol. 93, ser. A, no. 1, pp. 1–6, 1971; see also Erosionprobleme von Hochleistungsdampfturbinen, *Brennst. Warne-Kraft*, vol. 29, no. 7, pp. 286–290, July, 1977.
6. Heymann, F. J.: On the Prediction of Erosion in Steam Turbines, paper no. 37, *Proc. Sixth Conf. on Steam Turbines of Large Output*, 1972, Pilsen, Czechoslovakia.
7. Krzyzanowski, J.: Eine Analyse der Kollisionseffekte der Wassertropfen mit Laufschaufeln von Dampfturbinen und einige Versuchsergebnisse, *Trans. IFFM*, vols. 42–44, pp. 255–282, Polish Acad. Sci., Gdansk, Poland, 1969.
8. Kirillov, I. I., and I. P. Faddeev: Erosion of Blades of Turbines Operating on Wet Steam, *Thermal Engineering*, vol. 18, pp. 74–78, 1971; see also *Teploenergetika*, vol. 18, pp. 50–53, 1971.
9. Saric, W. S., and B. W. Marshall: An Experimental Investigation of the Stability of a Thin Liquid Layer Adjacent to a Supersonic Stream, *AIAA J.*, vol. 9, p. 1546, 1971.
10. Nayfeh, A. H., and W. S. Saric: Non-Linear Stability of a Liquid Adjacent to a Supersonic Stream, *J. Fluid Mech.*, vol. 58, p. 39, 1973.
11. Bordner, G. L., A. H. Nayfeh, and W. S. Saric: "Stability of Liquid Films Adjacent to Compressible Streams," Virginia Polytechnic Institute Rept. E-73-3, 1973.
12. Ryley, D. J., and K. A. Tubman: Spontaneous Condensation in High-Pressure Expanding Steam, *Proc. Second Conference on Large Steam Turbines*, IFFM, 1971, Polish Acad. Sci., Gdansk, Poland.
13. Ryley, D. J., and M. J. Holmes: Sampling of High Quality Wet Steam from Steam Mains Operating at 11.4 Bar Pressure, *Proc. Inst. Mech. Engr.*, vol. 187, no. 31/73, pp. 381–393, 1973.
14. Valha, J., and D. J. Ryley: Optical Studies of Nucleation in High Pressure Expanding Steam, *Proc. Symp. on Condensing Flows*, FED Spring Conf., ASME, June 15–17, 1977, pp. 27–42, New Haven, Conn.
15. Gyarmathy, G., H. P. Burkhard, F. Lesch, and A. Siegenthaler: Spontaneous Condensation of Steam at High Pressure; First Experimental Results, Conf. Publ. no. 3, *Proc. Inst. Mech. Engr.*, 1973, pp. 182–186.
16. Ryley, D. J., and P. D. Patel: Condensation on the Surface of a Low Pressure Steam Turbine Suction Blade, *Proc. Inst. Mech. Engr.*, vol. 186, no. 59/73, pp. 699–708, 1973.
17. Ryley, D. J., W. J. Ralph, and K. A. Tubman: The Collision Behavior of Water Drops within a Low Pressure Steam Atmosphere, *J. Mech. Sci.*, vol. 12, pp. 589–596, 1970.
18. Puzyrewski, R., and T. Krol: "Numerical Analysis of Hertz-Knudsen Model of Condensation upon Small Droplets in Water Vapor," Tech. Rept. of IFFM, Polish Acad. Sci., Gdansk, Poland, 1970.
19. Valha, J.: Liquid Phase Movement in Last Stages of Large Condensing Steam Turbines, paper 33, *Proc. Conf. Large Steam Turbines*, Inst. Mech. Engr., 1970.
20. Ryley, D. J., and J. Small: Re-Entrainment of Deposited Liquid from Simulated Steam Turbine Fixed Blades, Conf. Publ. no. 3, *Proc. Inst. Mech. Engr.*, 1973, pp. 9–18.
21. Ryley, D. J., and B. N. Bennett-Cowell: The Collision Behavior of Steam-Borne Water Drops, *Int. J. Mech. Sci.*, vol. 9, pp. 817–833, 1967.
22. Miles, J. W.: On the Generation of Surface Waves by Shear Flows, *J. Fluid Mech.*, vol. 3, p. 185, 1957.
23. Benjamin, T. B.: Shearing Flow Over a Wavy Boundary, *J. Fluid Mech.*, vol. 6, p. 161, 1959.
24. Hewitt, G. F., and P. M. C. Lacey: The Breakdown of the Liquid Film in Annular Two-Phase Flow, *Int. J. Heat and Mass Transfer*, vol. 8, p. 781, 1965.
25. Murgatroyd, W.: The Role of Shear and Form Forces in the Stability of a Dry Patch in Two-Phase Film Flow, *Int. J. Heat and Mass Transfer*, vol. 8, pp. 297–301, 1965.
26. Kim, W., S. Krezeczkowski, and F. G. Hammitt: Investigation of Behavior of Thin Liquid Films with Co-Current Steam Flow, *Proc. Two-Phase Flow and Heat Transfer Symposium-Workshop*,

Oct. 18–20, 1976, CERI, University of Miami, Miami, Florida; also available as DRDA Rept. UMICH 014571-3-T, July 1976, University of Michigan, Ann Arbor, Michigan.

27. Wurz, D.: Flow Behavior of Thin Water Films Under the Effect of Co-Current Air Flow of Moderate to High Subsonic Velocities, *Proc. Third Int'l Conf. on Rain Erosion and Associated Phenomena*, Aug. 11–12, 1970, pp. 727–750 (edited by A. A. Fyall of the Royal Aircraft Est., Farnborough, England).

28. Collier, J. G., and G. F. Hewitt: "Film Thickness Measurements," ASME paper no. 64-WA/HT-41, 1964.

29. Hanratty, T. J., and C. A. Thorsness: Turbulent Flow over Wavy Surfaces, *Proc. Symp. on Turbulent Shear Flow*, April, 1977, Pennsylvania State University, Philadelphia.

30. Kordyban, E. S.: Interfacial Shear in Two-Phase Wavy Flow in Closed Horizontal Channels, *Trans. ASME, J. Fluids Engr.*, vol. 96, ser. I, p. 97, 1974.

31. Kordyban, E. S.: "Long-Wave Disturbances in Two-Phase Wavy Flow," ASME paper no. 75-WA/HT-28, 1975.

32. Hewitt, G. F., R. D. King, and P. C. Lovegrove: Techniques for Liquid Film and Pressure Drop Studies in Annular Two-Phase Flow, *Chem. Engr. Prog.*, vol. 45, pp. 73–83, 1964.

33. Puzyrewski, R., and R. Jasinski: Measurement of the Thickness of Thin Water Films by Resistance Method, *Trans. IFFM, Polish Acad. Sci.*, vol. 26, pp. 73–83, 1965.

34. Hammitt, F. G., and J. Mikielewicz: Steam Tunnel Program and Liquid Film Thickness Gauge Development at University of Michigan, *Proc. Sixth Conf. on Steam Turbines of Large Output*, July, 1975, p. 134, Pilsen, Czechoslovakia; also available as DRDA Rept. UMICH 012449-8-T, 1975, University of Michigan, Ann Arbor, Michigan.

35. Mikielewicz, J., and F. G. Hammitt: Generalized Characteristics of Electrical Conductance Film Thickness Gauges, *Trans. IFFM, Polish Acad. Sci.*, vol. 162, p. 848, Gdansk, Poland, 1976.

36. Hammitt, F. G., J.-B. Hwang, and W. Kim: Liquid Film Thickness Measurements in University of Michigan Wet Steam Tunnel, *1976 ASME Cavitation Forum*, March, 1976, pp. 37–39.

37. Saric, W. S., A. H. Nayfeh, and S. G. Lekoudis: "Experiments on the Stability of Liquid Films Adjacent to Supersonic Boundary Layers," Virginia Polytechnic Institute Rept. VPI-E-75-21, Dept. of Engr. Sci. and Mech., February, 1976.

38. Marshall, B. W., and W. G. Tiederman: A Capacitance Depth Gage for Thin Liquid Films, *Rev. Scientific Instruments*, vol. 43, no. 3, pp. 544–547, March, 1972.

39. Solesio, J. N.: "Instabilities des Films Liquides Isothermes," CENG Rept. CEA-R-4835. Grenoble, France, 1977.

40. Cohen, L. S., and T. J. Hanratty: Effect of Waves at a Gas-Liquid Interface on a Turbulent Gas Flow, *J. Fluid Mech.*, vol. 31, pt. 3, p. 467, 1968.

41. Kim, W., F. G. Hammitt, S. Blome, and H. Hamed: Thin Shear Driven Water Film Wavelet Characteristics, *1977 ASME Cavitation Forum*, pp. 31–33, 1977.

42. Mitchell, J. E., and T. J. Hanratty: Study of Turbulence at a Wall Using an Electrochemical Wall-Shear Stress Meter, *J. Fluid Mech.*, vol. 26, p. 199, 1966.

43. Tailby, S. R., and S. Portalski: *Trans. Inst. Chem. Engrs.* (*London*), vol. 40, p. 114, 1962.

44. Son, J. S., and T. J. Hanratty: Velocity Gradients at the Wall for Flow Around a Cylinder at Reynolds Number from 5×10^3 to 10^5, *J. Fluid Mech.*, vol. 35, p. 353, 1969.

45. Kim, W.: "Study of Liquid Films, Fingers and Droplet Motion for Steam Turbine Blading Erosion Problem," Ph.D. Thesis, Mech. Engr. Dept., University of Michigan, Sept. 1978; also available as DRDA Rept. No. UMICH 014571-4-T. See also Kim, W., F. G. Hammitt: Investigation of Liquid Finger Behavior at Trailing Edge of a Simulated Steam Turbine Blade, *Proc. 2nd Multiphase Flow and Heat Transfer Symposium-Workshop*, CERI, Univ. Miami, Miami Beach, Florida, 16–18 April, 1979; also available as DRDA Rept. No. UMICH 014571-5-T, Sept. 1978, University of Michigan, Ann Arbor, Michigan.

46. Taitel, Y., and A. E. Duckler: A Model for Prediction Flow Regime Transitions in Horizontal and Near-Horizontal Gas-Liquid Flow, *AIChE J.*, vol. 22, pp. 47–55, 1976.

47. Baker, O.: Simultaneous Flow of Oil and Gas, *Oil Gas J.*, vol. 53, p. 185, July, 1954.

48. Nayfeh, A. H., and W. S. Saric: Stability of a Liquid Film, *AIAA J.*, vol. 9, p. 750, 1971.

49. Sabersky, R. H., A. J. Acosta, and E. G. Hauptmann: "Fluid Flow, A First Course in Fluid Mechanics," Macmillan, New York, 1971, p. 476.
50. Hartley, D. E., and W. Murgatroyd: Criteria for Break-up of Thin Liquid Layers Flowing Isothermally over a Solid Surface, *Int. J. Heat and Mass Transfer*, vol. 7, p. 1003, 1964.
51. Brauer, H.: Stromung und Warmeuhergang bie Rieselfilmen, *V.D.I. (Ver. Deut. Ingr.)-Forschungsheft*, p. 457, 1956.
52. Miles, J. W.: The Hydrodynamic Stability of a Thin Film of Liquid in Uniform Shearing Motion, *J. Fluid Mech.*, vol. 8, p. 38, 1960.
53. Mikielewicz, J., and J. R. Moszynski: Minimum Thickness of a Liquid Film Flowing Vertically Down a Solid Surface, *Int. J. Heat and Mass Transfer*, vol. 19, no. 7, pp. 771–776, 1976.
54. Mikielewicz, J., and J. R. Moszynski: Breakdown of a Shear Driven Liquid Film, *Trans. IFFM*, vol. 66, pp. 3–11, Polish Acad. Sci., Gdansk, Poland, 1975.
55. Anshus, B.: On the Asymptotic Solution to the Falling Film Stability Problem, *Ind. Eng. Chem. Fundam.*, vol. 11, no. 4, p. 502, 1972.
56. Krantz, W. B., and S. L. Goren: Stability of Thin Liquid Films Flowing Down a Plane, *Ind. Eng. Chem. Fundam.*, vol. 10, no. 1, p. 91, 1971.
57. Bankoff, S. G.: Stability of Liquid Flowing Down a Heated Inclined Plane, *Int. J. Heat and Mass Transfer*, vol. 14, p. 377, 1971.
58. Zuber, N., and F. W. Staub: Stability of Dry Patches Forming in Liquid Films Flowing over Heated Surfaces, *Int. J. Heat and Mass Transfer*, vol. 9, p. 897, 1966.
59. Orell, A., and S. G. Bankoff: Formation of a Dry Spot in Horizontal Liquid Film Heated from Below, *Int. J. Heat and Mass Transfer*, vol. 14, p. 1835, 1971.
60. Ryley, D. J., and B. H. Khoshaim: A New Method of Determining the Contact Angle Made by a Sessile Drop upon a Horizontal Surface, *J. Colloid. Chem.*, vol. 2, pp. 243–251, 1977.
61. Longson, B.: A New Method of Measuring the Contact Angle Between Liquid Sodium and a Solid," Risley Engr. and Mat. Lab. Rept. TRG-Memo-7096 (R), UKAEA, January, 1976.
62. Puzyrewski, R., and S. Krzeczkowski: Some Results of Investigations on Water-Film Break-up and Motion of Water Drops in Aerodynamic Trail, *Trans. IFFM*, vols. 29–31, Polish Acad. Sci., Gdansk, Poland, 1966.
63. Faddeev, J. P.: Structure of Erosion Inducing Streams of Droplets in Axial Clearance of Low-Pressure Part of the Turbine, *Proc. Third Conf. on Steam Turbines of Great Output, IFFM*, 1975, Polish Acad. Sci., Gdansk, Poland.
64. Krzeczkowski, S., W. Kim, F. G. Hammitt, and J.-B. Hwang: Investigations of Secondary Liquid Phase Structure in Steam Wake, ASME Paper No. 78-WA/FE-13, 1978; also available as University of Michigan ORA Rept. UMICH 014571-1-T, 1978.
65. Weigle, B., and H. Severin: Investigation of Relationship Between Gas Velocity Droplet Stream Structure, and Erosion Rate-Time Structure (in Polish), *IFFM Bull., Nr. arch.*, vol. 273/71, Polish Acad. Sci., Gdansk, Poland, 1971.
66. Krzeczkowski, S., and F. G. Hammitt: Experimental Investigation of Liquid Droplet Break-up Duration, *1977 ASME Cavitation Forum*, pp. 34–38; also available as University of Michigan ORA Rept. 014571-2-I, September, 1976, Ann Arbor, Michigan.
67. Valha, J.: Liquid Film Disintegration on the Trailing Edges of Swept Bodies, *Strojnicky Casopis*, vol. XXI, no. 3, Prague, Czechoslovakia, 1970.
68. Valha, J.: Rozpad Kapalinovych Filmuo na Odtokove Hrane Profilu pri Vysokych Rychlostech, *Proc. Fourth Conf. on Steam Turbines of Large Output*, 1972, Pilsen, Czechoslovakia.
69. Woodmansee, D. E., and T. J. Hanratty: Mechanism of Removal of Droplets from Liquid Surface by Parallel Air Flow, *Chem. Engr. Sci.*, vol. 24, pp. 299–307, 1969; see also Ph.D. thesis, University of Illinois, 1968, Champagne-Urbana, Illinois.
70. Kinney, G. R., A. E. Abramson, and J. L. Sloop: "Internal Liquid Film Cooling Experiments with Air Stream Temperatures to 2000°F," NACA Tech. Rept. 1087, 1952.
71. Tatnall, G. J.: Personal communications to F. G. Hammitt, Naval Air Devel. Center, Warminster, Pa., 1970.

EIGHT

WORLD RESEARCH LABORATORIES IN CAVITATION AND DROPLET-IMPINGEMENT RESEARCH

8-1 INTRODUCTION

This chapter lists and briefly discusses those laboratories of the world which at the time of writing were known to the author to be doing important research work in the fields of cavitation or liquid impingement. A brief discussion of the type of work, the most important investigators, the type of facilities and instrumentation available, and the limiting parameters of these facilities is included where the information is available to the writer. Cavitation is considered first and then liquid impingement. Under each subject the discussion is subdivided into countries. The listing within the countries is essentially random, although the more prominent tend to be towards the beginning of the list as they come first to mind. Somewhat similar information for a somewhat earlier time is presented in Ref. 1 and also in the various reports listed in the References. The present listing is by no means exhaustive or complete.

8-2 CAVITATION

8-2-1 United States

Pennsylvania State University (Penn State), U.S. Naval Ordnance Research Laboratory (John Garfield water tunnel), State College, Pennsylvania   Three major water tunnels are available at this facility [2, 55]. These include the 48-in (122-cm)

(John Garfield), 12-in (30-cm), and a "high-speed" tunnel. The last is a small tunnel with a $1\frac{1}{2}$-in (4-cm) diameter test section, but with maximum test-section velocity of ~ 92 m/s, maximum pressure of ~ 80 bar, and temperature of 150°C (Figs. 8-1 and 8-2). To date it has been operated with both water and Freon. Research has been mostly associated with cavitation-inception measurements, though the very high velocity capability of the small tunnel has also allowed damage research. The large tunnels have been used mostly for relatively standard profile tests, etc., for the U.S. Navy.

California Institute of Technology (CIT), Pasadena, California [1] The CIT high-speed water tunnel (completed 1947) was the forerunner of many of the present-day highly sophisticated tunnels. It was designed and operated primarily by the late Prof. R. T. Knapp and his colleagues. The test section is 14 in (36 cm) in diameter by 4 ft (120 cm) in length, and the maximum water velocity is about

Figure 8-1 Ultra-high-speed cavitation tunnel, 1.5 inch, Penn State University [2, 55].

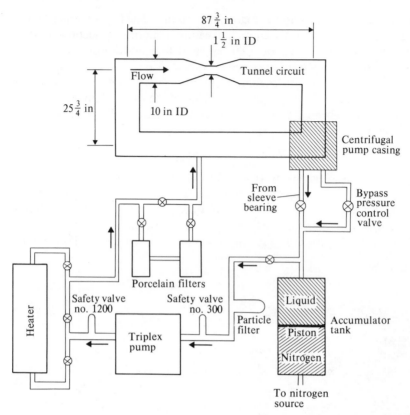

Figure 8-2 Schematic of Penn State University ultra-high-speed tunnel bypass circuit [2, 55].

30 m/s. The tunnel includes a large air "resorber" and methods for measuring air content.

A magnetostriction vibratory damage facility also exists in this laboratory, operated and designed by Prof. M. S. Pleaset.

University of Michigan (U-M), Cavitation and Multiphase Flow Laboratory, Mechanical Engineering Department, Ann Arbor, Michigan [3] A small circulating high-speed tunnel as well as vibratory cavitation and jet impact damage facilities are available. An older, smaller tunnel using water, mercury, and Freon has been disassembled. The maximum test-section velocity is ∼70 m/s. Only the high-speed Penn State tunnel is faster in the United States to the author's knowledge. However, a somewhat similar new unit exists at Tokyo University as well as at Hydronautics (discussed later). The test fluid in the present U-M tunnel has been limited to water† to ∼90°C. The tunnel includes four working loops in parallel, each with a ½-in (1.3-cm) diameter throat. If desired, all the

† Other fluids such as Freons are possible.

liquid can be passed through a single throat of up to 4.5 cm diameter. The tunnel has been used both for cavitation inception and for damage studies. It has been used with a Coulter-counter system for measurement of gas nuclei spectra and a Van Slyke apparatus for total gas measurement. A deaerator is available and an ultra-high-speed motion picture facility ($> 10^6$ f/s)† has been used with this tunnel.

Vibratory cavitation facilities are also available for study of nucleation in various fluids (including high-temperature sodium) and for damage studies. These facilities allow fluid temperatures to about 800°C. The facility has been used previously with both molten sodium and lead-bismuth alloy to this temperature, and pressures to ~35 bar. Higher fluid pressures could be accommodated with a redesigned vessel, using essentially the same vibratory system and sealing arrangement.

University of Minnesota, St. Anthony Falls Hydraulic Laboratory, Minneapolis, Minnesota   Three water tunnels exist at this facility [4–6]. One involves a vertical free jet (0.25 m diameter) and is nonrecirculating. The other two are recirculating with test-section diameters of 0.15 and 1.07 m, respectively. The test-section velocity of the free jet and 0.25-m tunnels is about 15 m/s, and the flow rate of the 1.07-m tunnel is 1.7 m³/s.

A large volume of cavitation research and testing continuously originates from this laboratory. A particularly important contribution has been the development of ultrasonic methods for the continuous measurement of gas nuclei spectra within the water tunnel. This measurement capability seems necessary for further progress in the correlation of cavitation nucleation thresholds with other flow parameters. Some cavitation and impact-damage work using a magnetostriction vibratory facility for cavitation damage and a rotating-wheel facility for liquid-impingement damage has been done here.

Colorado State University, Engineering Research Center, Fort Collins, Colorado   The major cavitation facility at this location consists of a once-through 24-in diameter pipe connecting a mountain reservoir above the laboratory with a tailrace at the laboratory [4, 5]. The elevation head is ~105 m. With pump assistance, the total head available is ~190 m for a flow rate of 5 ft³/(0.14 m³/s). The facility is used for both the study of the cavitation performance of valves, orifices, etc., of different sizes, and for cavitation-damage studies downstream of such flow restrictions as valves, orifices, etc.

Hydronautics, Inc., Laurel, Maryland   Three cavitation tunnels are listed at this location [4–6]: a large tunnel of 32,000 gal/min (2.0 m³/s) capability, a smaller tunnel at 2200 gal/min (~0.13 m³/s), and a small high-speed tunnel capable of 200 ft/s (61 m/s). In addition, a vibratory facility for liquid-metal cavitation-damage work was developed and used several years ago.

† Frames per second = Hz.

National Bureau of Standards, U.S. Department of Commerce, Cryogenics Division, Boulder, Colorado Work has been done at this laboratory to measure incipient and desinent (as well as developed) cavitation numbers for several hydrodynamic bodies in both liquid hydrogen and liquid nitrogen. The aim is to extend the capability of design and prediction for pumps and other liquid handling components to account for size, geometry, fluid, fluid velocity, temperature, etc., effects, i.e., "scale effects" in general. In the past their results have been compared with results with water from the NASA Lewis Laboratory, using precisely the same geometries, etc. The following section gives further details.

National Aeronautics and Space Administration (NASA), Lewis Research Center, Cleveland, Ohio Flowing system cavitation experimentation was performed at this location in water using different venturi geometries to study effects of size, velocity, temperature, geometry, etc., on cavitation inception and related head losses for more developed cavitation. This program was closely coordinated with that at National Bureau of Standards, Boulder, Colorado, previously described.

Previous flow work at NASA-Lewis related to cavitation involving liquid-sodium work was coordinated with work supported by NASA at CANEL† (previously an operating division of Pratt and Whitney Aircraft) and at Hydronautics.

A vibratory cavitation-damage facility also existed at NASA-Lewis in which tests in liquid sodium, mercury, and water on various specialized materials were performed for the space program.

Oak Ridge National Laboratory, U.S. Atomic Energy Commission, Oak Ridge, Tennessee Long-term tests of liquid metal and molten-salt pump cavitation were made at this location in support of the Molten-Salt Nuclear Reactor Program. These pump tests have been completed, but other liquid-metal cavitation work, such as with an electromagnetic pump, has been reported from time to time.

U.S. Naval Ship Research and Development Center (USNSRDC), Washington, D.C. There are three major water tunnels listed [4, 5] at this location: 30 cm, 61 cm, and 91 cm; their test-section velocities are 6.5, 17, and 25.7 m/s, respectively. Cavitation performance research has been done in some or all of these. Also, there has been work reported on gas nuclei spectra measurements, particularly using a holographic technique.

In addition, work concerned with the development of coatings and overlay materials to provide corrosion, erosion, and antifouling protection in cavitating environments has been done at the Annapolis Laboratory of USNSRDC,‡ using rotating-disc, venturi, and vibratory equipment. Field trials have also been included on propellors and hydrofoils.

U.S. Naval Undersea Research and Development Center, Pasadena, California A closed, recirculating tunnel with a 12-in (30-cm) diameter semi-open jet is

---

† Connecticut Advanced Nuclear Engineering Laboratory, Middletown, Connecticut.
‡ Formerly done at the Naval Applied Science Laboratory at Brooklyn Navy Yard.

reported [4, 5] at this location with a test-section velocity ~ 12 m/s. Work upon the effects of macromolecules upon cavitation inception is reported. Dilute aqueous polymer solutions were tested in a blowdown tunnel, and it was found that the long-chain polymers inhibited cavitation inception to some extent.

U.S. Department of Interior, Bureau of Reclamation, Division of Research, Denver, Colorado   Large water-circulating facilities are available in this laboratory [4, 5]. Cavitation research is reported to determine the velocity-head relations for cavitation inception on concrete surfaces.

Harvard University Acoustics Laboratory, Cambridge, Massachusetts   Theoretical and experimental work has been reported from this location relating to ultrasonic cavitation and cavitation bubble-dynamics studies.

University of Rochester, Rochester, New York   Theoretical and experimental work relating to ultrasonic cavitation and cavitation bubble dynamics has been reported from here, under the direction of Prof. H. G. Flynn.

University of Houston, Houston, Texas   Theoretical and experimental work relating to ultrasonic cavitation, underwater sound transmission, and cavitation bubble dynamics has been reported from this laboratory, under the direction of Prof. R. D. Finch.

State University of New York (Stony Brook)   Continuing work with a vibratory facility, especially involving micromaterial effects, was recently reported by Prof. C. Preece† from this location.

Texas A&M University, Center for Dredging Studies, Hydromechanics Laboratories, Department of Civil Engineering, College Station, Texas   The Center for Dredging Studies was established in 1968 and the Coastal Engineering Laboratory at same location in 1969. Dredging studies include work on cavitating slurries.

8-2-2 United Kingdom

National Engineering Laboratory (NEL), East Kilbride, Glasgow, Scotland   There are several large water tunnels at NEL capable of testing large hydraulic turbomachines and performing other cavitation tests [4, 7]. In addition, much highly significant cavitation and impingement-damage research, utilizing magnetostriction vibratory, impact wheel, and venturi facilities has been performed here.

Imperial College of Science and Technology, University of London, Exhibition Road, London   A recirculating water facility of 250 kW and 0.55 m³/s (~ 9000 gal/min) is reported here [4]. Research activities are reported to include cavitation.

† Now at the Bell Telephone Research Laboratories, Murray Hill, New Jersey.

Southampton University, Mechanical Engineering Department, Southampton
Work on cavitation damage to bearings and cavitation inception in a small-tunnel facility has been reported from here, under the direction of Prof. S. P. Hutton.

National Physical Laboratory, Teddington, England Two large tunnels exist at this location (25 and 19 m/s, respectively) with 3.3 and 8.3 m diameter test sections [4, 8]. An important symposium on cavitation was held here in 1955 and much significant work was reported at that time.

University College, Mechanical Engineering Department, Cardiff This group has done cavitation-damage work with a vibratory facility.

8-2-3 Canada

University of British Columbia, Hydraulics Laboratory, Department of Civil Engineering, Vancouver, British Columbia Work on cavitation damage to concrete as it is affected by air entrainment is reported from this location [4, 5]. Work is done in an open flume, 7.5 cm wide, with velocities to 30 m/s. Liquid impingement studies relating to jet-cutting of materials are also reported from this university, under Prof. N. C. Franz.

McGill University, Department of Civil Engineering and Applied Mechanics, Montreal A study of cavitation at high-head sluice gates is reported. Work has been done in both a 28 by 28 cm and 7.5 by 7.5 cm water-tunnel section. The work included the measurement of pressure fluctuations on the wall behind an obstacle which produces cavitation.

Concordia University, Montreal Small tunnel cavitation sigma studies have recently been performed at this location.

8-2-4 France

Electricité de France Research Laboratory, Chatou There are large facilities at this location for the testing of hydraulic turbines and pumps. In addition there are also cavitation and impingement rotating discs for erosion studies, as well as spark-bubble apparata for basic cavitation bubble-dynamics studies. Much of this work is reported in the journal published by this laboratory [9].

Naval Towing Tank Laboratory, Bassin d'Essais des Carènes, Paris Two propellor-testing cavitation tunnels are reported from this laboratory, as well as much other related apparata. Maximum test-section velocity is 24 m/s, with a 0.8-m diameter test section.

SOGREAH, Grenoble   Six cavitation tunnels are listed [10] at this location, two of which are for turbine testing. The highest-velocity tunnel provides 42 m/s and the lowest 18 m/s. Hence all are very suitable for cavitation work. Some damage testing has been reported from this laboratory.

University of Grenoble, Grenoble   A relatively large cavitation tunnel is available here. This has been used particularly for supercavitation studies with submerged hydrofoils. No damage work has been reported.

CEA-Cadarache   The cavitation work here [11, 12] is directed toward sodium cavitation, as applied to the fast-breeder reactor development program in France. Probably the world's largest sodium cavitation tunnel exists here, for testing both cavitation performance and damage effects. A water tunnel with identical flow performance and geometry is also being used, so that direct comparisons between sodium and water cavitation can be made [11, 12].

École Nationale Supérieure de Techniques Avancées (ENSTA), Paris   A relatively small cavitation tunnel exists at this location as well as bubble chamber apparata for the basic study of cavitation bubble dynamics, developed primarily by Darroze and Chahine [13].

CEAT-Poitiers   A blowdown tunnel is reported [10] at this location for cavitation tests upon foils, etc.

8-2-5 Netherlands

Laboratory of Mechanics and Electronics, Kon. Machinefabriek Stork, Hengelo
Cavitation work, particularly on pumps, is reported from this laboratory. Closed water tunnels of 75 and 750 kW capacity exist here [10].

Apeldoorn-Delft Technical University   A multipurpose, closed, recirculating tunnel is reported here with a maximum water velocity of 20 m/s [10].

Technical University, Delft

Laboratory of Fluid Mechanics   A closed, recirculating tunnel is reported in this laboratory with a test-section velocity of about 11 m/s.

Laboratory of Metallurgy   A research program is indicated here involving cavitation-damage tests with a rotating-disk apparatus.

Naval Ship Model Basin, Wageningen   Four closed, vertical, recirculating water tunnels are reported [10]. Test-section velocities are 11, 7, 7, and 65 m/s, respectively. The high-speed unit has a test-section diameter of 4 cm, but the others are much larger. Presumably only the last would be well suited to cavitation research, since the test-section velocities of the others are probably too small, with the possible exception of the first.

Metal Research Institute TNO, Tribology Department, Delft   Considerable work has been reported [14] from this laboratory with cavitation-damage studies using a vibratory-type facility upon various metallic alloys. The emphasis has been primarily metallurgical rather than fluid dynamical.

Netherlands Ship Model Basin, Wageningen   A program on cavitation erosion of model ship propellors is reported from this laboratory [15].

Delft Hydraulics Laboratory, Delft Technical University   This laboratory includes the Civil Engineering Fluid Mechanics Laboratory and a laboratory for hydraulic machinery in the Mechanical Engineering Department. The laboratory is building a large testing facility for hydraulic machinery, including cavitation, under the direction of J. Wijdieks.

8-2-6 Norway

Trondheim, S.M.T.   Two tunnels are reported [10] at this location, with only the higher-velocity unit presently in service. The maximum velocity is 55 m/s, and hence it is well suited for cavitation-damage testing over a wide range of parameters.

8-2-7 Sweden

Göteborg   Two tunnels are reported here [10] with test-section velocities of 11 and 23 m/s, respectively. Hence both are suitable for cavitation testing.

8-2-8 West Germany

Institute of Technology, Laboratory of Hydraulic and Hydrologic Research, Darmstadt   A closed recirculating tunnel is reported [4, 10] at this location with a maximum test-section velocity of 25 m/s. The tunnel is reportedly used for the measurement and photography of cavitation incidence.

Hamburg Shipbuilding Experimental Station   Three closed, recirculating water tunnels of large capacity are reported at this laboratory [4, 10]. However, only the largest of these (75-cm diameter test section) has a large enough test-section velocity to be used for cavitation research (19.5 m/s).

Berlin Technical University Two free-surface, high-speed, recirculating, variable-depth tunnels are reported here [4, 10]. The highest velocity is 18 m/s, which seems adequate for cavitation research.

Kiel Staatliche Engineering School A closed, recirculating tunnel is reported at this location with a maximum velocity of 10 m/s. Hence it seems only marginal for cavitation research.

Institute of Hydraulics and Hydrology (Oskar-v-Miller Institute), Technical University, Munich A relatively large recirculating tunnel exists at this location for the testing of hydraulic pumps and turbines, under the direction of Prof. J. C. W. Raabe. The test-section area is 0.4 m², power is ~90 kW, and velocity is 6 m/s. Also in the same institute, important work is reported on the development of a method using laser-light diffraction to measure the spectra of gas micro-bubbles in a liquid, for prediction of cavitation thresholds, etc. It is attempted to develop a commercial instrument package of this type, under the direction of Dr. A. Keller at the Garmisch-Partenkirchen laboratory. This work is discussed in Chap. 3.

Daimler-Benz Company, Stuttgart Vibratory cavitation-damage tests in water have been recently reported from this location.

8-2-9 Italy

CEIMM, Rome A closed, recirculating tunnel with vertical test section is reported here [10]. It is quite large, with about a 4-m² test-section area and a test-section velocity of 14 m/s. Thus cavitation work is probably possible. Pump power is 100 kW.

University of Genoa, Hydraulic Institute, Genoa Cavitation research at this institute [16] has involved primarily pump and cavitation performance, including especially the acoustic manifestations.

8-2-10 Japan

Tohoku University, Institute of High Speed Mechanics, Sendai Four closed, recirculating tunnels with vertical test sections exist at this location [4, 10, 17]. Much cavitation flow testing of hydrofoil cascades,† etc., has been done in this laboratory, following the procedures developed there by F. Numachi (now emeritus). Numachi's group has contributed very significantly to our knowledge of cavitating cascades. One of the tunnels has a test-section velocity of 33 m/s, and hence might be used for cavitation-damage research as well. This is the most powerful of the units with a 520-kW pump motor.

† See various Numachi references.

Tokyo University   A high-speed tunnel is reported at this location with a maximum test-section velocity of 80 m/s in a 3.0-cm diameter test section. Pump motor power is 70 kW. Such a tunnel appears most suitable for cavitation-damage studies, and may well have been built primarily for that purpose. It is no doubt also useful for cavitation inception and other performance work.

Hitachi Hydraulic Research Laboratory, Hitachi-chi, Ibaraki-ken   A closed, recirculating tunnel with closed test section exists at this location [4]. Maximum test-section velocity is 30 m/s, so that cavitation research for both performance and damage should be easily possible. Use of the tunnel for observation and measure of cavitation incidence and desinence is reported.

Hydraulics Department, Technical Laboratory No. 2, Central Research Institute of Electric Power Industry, Tokyo   A cavitation tunnel is reported [4] at this location with a flow rate of 0.1 m^3/s (1580 gal/min). The scope of activities includes cavitation in hydraulic structures.

Ebara Manufacturing Company, Research Laboratory, Haneda Plant, Tokyo   Test facilities with total power of 800 kW exist at this location for the testing of pumps and turbines [4]. Their scope includes cavitation in pumps and turbines.

Hiroshima University, Faculty of Engineering, Sendamachi, Hiroshima   Cavitation-damage experimental studies have been carried out using both a vibratory apparatus and a tunnel (cavitation and the pulses therefrom were observed on a sphere suspended in the test section) with a velocity of 15 m/s. Tests with fluids other than water are reported [4] for both types of facilities.

Meiji University, Faculty of Engineering, Kawasaki   Vibratory cavitation-damage tests in water have been reported from this location recently [18].

Fukui University   Vibratory damage work by Endo and Okada [56] reported here.

8-2-11 India

Indian Institute of Science, Civil and Hydraulic Engineering Department, Bangalore [19]   A relatively new 15-in closed, recirculating water tunnel exists at this location [4, 19]. Total water flow capacity is 0.5 m^3/s (8000 gal/min) with a pump power of 450 kW. The laboratory is primarily active in cavitation damage using venturi systems as well as a rotating-disk and a vibratory facility.

Hydraulics and Water Resources Department, College of Engineering, Madras   A water tunnel with a 20-cm diameter test section is reported at this location [4]. Flow rate is about 0.1 m^3/s (1600 gal/min). Their activities include cavitation research.

Central Water and Power Research Station, Khadakwasla, Poona   A water tunnel with a flow rate of 17 m^3/s and 2000-kW pump exists in this laboratory [4], along with much other specialized equipment, some of which pertains to cavitation research. The study of cavitation in gates, valves, turbines, and pumps is within their scope of activities.

Irrigation Research Station, Poona   A water tunnel is reported in this laboratory [4] with a 280-kW pump and a flow rate of about 0.7 m³/s. Cavitation studies are included in their scope of activities.

8-2-12 Soviet Union

The coverage here of cavitation research facilities in the Soviet Union and Eastern Europe is necessarily limited, due to the writer's lack of full information.

B. E. Vedeneev All-Union Scientific Research Institute of Hydrotechnics, Kriloff No. 1 and No. 2 tunnels, Leningrad   These are both large tunnels, with reported test-section velocities of 10 and 13 m/s, respectively, so that cavitation research is a possibility [4].

Institute of Problems of Mechanics, Moscow   This is the institute of Prof. K. K. Shal'nev, who has produced an important part of the Russian cavitation research literature over the past several decades. In this laboratory there are two relatively small, closed, recirculating tunnels, wherein cavitation-damage studies have been conducted in water in "Shal'nev-type" venturis, i.e., a transverse cylindrical pin across a rectangular, constant-area test section. Work has included the effects of magnetic and electric fields. The work of this group is discussed elsewhere in this book.

Ural Polytechnic Institute imeni S. M. Kirov, Sverdlovsk, Vtuzgorodak   A considerable amount of cavitation-damage research has been reported from this laboratory [20] with particular emphasis upon the development of more resistant steels, particularly for water applications. The experimentation has been done using both vibratory and rotating-disc facilities. Work has also been done in molten lead using a vibratory apparatus.

Scientific Research Center of the Institute "Hydroproject," Moscow   This laboratory is under the direction of L. A. Zolotov. Cavitation research in machinery and other hydraulic structures is conducted here.

8-2-13 Hungary

Technical University, Department of Hydraulic Machinery, Budapest   A very considerable amount of cavitation research, involving both cavitation-damage and performance work, has been published by this laboratory over the past decade, particularly under the guidance of Prof. J. J. Varga [21–23]. This has involved a relatively small tunnel using a "Shal'nev-type" damage venturi, vibratory facilities, and direct pump tests. Work has included attempts to correlate cavitation damage with the acoustic signal [22, 23].

8-2-14 Romania

Timisoara Technical Institute   Considerable cavitation research has originated in this location, particularly under the guidance of Dr. I. Anton [24, 25]. This has included work with a new fairly large water tunnel, vibratory facilities, and other miscellaneous cavitation facilities such as a bubble spark-chamber for photographic studies of individual bubble growth and collapse.

8-2-15 Poland

Institute of Fluid Flow Machines, Polish Academy of Science, Gdansk   A relatively new propellor tunnel, vibratory rotating disc, and venturi-damage facilities exist at this location. This work has been under the supervision of Dr. K. Steller [26–28].

8-2-16 Czechoslovakia

Important work in this country has been done by Dr. M. Nechleba at Brno, particularly on turbomachines [29], and by S. Nemecek at Libercec on cavitation damage [30] using a vibratory facility. Work in Czechoslovakia is well described by Noskievic [31].

8-3 JET AND DROPLET LIQUID IMPACT

8-3-1 United States

Bell Aerosystems Company, Buffalo, New York [32, 33]   The largest and highest velocity rotating-arm facility in the world (to the author's knowledge) for the testing of materials for resistance to high-velocity droplet impingement exists at this facility for velocity up to air Mach 3 (~ 1000 m/s). It includes provision for reduced-pressure testing (to simulate altitude) and for testing with dust impingement as well as water-droplet impingement.

Wright-Patterson Air Force Base (WPAFB), Materials Laboratory, Ohio   A propellor arm at the Wright-Patterson Air Force Base [34] of Mach 1.2 capability (~ 400 m/s) is shown as Fig. 6-37. This group has also used a subsonic (~ 220 m/s) propellor arm at Cincinnati University, Cincinnati, Ohio, for some of their tests, and have supported the large supersonic rotating arm at Bell [32, 33] previously mentioned. The subsonic arm at B. F. Goodrich, Cleveland, Ohio, has also been used by this group.

The group at WPAFB also has conducted numerous rain-impact erosion tests at impact speeds up to air Mach 5, using a rocket-sled device on a ~ 15-km† track at Holloman Air Force Base, New Mexico [35–37]. The objective of these, and also the high-speed propellor-arm tests previously mentioned, is the development of rain-erosion resistant materials for high-speed aircraft and missile components such as helicopter blade tips, radomes, nose cones, etc.

B. F. Goodrich Company, Cleveland, Ohio   This company is stated to have a subsonic propellor arm and also a vibratory cavitation device for testing of rubberized materials and coatings for both cavitation and droplet-impingement resistance.

University of Minnesota, St. Anthony Falls Hydraulic Laboratory, Minneapolis, Minnesota   A rotating-wheel droplet-impact device for the study of spherical liquid droplet impact (a single drop per wheel revolution) has been developed and used here. Tests on various metallic materials have been used and compared with the resistance of the same materials to cavitation in their magnetostriction vibratory facility [38] under direction of J. F. Ripken.

University of Michigan, Cavitation and Multiphase Flow Laboratory, Mechanical Engineering Department, Ann Arbor, Michigan   Liquid-slug impact tests on metallic and other materials have been performed in this laboratory using both a single-shot water-gun device and a somewhat similar automated gun device, described in Chap. 6 on droplet impact from a device pioneered by Kenyon [39] (see Fig. 6-39). Both are capable of velocities up to about 600 m/s, and project an elongated slug of liquid of about 1 mm diameter. The single-shot water gun

(a)   (b)

Figure 8-3 University of Michigan propellor arm. (a) Arm tip and test specimen. (b) Arm and fiberglass housing [41].

† Recently increased from 11 km [34].

is a momentum-exchange type of apparatus of a concept first developed by Bowden and Brunton [40] at Cambridge University (see Fig. 6-38). More recently, a propellor-arm device for working velocity ~1500 ft/s (450 m/s) has been built here [41], shown in Fig. 8-3.

Hydronautics, Inc., Laurel, Maryland A rotating-wheel device (30 cm diameter) wherein the specimens are whirled at high rpm through a relatively low-velocity jet exists at this location. Impact velocity is ~230 m/s. Some relatively recent test results have been reported [42]. Figure 8-4 shows this apparatus, developed by A. Conn.

Raytheon, Boston, Massachusetts A subsonic propellor-arm and air-tunnel apparatus with droplet injection has been used in this location for droplet-impact studies.

Rotating disk

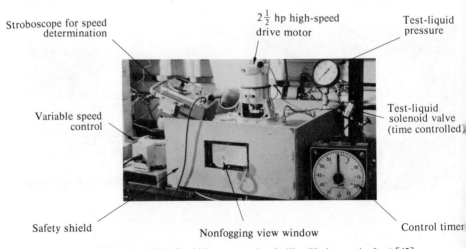

Stroboscope for speed determination

$2\frac{1}{2}$ hp high-speed drive motor

Test-liquid pressure

Variable speed control

Test-liquid solenoid valve (time controlled)

Safety shield

Nonfogging view window

Control timer

Figure 8-4 General view of multiple liquid-impact erosion facility (Hydronautics Inc.) [42].

8-3-2 Canada

Concordia University, Montreal   A high-velocity continuous-flow liquid jet facility exists at this location for research into useful applications of liquid jet cutting. A maximum pressure of ~ 3500 bar is available. Such applications include the cutting of wood and other materials, coal-mining assist, earth-tunneling, etc. These applications, and the present state of the art in relation to them, have been covered in various recent symposia [43–45].

University of British Columbia, Faculty of Forestry, Vancouver   Research has been reported from this location on useful high-velocity liquid jet impact by Dr. Franz [46], using a continuous high-velocity jet device somewhat similar to that mentioned above at Concordia.

8-3-3 United Kingdom

National Engineering Laboratory (NEL), East Kilbride, Glasgow, Scotland   A rotating-wheel impact apparatus has been developed by Dr. Hobbs wherein the specimens are attached to the periphery of the wheel and whirled through a relatively low-velocity jet of water at velocities of about 100 m/s. Facilities of about this capability were developed before World War II, and are fairly numerous in Europe.

Central Electricity Generating Board (CEGB), Marchwood Engineering Laboratory   A comparison of liquid-droplet erosion results on three steam-turbine blading materials in four separate rotating-wheel facilities in the United Kingdom of much higher velocity than that at NEL (discussed above) are reported by this group [47]. Two of these facilities are at the Marchwood Laboratory of CEGB, and the other two will be discussed in the two items immediately following. The machines at CEGB have velocity capacities of 430 and 310 m/s, respectively.

English Electric Turbine Laboratory [47]   This facility was used to compare tests on three turbine materials with the CEGB facilities described above and the C. A. Parsons facility to be described next. The English Electric facility has been used at a velocity of 430 m/s in these tests, though it may be capable of a still higher speed, since it involves two counter-rotating wheels, one carrying the specimen and the other releasing the water drops to be impacted.

C. A. Parsons Facility, Heaton Research and Development Laboratories, Newcastle-upon-Tyne [47]   This test facility was also used at 350 m/s for the above comparative tests, but is capable of ~ 610 m/s.

Associated Electrical Industries (AEI), Birmingham, England   A unique type of liquid-impact facility (Fig. 6-39) was developed and utilized for some years at this laboratory by Dr. P. Kenyon [39]. This same type of facility is now in use in the U-M laboratory (described in Chap. 6). It is an automated water gun capable of projecting about 30 liquid slugs per minute. These are about 1 mm in diameter and ~1 cm long. Maximum velocity is about 600 m/s. The characteristics of the liquid slugs (shape and size) are very similar to those from the Bowden-Brunton momentum-exchange gun, which is discussed next.

University of Cambridge, Department of Mechanical Engineering, Cambridge, England [40]   Work has been conducted in this laboratory over the past decade, initially under the direction of Dr. Bowden (deceased) and then carried on to the present by Dr. J. M. Brunton and his colleagues, upon single droplet-impact effects, using an air-gun-driven momentum-exchange device capable of propelling single liquid slugs against stationary targets at velocities up to ~1000 m/s. The droplets appear to have a somewhat enlarged head which is roughly hemispherical, followed by a relatively long tail of reduced diameter. Since the very high pressures during impact are induced only during the initial portion of impact, the existence of the relatively low-velocity tail may not affect the erosion process significantly. This laboratory is also equipped with excellent ultra-high-speed photographic equipment (10^6 to 10^7 Hz motion picture camera).

Royal Aircraft Establishment (RAE), Farnborough   Very significant basic research upon the impact of single spherical drops upon solid materials has been done by A. A. Fyall and his colleagues at RAE, Farnborough [48]. They have developed an apparatus wherein single liquid drops are suspended in the path of a projectile from a gun device. Since the droplet is stationary before impact, it remains highly spherical up to the moment of impact. Since they are also equipped with excellent ultra-high-speed motion picture equipment (10^6 to 10^7 Hz) they have obtained and published excellent and highly informative pictures of such collisions. Also, a relatively small subsonic propellor-arm facility exists here.

8-3-4 West Germany [49, 50]

Dornier-System GmbH, Friedrichshafen   A considerable amount of rain-erosion-oriented liquid-impact research has been accomplished at this location during the past several years. This has included the development of two propellor-arm facilities for approximately Mach 1 and Mach 2 capability. In addition, a Brunton-type momentum-exchange gun device has been developed using a small caliber rifle for driver, rather than an air gun as used in the other laboratories using this type of device, and producing a velocity of ~1000 m/s. In addition, the several Meersburg conferences [49–51] have been sponsored by this group.

8-3-5 France [49, 50]

ONERA Laboratories at Modane   Some rain-erosion work has been reported from this facility wherein a droplet-laden wind was impacted against relatively large models of aircraft components in the large wind tunnel.

Sud-Aviation, Marseilles   A whirling-arm facility exists at this location with a velocity capability of about Mach 1. It has been used both for liquid-droplet and also solid-particle impingement. Primary applications appear to be rain and dust erosion of helicopter blades.

Electricité de France, Chatou Laboratory   A rotating-wheel facility exists in this laboratory wherein test specimens are whirled through relatively low-velocity jets [9, 52]. The number of jets can be varied, thus obtaining differing frequencies of impact for the same impact velocity. Otherwise the facility is similar to that at NEL, previously described, with the maximum velocity capability ~ 100 m/s.

Société Rateau, La Corneuve   A rotating-disc facility with a maximum velocity capability of about 600 m/s exists at this location, and has been used primarily to test steam-turbine blading steels and other alloys such as titanium, etc.

8-3-6 Russia and Eastern Europe

A certain amount of work has been reported [20] on liquid impingement for relatively low-velocity rotating-wheel devices (such as those of EDF, Chatou, or NEL) with a maximum impact velocity of about 100 m/s. It has been generally assumed in the Russian literature that this type of test is equivalent to a cavitation test, so that data obtained from such devices is assumed pertinent to cavitation-erosion applications. Such facilities have been reported at Kharkov Polytechnic Institute, Traktorniy Institute in Chekov, and also in Prof. Varga's laboratory at the Technical University of Budapest [21–23]. A newer and higher velocity facility is also reported at the Moscow Engineering Institute (MEI) in Moscow.

General bibliographical items [53, 54] contain much pertinent information, although these were not directly cited in the discussions.

REFERENCES

1. Knapp, R. T., J. W. Daily, and F. G. Hammitt: "Cavitation," McGraw-Hill, 1970.
2. Holl, J. W.: Cavitation Research Facilities at the Ordnance Research Laboratory of the Pennsylvania State University, in J. W. Holl and G. M. Wood (eds.), *Symp. on Cavitation Research Facilities and Techniques*, 1964, pp. 11–18, ASME.
3. Hammitt, F. G.: Cavitation Damage and Performance Research Techniques, in J. W. Holl and G. M. Wood (eds.), *Symp. on Cavitation Research Facilities and Techniques*, 1964, pp. 175–184, ASME.

4. "Directory of Hydraulic Research Institutes and Laboratories," vols. I and II, IAHR, April, 1971.
5. Kulin, G., and P. H. Gurewitz (eds): "Hydraulic Research in the United States, 1970 (Including Canadian Laboratories)," U.S. Dept. of Commerce, National Bureau of Standards, NBS Special Publ. No. 346, SD No. C13.10:346, issued March, 1971, U.S. Government Printing Office, Washington, D.C.
6. Holl, J. W., and G. M. Wood (eds.): *Symp. on Cavitation Research Facilities and Techniques,* 1964, ASME.
7. Nixon, R. A., and W. D. Carney: "Scale Effects in Centrifugal Cooling Water Pumps for Thermal Power Stations," NEL Rept. 505, East Kilbride, Glasgow, Scotland, 1972; see also *Fluid Mech.,* No. 473 and 474 for lists of NEL publications and various brochures describing NEL facilities, 1971.
8. *Proc. 1955 NPL Symp. on Cavitation in Hydrodynamics,* HMSO, London, 1956.
9. *Bull. Directions des Etudes et Recherches,* ser. A, Electricité de France, Chatou, 1970–1979.
10. Report of Cavitation Committee, *Proc. Twelfth Int'l Towing Tank Conf.,* Sept. 22–30, 1969, Rome.
11. Bisci, R., and P. Courbière: Analogy of Water as Compared to Sodium in Cavitation, *Proc. Specialists' Meeting on Cavitation in Sodium and Studies of Analogy with Water as Compared to Sodium,* April 12–16, 1976, pp. 62–67, CEA Cadarache, France.
12. Ardellier, A., and J. C. Duquesne: "Etude Experimentale de la Cavitation dans un Ecoulement de Sodium à Travers des Diaphragmes et une Tuyère, Tech. note SDER/73/163, Dept. of Fast Reactors, CEA Cadarache, France, 1973.
13. Chahine, G. L.: Interaction Between an Oscillating Bubble and a Free Surface, *Trans. ASME, J. Fluids Engr.,* 1978; see also *J. de Mécanique,* vol. 15, no. 2, pp. 287–306, 1976, France.
14. Tichler, J. W., A. W. J. DeGee, and H. C. VanElst: Applied Cavitation Erosion Testing, *ASTM STP* 567, pp. 18–29, 1973.
15. Van der Neulen, J. H. J.: Cavitation Erosion of a Ship Model Propellor, *ASTM STP* 474, pp. 162–181, October, 1970.
16. Frederici, G., E. Raiteri, and F. Siccardi: Cavitation Detection and Control in Pumps, *Proc. Fifth Conf. Fluid Mach.,* September, 1975, pp. 273–284. Budapest, Hungary.
17. *Proc. Inst. High Speed Mechanics,* Sendai, Japan.
18. Hirotsu, M.: Cavitation Damage Mechanism and its Correlation to Physical Properties of Material, in J. W. Holl and G. M. Wood (eds.), *Symp. on Cavitation Research Facilities and Techniques,* 1964, pp. 48–66, ASME.
19. Indian Institute of Science, Dept. of Civil and Hydraulic Engr., Annual Rept. Publ. no. 31, Bangalore, India, 1968.
20. Bogachev, I. N., and R. I. Mints: "Cavitational Erosion and Means for its Prevention," Izdateistvo Mashinostroenie, Moscow, 1964, trans. from Russian by A. Wald, IPST Cat. no. 1428, Israel Program for Scientific Translations Ltd., 1966.
21. Varga, J. J., T. Bata, and Gy. Sebestyen: Investigation of Cavitation Erosion in Rotary Disc Equipment and Some Results, in L. Kisbocskoi and A. Szabo (eds.), *Proc. Third Conf. on Fluid Mechanics and Fluid Machinery,* 1969, pp. 694–704, Publishing House of Hungarian Academy of Sciences.
22. Varga, J. J., Gy. Sebestyen, and T. Bata: Investigation of Erosion and Acoustic Characteristics of Cavitation in Rotary Disk Equipment, *Proc. Fourth Conf. on Fluid Mechanics and Fluid Machinery,* 1972, pp. 1431–1444, Publishing House of Hungarian Academy of Sciences.
23. Varga, J. J., Gy. Sebestyen, and A. Fay: Detection of Cavitation by Acoustic and Vibration Measurement Methods, *La Houille Blanche,* no. 2, pp. 137–149, 1969.
24. Anton, I., and M. Popoviciu: The Behaviour of Hemispherical Bubbles Generated by Electric Sparks, *Proc. Fourth Conf. on Fluid Mechanics and Fluid Machinery,* 1972, pp. 89–102, Publishing House of Hungarian Academy of Sciences.
25. Anton, I., and L. Vekas: The Influence of the Geometrical Shape of Solid Surface Micro-irregularities on the Cavitation Bubble Nucleation, *Proc. Fourth Conf. on Fluid Mechanics and Fluid Machinery,* 1972, pp. 103–114, Publishing House of Hungarian Academy of Sciences.
26. Steller, K.: Correlation Between Different Symptoms of Cavitation Observed During the Tests of a

Kaplan Turbine and a Rotodynamic Pump, *Proc. Fourth Conf. on Fluid Mechanics and Fluid Machinery*, 1972, pp. 1289–1316, Publishing House of Hungarian Academy of Sciences.

27. Steller, K., and Z. Reymann: Cavitation Erosion with Different Surface Finish of Materials, *Proc. Fourth Conf. on Fluid Mechanics and Fluid Machinery*, 1972, pp. 1317–1348, Publishing House of Hungarian Academy of Sciences.

28. Steller, K., T. Krzysztofowicz, and Z. Reymann: Effects of Cavitation on Materials in Field and Laboratory Conditions, *ASTM STP* 567, pp. 152–170, 1974.

29. Nechleba, M.: "Entwicklungsperspektiven der Wasserturbinen," Die Schwerindustrie der Tschechoslowakei, 1960.

30. Nemecek, S.: "Urcovani Kavitacni Odolnosti Materialu. I. Dilci Zprava Vedeckovyzkumneho Ukolu 13–13a," VSST, Liberec, 1962.

31. Noskievic, J.: "Kavitace" (in Czech.), Academia Praha, 1969.

32. Wahl, N. E.: "Investigation of the Phenomena of Rain Erosion at Subsonic and Supersonic Speeds," Air Force Materials Lab., Tech. Rept. AFML-TR-65-330, Wright-Patterson Air Force Base, Ohio, October, 1965.

33. Wahl, N. E.: Design and Operation of Mach 3 Rotating Arm Erosion Testing Apparatus, *Proc. Third Int'l Conf. on Rain Erosion and Assoc. Phenomena*, Aug. 11–13, 1970, pp. 13–42, Royal Aircraft Est., Farnborough, England.

34. Schmitt Jr., G. F.: "Influence of Materials Construction Variables on the Rain Erosion Performance of Carbon-Carbon Composites, Heat Shield and Radome Materials," Air Force Materials Lab., Tech. Rept. AFML-TR-76-203, Wright-Patterson Air Force Base, Ohio, February, 1977.

35. Hurley, C. J., and G. F. Schmitt: "Development and Calibration of a Mach 1.2 Rain Erosion Test Apparatus," Air Force Materials Lab., Tech. Rept. AFML-TR-70-240, Wright-Patterson Air Force Base, October, 1970.

36. Schmitt Jr., G. F., W. G. Reinecke, and G. D. Waldman: Influence of Velocity, Impingement Angle, Heating, and Aerodynamic Shock Layers on Erosion of Materials at Velocities of 5500 ft/sec (1700 m/s), *ASTM STP* 567, pp. 219–238, 1974.

37. Letson, K. N., and P. A. Ormsby: "Rain Erosion Testing of Slip Cast Fused Silica at Mach 5," ASME paper no. 76-ENAS-6, 1976.

38. Ripken, J. F.: A Test Rig for Studying Impingement and Cavitation Damage, in "Erosion by Cavitation or Impingement," Amer. Soc. for Testing and Materials, ASTM Special Tech. Publ. 408, 1967.

39. Kenyon, H. F.: "Erosion by Water Jet Impacts, Parts I and II," Assoc. Elec. Industries Rept. T.P. IR 5587, January, 1967.

40. Bowden, F. P., and J. H. Brunton: Deformation of Solids by Liquid Impact at Supersonic Speeds, *Proc. Roy. Soc.* (*London*), ser. A, vol. 263, pp. 433–450, 1969; see also ser. A, vol. 260, no. 1110, pp. 79–85, 1966.

41. Hammitt, F. G., J. B. Hwang, et al.: Experimental and Theoretical Research on Liquid Droplet Impact, *Proc. Fourth Int'l Conf. on Rain Erosion and Assoc. Phenomena*, May 8–10, 1974, pp. 319–345, Meersburg, Germany (edited by A. A. Fyall and R. B. King at the Royal Aircraft Est., Farnborough, England).

42. Conn, A. F.: Report of Cavitation Committee, Appendix 4—Mechanisms of Cavitation Damage, *Proc. Fourteenth Int'l Towing Tank Conference*, 1975, Ottowa, Canada.

43. *Third Int'l Symp. Jet-Cutting Tech.*, May 11–13, 1976, Chicago, Ill., sponsored by BHRA, Cranfield, Bedford, England.

44. *Fourth Int'l Symp. on Jet-Cutting*, April 12–14, 1978, Canterbury, England, sponsored by BHRA, Cranfield, Bedford, England.

45. *ASTM Symp. on Erosion: Prevention and Useful Applications*, Oct. 24–26, 1977, Vail, Colorado.

46. Franz, N. C.: Preprint delivered at *First Nat'l Symp. on Jet Cutting*, April, 1972, Coventry, England, sponsored by BHRA, Cranfield, Bedford, England.

47. Elliott, D. E., J. B. Marriott, and A. Smith: Comparison of Erosion Resistance of Standard Steam Turbine Blade and Shield Materials on Four Test Rigs, *ASTM STP* 474, pp. 127–161, October, 1970.

48. Fyall, A. A.: Single Impact Studies of Rain Erosion, *Shell Aviation News*, vol. 374, pp. 16–23, 1969.
49. *Proc. First Meersburg Conf. on Rain Erosion and Allied Phenomena*, 1964, Meersburg, Germany (edited by A. A. Fyall et al. and available in English through the Royal Aircraft Est., Farnborough, England).
50. *Proc. Second Meersburg Conf. on Rain Erosion and Allied Phenomena*, August, 1967, Meersburg, Germany (edited by A. A. Fyall et al. and available in English through the Royal Aircraft Est., Farnborough, England).
51. *Proc. Fourth Meersburg Conf. on Rain Erosion and Allied Phenomena*, May, 1974, Meersburg, Germany (edited by A. A. Fyall, et al. and available in English through the Royal Aircraft Est., Farnborough, England).
52. Canavelis, R.: Jet Impact and Cavitation Damage, *Trans. ASME, J. Basic Engr.*, vol. 90, ser D, no. 3, pp. 355–367, September, 1968.
53. Erosion by Cavitation or Impingement, *ASTM STP* 408, March, 1967.
54. Characterization and Determination of Erosion Resistance, *ASTM STP* 474, October, 1970.
55. Weir, D. S., M. L. Billet, and J. W. Holl: "The 1.5 inch Ultra-High-Speed Cavitation Tunnel at the Applied Research Laboratory of the Pennsylvania State University," Applied Res. Lab. Rept. TM 75-188, Pennsylvania State University, July, 1975.
56. Endo, K., T. Okada, and M. Nakashima: A Study of Erosion Between Two Parallel Surfaces Oscillating at Close Proximity in Liquids, *Trans. ASME*, vol. 89, ser. F, no. 4, July, 1967.

AUTHOR INDEX

SUBJECT INDEX

Acoustic effects, 208, 257
Aircraft and space applications, 23, 231–232, 302, 307, 331, 334–335
  rain erosion, 351–353, 361–362, 384–386, 404–405
  transpiration cooling; missile nose-cones, 366, 378, 384–386
ASTM, Committee G-2 on Erosion & Wear, 22, 23

BHRA, 411
Bismuth (molten), 195–196
Boiling, 44, 56, 57, 84, 120, 123, 126, 127, 146, 180, 183, 185, 192, 207–208, 275, 371, 377
Bubble collapse and growth (*see also* bubble growth and collapse):
  asymmetrical collapse, 153, 157–160, 162–182
    bubble collapse sorting mechanism, 213–214
    damage, 213–214
    double and multiple bubble, 164–171, 178, 180, 182
    free surface, 164, 182, 213–214
    gravity, 163, 176–177, 182
    microjet, 163–171, 176–188, 213–214, 228, 244, 283
    pressure gradient, 163, 174–177, 182
    resilient surface, 164–167, 177–180, 182, 213–214
    thermodynamic effect, 189
    velocity (external), 163, 170–171, 176–177, 182
    wall proximity, 163–171, 176, 182
      roughness, 171
  besant analyses, 136–140, 145
    growth applications of equations, 145

Bubble collapse and growth (*continued*)
  bubble rebound and entrapped gas, 30, 155–160, 172, 175, 180–182, 213–214, 244–245, 248
  Rayleigh bubble collapse (and growth) analysis, 10–11, 138–147
    collapse time, 141–144, 167
    growth application of equations, 145–146
    internal gas effects, 144–146
    surface tension effects, 144–146
  real-fluid parameter effects, 146–162, 180
    compressibility, 146, 152–161, 180
      equation of state (Tait), 154, 309
    density, 148–149, 194
    gas content, 30, 155–161, 188–194
    surface tension, 146, 148–149, 151–152, 159–162
  thermodynamic effects, 146, 183–211, 251–253
    Bonnin B-factors, 183–185
    fixed cavity theories, 197–207
    Florscheutz, Chao & Wittke analyses, 185
    Garcia, Hammitt & Bhatt studies, 193–198
    Hickling, Plesset, Zwick & Hsieh analyses, 185–186
    Mitchell & Hammitt analysis, 186–193
    NASA (Ruggeri et al), Bureau of Standards, (Hord et al), 198–204
    Penn State studies (Holl et al) entrainment model, 204–209
    sodium damage results, 209–211
    Stepanoff's B-factor, 183
  viscosity, 146–152, 159–160
    dissipation function, 151
Bubble growth and collapse, 9, 136–219
  single bubble studies, 9